OPTICAL FILTER DESIGN AND ANALYSIS: A SIGNAL PROCESSING APPROACH • *Christi K. Madsen and Jian H. Zhao*

OPTOELECTRONIC PACKAGING • *A. R. Mickelson, N. R. Basavanhally, and Y. C. Lee (eds.)*

OPTICAL CHARACTER RECOGNITION • *Shunji Mori, Hirobumi Nishida, and Hiromitsu Yamada*

ANTENNAS FOR RADAR AND COMMUNICATIONS: A POLARIMETRIC APPROACH • *Harold Mott*

INTEGRATED ACTIVE ANTENNAS AND SPATIAL POWER COMBINING • *Julio A. Navarro and Kai Chang*

ANALYSIS METHODS FOR RF, MICROWAVE, AND MILLIMETER-WAVE PLANAR TRANSMISSION LINE STRUCTURES • *Cam Nguyen*

FREQUENCY CONTROL OF SEMICONDUCTOR LASERS • *Motoichi Ohtsu (ed.)*

SOLAR CELLS AND THEIR APPLICATIONS • *Larry D. Partain (ed.)*

ANALYSIS OF MULTICONDUCTOR TRANSMISSION LINES • *Clayton R. Paul*

INTRODUCTION TO ELECTROMAGNETIC COMPATIBILITY • *Clayton R. Paul*

ELECTROMAGNETIC OPTIMIZATION BY GENETIC ALGORITHMS • *Yahya Rahmat-Samii and Eric Michielssen (eds.)*

INTRODUCTION TO HIGH-SPEED ELECTRONICS AND OPTOELECTRONICS • *Leonard M. Riaziat*

NEW FRONTIERS IN MEDICAL DEVICE TECHNOLOGY • *Arye Rosen and Harel Rosen (eds.)*

ELECTROMAGNETIC PROPAGATION IN MULTI-MODE RANDOM MEDIA • *Harrison E. Rowe*

ELECTROMAGNETIC PROPAGATION IN ONE-DIMENSIONAL RANDOM MEDIA • *Harrison E. Rowe*

NONLINEAR OPTICS • *E. G. Sauter*

ELECTROMAGNETIC FIELDS IN UNCONVENTIONAL MATERIALS AND STRUCTURES • *Onkar N. Singh and Akhlesh Lakhtakia (eds.)*

FUNDAMENTALS OF GLOBAL POSITIONING SYSTEM RECEIVERS: A SOFTWARE APPROACH • *James Bao-yen Tsui*

InP-BASED MATERIALS AND DEVICES: PHYSICS AND TECHNOLOGY • *Osamu Wada and Hideki Hasegawa (eds.)*

DESIGN OF NONPLANAR MICROSTRIP ANTENNAS AND TRANSMISSION LINES • *Kin-Lu Wong*

FREQUENCY SELECTIVE SURFACE AND GRID ARRAY • *T. K. Wu (ed.)*

ACTIVE AND QUASI-OPTICAL ARRAYS FOR SOLID-STATE POWER COMBINING • *Robert A. York and Zoya B. Popović (eds.)*

OPTICAL SIGNAL PROCESSING, COMPUTING AND NEURAL NETWORKS • *Francis T. S. Yu and Suganda Jutamulia*

SiGe, GaAs, AND InP HETEROJUNCTION BIPOLAR TRANSISTORS • *Jiann Yuan*

ELECTRODYNAMICS OF SOLIDS AND MICROWAVE SUPERCONDUCTIVITY • *Shu-Ang Zhou*

Analysis Methods for RF, Microwave, and Millimeter-Wave Planar Transmission Line Structures

Analysis Methods for RF, Microwave, and Millimeter-Wave Planar Transmission Line Structures

CAM NGUYEN
Texas A&M University

A Wiley-Interscience Publication
JOHN WILEY & SONS, INC.
New York / Chichester / Weinheim / Brisbane / Singapore / Toronto

This book is printed on acid-free paper. ♾

Published simultaneously in Canada.

Library of Congress Cataloging-in-Publication Data:

Nguyen, Cam

Analysis methods for RF, microwave, and millimeter-wave planar transmission line structures/Cam Nguyen.

p.cm.—(Wiley series in microwave and optical engineering)

"Wiley-Interscience publication."

Includes index.

ISBN 0-471-01750-7 (cloth : alk. paper)

1. Electric circuit analysis. 2. Microwave transmission lines. 3. Strip transmission lines. 4. Microwave integrated circuits. 5. Electric circuit analysis. I. Series.

TK7876.N48 2000
621.381′31–dc21 99-086737

10 9 8 7 6 5 4 3 2 1

To my wife, Ngọc-Diệp,
and my children, Christine (Nhã-Uyên) and Andrew (An)

Contents

Preface

RF integrated circuit (RFICs) and microwave integrated circuits (MICs), both hybrid and monolithic, have advanced rapidly in the last two decades. This progress has been achieved not only because of the advance of solid-state devices, but also due to the progression of planar transmission lines. Many milestones have been achieved: one of them being the development of various analysis methods for RF microwave and millimeter-wave passive structures, in general, and planar transmission lines, in particular. These methods have played an important role in providing accurate transmission line parameters for designing RFICs and MICs, as well as in investigating and developing new planar transmission lines.

The primary objective of this book is to present the Green's function, conformal-mapping, variational, spectral-domain, and mode-matching methods, which are some of the most useful and commonly used techniques for analyzing planar transmission lines. Information for these methods in the literature is at a level that is not very suitable for the majority of first-year graduate students and practicing RF and microwave engineers. The material in this book is self-contained and presented in a way that allows readers with only fundamental knowledge in electromagnetic theory to easily understand and implement the techniques. The book also includes problems at the end of each chapter, allowing readers to reinforce their knowledge and to practice their understanding. Some of these problems are relatively long and difficult, and thus are more suitable for class projects. The book can therefore serve not only as a textbook for first-year graduate students, but also as a reference book for practicing RF and microwave engineers. Another purpose of the book is to use these methods as means to present the principles of applying electromagnetic theory to the analysis of microwave boundary-value problems. This knowledge is essential for microwave students and engineers, as it allows them to modify and improve these methods, as well as to develop new techniques.

This book is based on the material of a graduate course on field theory for microwave passive structures offered at Texas A&M University. It is completely self-contained and requires readers to have only the fundamentals of electromagnetic theory, which is normally fulfilled by the first undergraduate course in electromagnetics.

I sincerely appreciate the patience of Professor Kai Chang, Editor of the Wiley Series in Microwave and Optical Engineering, and Mr. George Telecki, Executive Editor of Wiley-Interscience, during the writing of the manuscript for this book. I am also grateful to my former students who took the course and provided me with a purpose for writing this book. Finally, I wish to express my heartfelt thanks and deepest appreciation to my wife, Ngoc-Diep, for her constant encouragement and support, and my children, Christine and Andrew, for their understanding during the writing of this book.

CAM NGUYEN

College Station, Texas

Analysis Methods for RF, Microwave, and Millimeter-Wave Planar Transmission Line Structures

CHAPTER ONE

Introduction

Microwave integrated circuits (MICs) were introduced in the 1950s. Since then, they have played perhaps the most important role in advancing the radiofrequency (RF) and microwave technologies. The most noticeable and important milestone was possibly the emergence of *monolithic microwave integrated circuits* (MMICs). This progress of MICs would not have been possible without the advances of solid-state devices and planar transmission lines. Planar transmission lines refer to transmission lines that consist of conducting strips printed on surfaces of the transmission lines' substrates. These structures are the backbone of MICs, and represent an important and interesting research topic for many microwave engineers. Along with the advances of MICs and planar transmission lines, numerous analysis methods for microwave and millimeter-wave passive structures, in general, and planar transmission lines, in particular, have been developed in response to the need for accurate analysis and design of MICs. These analysis methods have in turn helped further investigation and development of new planar transmission lines.

This book presents the Green's function, conformal-mapping, variational, spectral-domain, and mode-matching methods. They are useful and commonly used techniques for analyzing microwave and millimeter-wave planar transmission lines, in particular, and passive structures, in general. Information for these methods in the literature is at a level that is not very suitable for the majority of first-year graduate students and practicing microwave engineers. This book attempts to present the materials in such a way as to allow students and engineers with basic knowledge in electromagnetic theory to understand and implement the techniques. The book also includes problems for each chapter so readers can reinforce and practice their knowledge.

1.1 PLANAR TRANSMISSION LINES AND MICROWAVE INTEGRATED CIRCUITS

Planar transmission lines are essential components of MICs. They have been used to realize many circuit functions, such as baluns, filters, hybrids, and couplers, as well as simply to carry signals. Figure 1.1 shows some commonly used planar

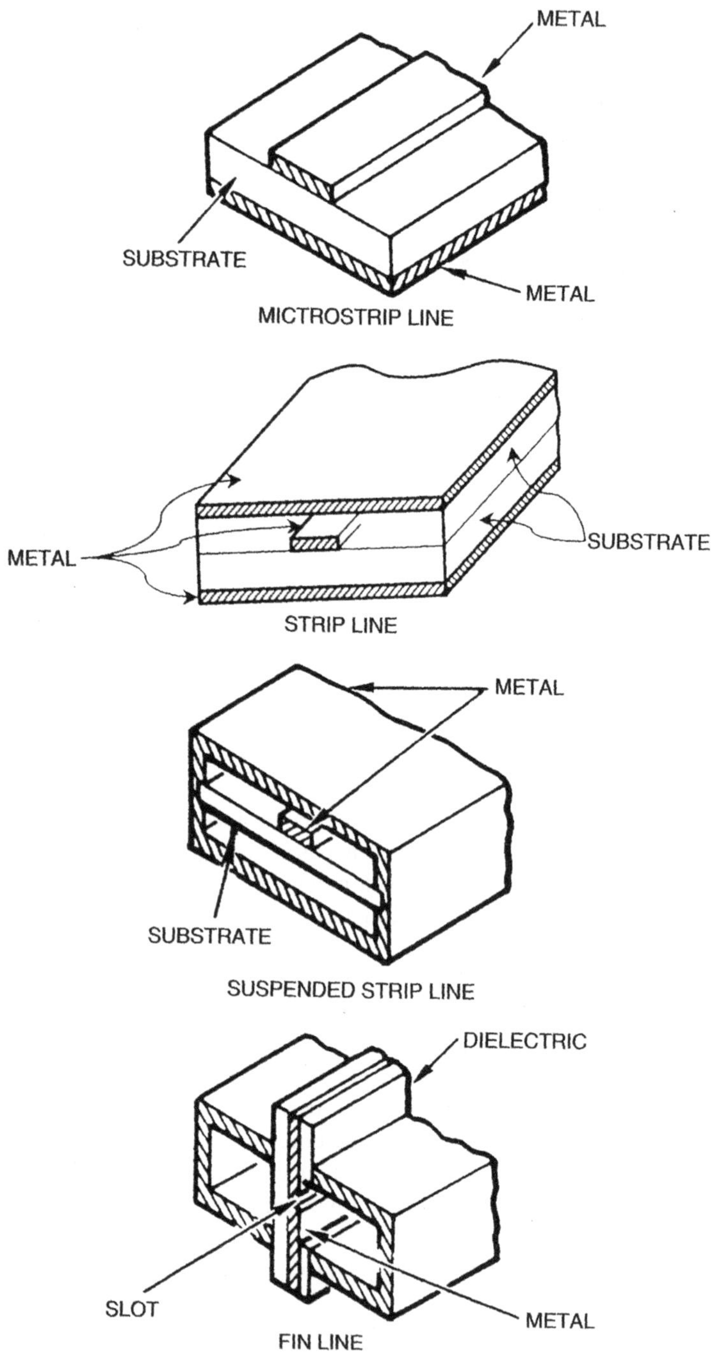

Figure 1.1 Common planar transmission lines.

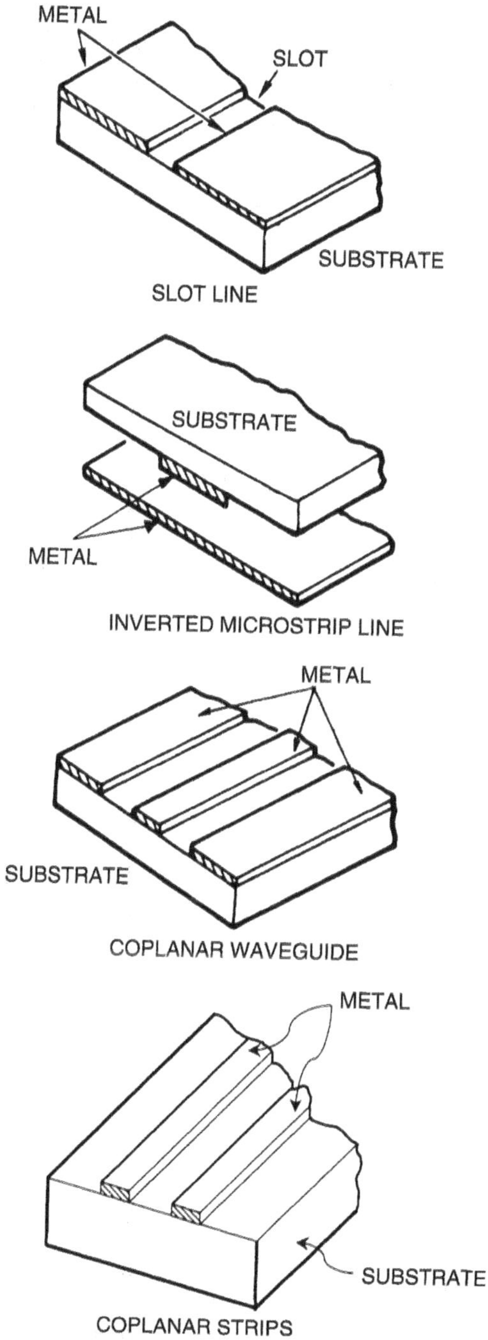

Figure 1.1 *(Continued)*

TABLE 1.1 Properties of Planar Transmission Lines Shown in Fig. 1.1

Transmission Line	Operating Frequency (GHz)	Characteristic Impedance Range (Ohm)	Dimension	Loss	Power Handling	Solid-State Device Mounting	Low-Cost Production
Microstrip line	$\leq$110 GHz	10–100	Small	High	Low	Fair	Good
Strip line	$\leq$60 GHz	20–150	Moderate	Low	Low	Moderate	Good
Suspended	$\leq$220 GHz	20–150	Moderate	Low	Low	Moderate strip line	Fair
Fin line	$\leq$220 GHz	20–400	Moderate	Moderate	Low	Easy	Fair
Slot line	$\leq$110 GHz	60–200	Small	High	Low	Easy	Good
Inverted microstrip line	$\leq$220 GHz	25–130	Small	Moderate	Low	Moderate	Fair
Coplanar waveguide	$\leq$110 GHz	40–150	Small	High	Low	Very easy	Good
Coplanar strips	$\leq$110 GHz	30–250	Small	High	Low	Easy	Good

transmission lines and Table 1.1 summarizes their properties. Each transmission line has its own unique advantages and disadvantages and, depending on circuit types, either an individual transmission line or a combination of them is needed to achieve desired circuit functions as well as optimum performances. The most viable planar transmission lines are perhaps the conventional microstrip line and coplanar waveguide (CPW), from which many other planar transmission lines have evolved. Multilayer planar transmission lines, such as that shown in Fig. 1.2, are especially attractive for MICs due to their flexibility and ability to realize complicated circuits, ultimately allowing very compact, high-density circuit integration. They also allow thin dielectric layers to be deposited on conductor-backed semiconducting substrates for achieving ultracompact MICs. Furthermore, multilayer transmission lines have significantly less cross talk and distortion via appropriate selection of dielectric layers.

There are two classes of MICs: hybrid and monolithic circuits. Hybrid MIC refers to a planar circuit in which only parts of the circuit are formed on surfaces of the circuit's substrates by some deposition schemes. A typical hybrid MIC has all the transmission lines deposited on the dielectric surfaces, except solid-state devices such as transistors and other passive components like capacitors. These solid-state devices and passive elements are discrete components and connected to the transmission lines by bonding, soldering, or conducting epoxy. The substrates of a hybrid MIC are generally low-loss insulators, used solely for supporting the circuit components and delivering the signals. Advantages of hybrid MICs include small size, light weight, easy fabrication, low cost, and high-volume production. In practice, hybrid MICs are normally referred to simply as MICs. Figures 1.3–1.7 show photographs of some hybrid MICs employing planar transmission lines.

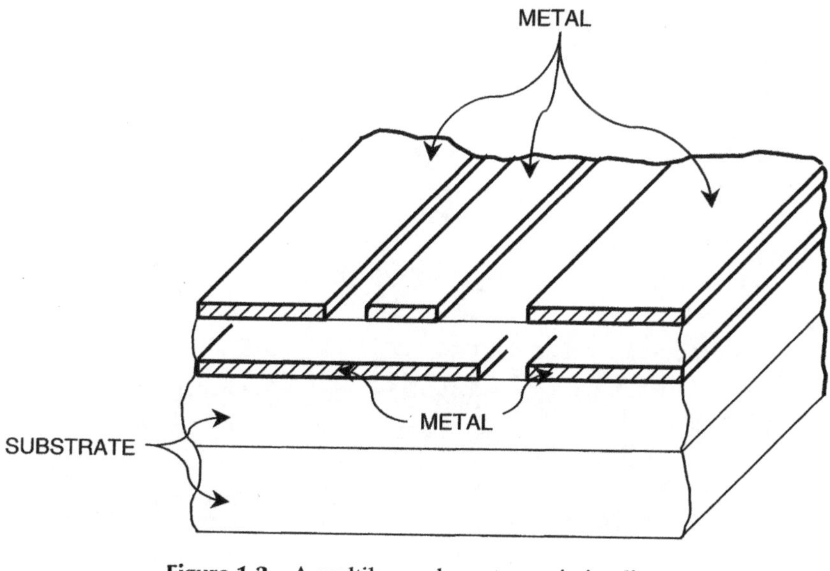

Figure 1.2 A multilayer planar transmission line.

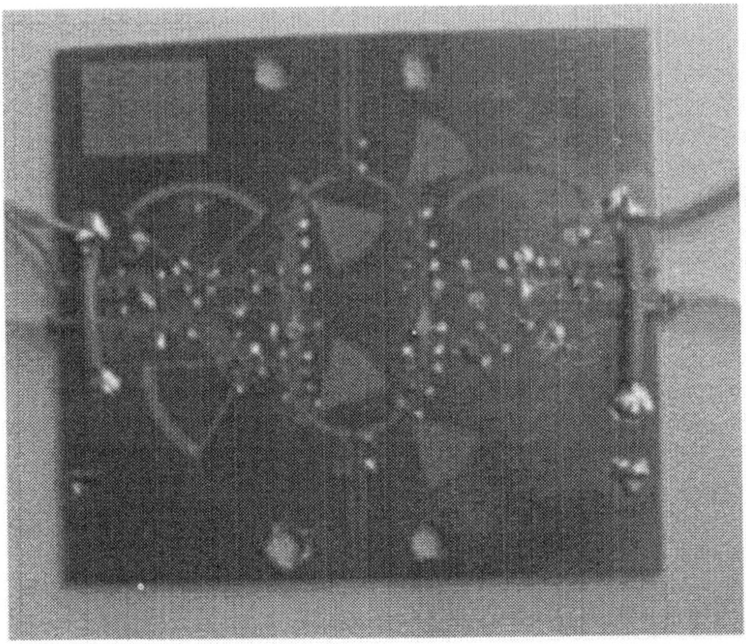

Figure 1.3 S-band (2–4 GHz) MIC push–pull field effect transistor (FET) amplifier using CPW and slot line.

Figure 1.4 W-band (75–110 GHz) MIC diode balanced mixer using fin line, CPW, and suspended strip line.

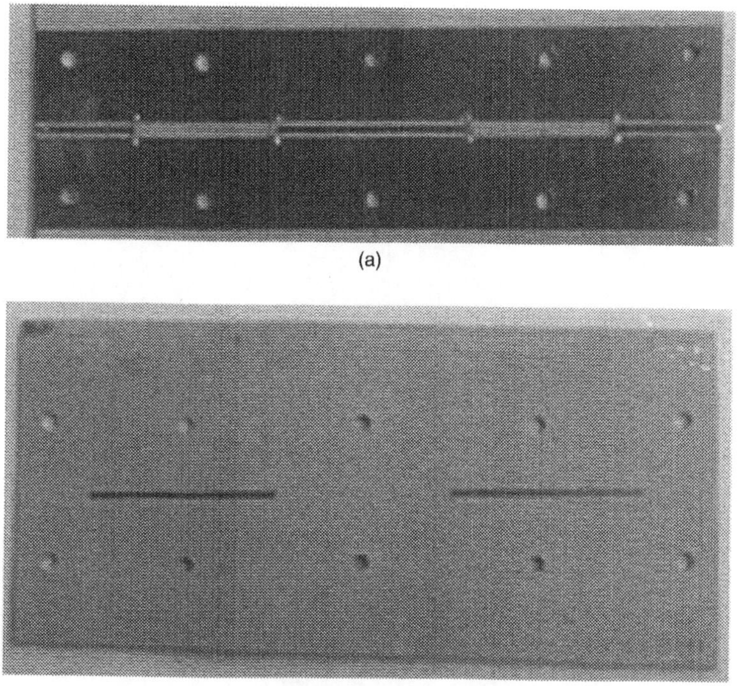

(a)

(b)

Figure 1.5 Top (a) and bottom (b) sides of an S-band MIC bandpass filter using multi-layer broadside-coupled CPW.

The monolithic MIC (MMIC) is a special class of MICs, in which all the circuit elements, including passive components and solid-state devices, are formed into the bulk or onto the surface of a semi-insulating semiconductor substrate by some deposition technique. In contrast to hybrid MICs, the substrates are used in MMICs not only as a signal-propagating medium and a supporting structure for passive components, but also as a material onto which semiconducting layers with good properties for realizing microwave solid-state devices are grown or deposited. Compared to hybrid MICs, the advantages of MMICs are lower-cost circuits through batch processing, improved reliability and reproducibility through minimization of wire bonds and discrete components, smaller size and weight, more circuit design flexibility, and multifunction performance on a single chip. MMICs are very important for microwave technology. Most microwave and millimeter-wave applications are expected eventually to employ all MMICs. Figure. 1.8 shows a photograph of a Ka-band (26.5–40 GHz) push–push MMIC oscillator.

1.2 ANALYSIS METHODS FOR PLANAR TRANSMISSION LINES

In using planar transmission lines in MICs, analysis methods are needed in order to determine the transmission lines' characteristics such as characteristic

Figure 1.6 A 5–20 GHz MIC balun using microstrip line.

impedance, effective dielectric constant, and loss. The design of MICs depends partly on accurate analysis of planar transmission lines.

The microwave technology is changing rapidly and, in connecting with it, useful analysis methods for microwave and millimeter-wave planar transmission lines, either completely brand new or modifications of existing techniques, appear constantly. In fact, microwave engineers are now faced with many different techniques and a vast amount of information, making the techniques difficult to understand and hence to implement in the short time normally encountered in an industrial setting. Each method has its own unique advantages and disadvantages for particular problems and needs. However, they are all based on Maxwell's

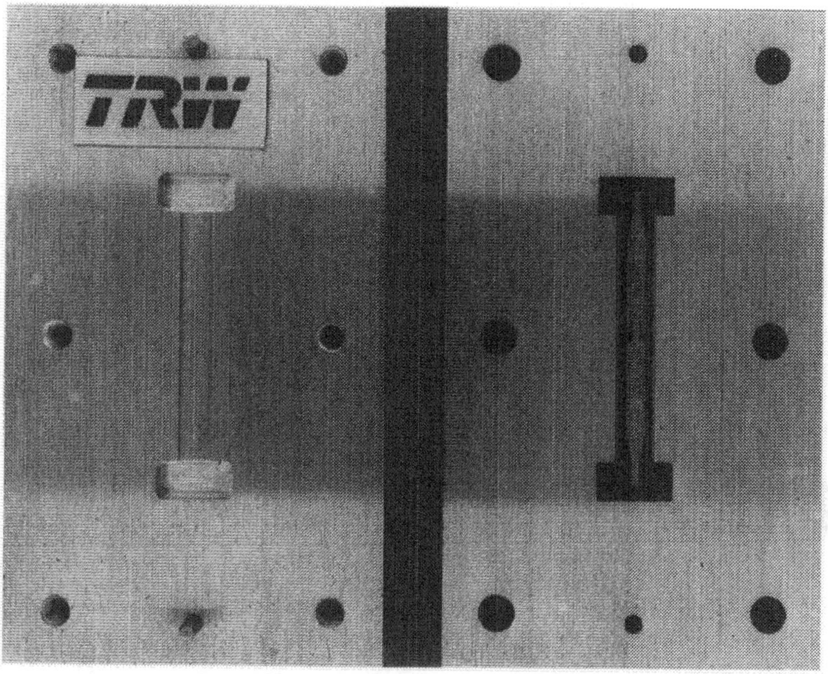

Figure 1.7 Ka-band MIC bandstop filter using suspended strip line.

equations, in general, and wave equations and boundary conditions, in particular. These are the fundamentals of these methods and, while techniques can change steadily, the fundamentals always remain the same. They, in fact, provide a foundation for the derivation, modification, and implementation of all current and future analysis methods. In this book, we describe particularly the details of the Green's function, conformal-mapping, variational, spectral-domain, and mode-matching methods. These methods not only represent some of the most useful and commonly used techniques for analyzing planar transmission lines, but also serve as means to present the fundamentals of applying electromagnetic theory to the analysis of microwave boundary-value problems. This knowledge would then allow readers to modify and improve these methods, or to develop new techniques.

1.3 ORGANIZATION OF THE BOOK

The book is organized into eight chapters and is self-contained. Chapter 2 gives the fundamentals of electromagnetic theory, which are needed for the formulation of the methods addressed in this book. Chapter 3 covers Green's functions used in various methods. Chapter 4 discusses the fundamentals of planar transmission lines and provides useful equations for commonly used transmission lines

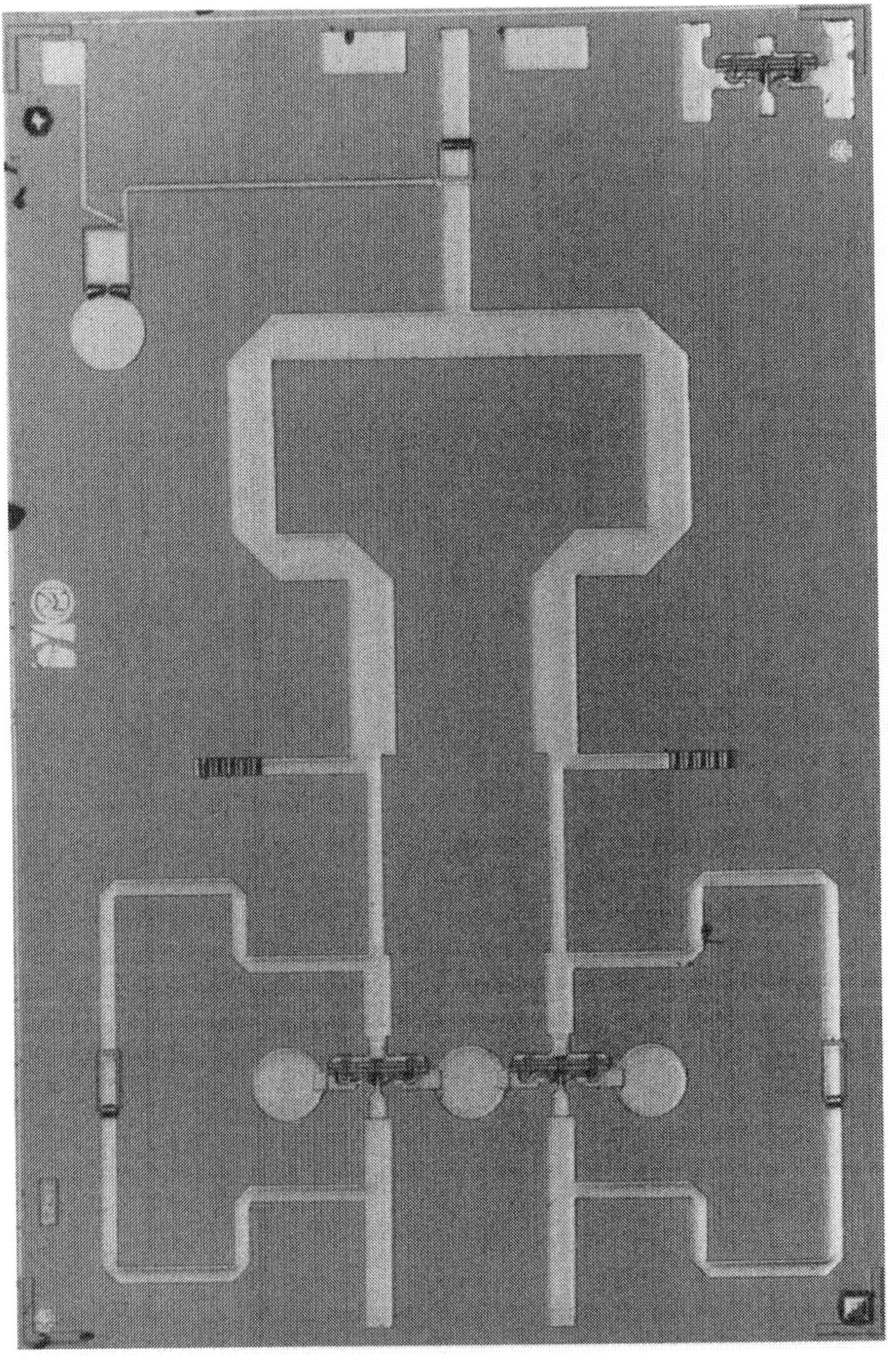

Figure 1.8 Ka-band MMIC push–push FET oscillator using microstrip line.

in MICs. Chapter 5 covers the principles of conformal mapping and demonstrates its use in analyzing planar transmission lines. Chapter 6 presents the variational methods in both the space and spectral domains and uses them to analyze planar transmission lines. Chapter 7 gives the foundation of the spectral-domain methods and then applies them in the analysis of planar transmission lines. Finally, Chapter 8 formulates the mode-matching method for both planar transmission lines and their discontinuities.

CHAPTER TWO

Fundamentals of Electromagnetic Theory

Electromagnetic theory forms the foundation for electrical engineering. Not only can it be used to explain many phenomena of electronic, it can be employed to design and analyze accurately many electronic circuits operating across the electromagnetic spectrum. While circuit theory may fail to explain adequately an electrical phenomenon or to accurately analyze and design an electronic circuit, electromagnetic theory, in general, will not. In this chapter we will review sufficient fundamentals of electromagnetics to allow readers to understand the methods we will present in subsequent chapters.

2.1 MAXWELL'S EQUATIONS

Perhaps the most important equations in electromagnetic theory are *Maxwell's equations*, which altogether create the foundation of electromagnetic theory. Maxwell's equations can be written in a differential or integral form. For general time-varying electromagnetic fields, they are given as follows:

Differential Form

$$\nabla \times \mathcal{E} = -\frac{\partial \mathcal{B}}{\partial t} \tag{2.1a}$$

$$\nabla \times \mathcal{H} = \mathcal{J} + \frac{\partial \mathcal{D}}{\partial t} \tag{2.1b}$$

$$\nabla \cdot \mathcal{D} = \rho \tag{2.1c}$$

$$\nabla \cdot \mathcal{B} = 0 \tag{2.1d}$$

Integral Form

$$\oint \boldsymbol{\mathcal{E}} \cdot d\boldsymbol{\ell} = -\frac{d}{dt} \iint \boldsymbol{\mathcal{B}} \cdot d\mathbf{S} \tag{2.2a}$$

$$\oint \boldsymbol{\mathcal{H}} \cdot d\boldsymbol{\ell} = \iint \boldsymbol{\mathcal{J}} \cdot d\mathbf{S} + \frac{d}{dt} \iint \boldsymbol{\mathcal{D}} \cdot d\mathbf{S} \tag{2.2b}$$

$$\oiint \boldsymbol{\mathcal{B}} \cdot d\mathbf{S} = 0 \tag{2.2c}$$

$$\oiint \boldsymbol{\mathcal{D}} \cdot d\mathbf{S} = \iiint \rho \, dv \tag{2.2d}$$

where

$\boldsymbol{\mathcal{E}}(x, y, z, t)$ is *electric field* or *electric field intensity* in volts/meter (V/m)
$\boldsymbol{\mathcal{H}}(x, y, z, t)$ is *magnetic field* or *magnetic field intensity* in amperes/meter (A/m)
$\boldsymbol{\mathcal{D}}(x, y, z, t)$ is *electric flux density* in coulombs/meter^2 (C/m^2)
$\boldsymbol{\mathcal{B}}(x, y, z, t)$ is *magnetic flux density* in webers/meter^2 (Wb/m^2)
$\boldsymbol{\mathcal{J}}(x, y, z, t)$ is *electric current density* in amperes/meter^2 (A/m^2)
$\rho(x, y, z, t)$ is *electric charge density* in coulombs/meter^3 (C/m^3)

The integral Maxwell equations can be derived from their differential forms by using the *Stokes* and *divergence theorems*. The parameter defined by

$$\boldsymbol{\mathcal{J}}_d = \frac{d\boldsymbol{\mathcal{D}}}{dt} \tag{2.3}$$

is known as the *displacement current density* (in A/m^2). The time (t) and location (e.g., x, y, and z) dependence are assumed for all these fields. These Maxwell equations are general and hold for fields with arbitrary time dependence in any electronic structure and at any location in the structure. They become simpler for special cases such as static or quasi-static fields, sinusoidal time-varying (or time-harmonic) fields, and source-free media.

Under the assumption of *static* or *quasi-static*, we let d/dt equal zero and write the differential-form Maxwell equations as

$$\nabla \times \hat{\mathbf{E}} = 0 \tag{2.4a}$$

$$\nabla \times \hat{\mathbf{H}} = \hat{\mathbf{J}} \tag{2.4b}$$

$$\nabla \cdot \hat{\mathbf{D}} = \rho \tag{2.4c}$$

$$\nabla \cdot \hat{\mathbf{B}} = 0 \tag{2.4d}$$

Note that the static quantities are denoted with a hat ($\hat{}$). These field quantities are independent of time. It should be noted that these equations are only valid with direct current (dc). In most engineering practices, however, they can also be used when the operating frequencies are not high.

For the case of time-harmonic fields, we can replace d/dt by $j\omega$ and obtain Maxwell's equations as

$$\nabla \times \mathbf{E} = -j\omega \mathbf{B} \tag{2.5a}$$

$$\nabla \times \mathbf{H} = \mathbf{J} + j\omega \mathbf{D} \tag{2.5b}$$

$$\nabla \times \mathbf{D} = \rho \tag{2.5c}$$

$$\nabla \cdot \mathbf{B} = 0 \tag{2.5d}$$

where the fields now represent *phasor fields*, which are functions of location only. These Maxwell equations are commonly known as the *time-harmonic Maxwell equations*. The phasor representation such as **E** and its corresponding *instantaneous* field quantity $\mathcal{E}$ are related, with reference to cos ωt, by

$$\mathcal{E}(x, y, z, t) = \text{Re}[\mathbf{E}(x, y, z)e^{j\omega t}] \tag{2.6}$$

The time-harmonic case is perhaps most commonly used in electrical engineering and will be considered in this book together with the static case.

Maxwell's equations under the source-free condition are obtained by letting $\rho = J = 0$. These equations are applicable to passive microwave structures such as transmission lines.

2.2 CONSTITUTIVE RELATIONS

In order to solve for field quantities using Maxwell's equations, three constitutive relations are needed. They basically describe the relations between the fields through the properties of the medium. Under the time-harmonic assumption, the (phasor) electric flux density **D** and electric field **E** in a simple medium are related by

$$\mathbf{D} = \varepsilon(\omega)\mathbf{E} = \varepsilon_0 \varepsilon_r(\omega)\mathbf{E} \tag{2.7}$$

where $\varepsilon_0 = 8.854 \times 10^{-12}$ F/m (farad/meter) and ε are the permittivity or dielectric constant of the vacuum and medium, respectively. Note that, in practice, free space is normally considered a vacuum. ε_r is called the relative permittivity or relative dielectric constant of the medium. The relation between the magnetic flux density **B** and magnetic field **H** in a simple medium is given as

$$\mathbf{B} = \mu(\omega)\mathbf{H} = \mu_0 \mu_r(\omega)\mathbf{H} \tag{2.8}$$

where $\mu_0 = 4\pi \times 10^{-7}$ H/m (henries/meter) and μ are the permeability of the vacuum and medium, respectively. μ_r is the relative permeability of the medium.

For time-harmonic fields and simple media, the current relates to the electric field by

$$\mathbf{J} = \sigma(\omega)\mathbf{E} \tag{2.9}$$

In general ε_r, μ_r, and σ are a function of the location and direction in the medium as well as the power level applied to the medium. Most substrates used for electronic circuits, however, are homogeneous, isotropic, and linear, having constant ε_r, μ_r, and σ. They are known as *simple materials (media)*. Furthermore, most electronic substrates are *nonmagnetic*, having a relative permeability of 1. In this book, we will consider only simple and nonmagnetic substrates. There are also other materials classified as *anisotropic* (or *nonisotropic*), such as sapphire, and *magnetic*, such as ferrite. For these materials, the relative dielectric constant and permeability are described by the *relative tensor dielectric constant* and the *permeability*, respectively. In general, the conductivity and relative dielectric constant and permeability are also dependent on frequency. Good nonmagnetic substrates, however, have relative dielectric constants almost constant up to high frequencies. Good conductors have almost constant conductivity from dc up to the infrared frequencies. Their permittivity and permeability are approximately equal to those of a vacuum.

2.3 CONTINUITY EQUATION

The continuity equation is obtained from the conservation of charge as

$$\nabla \cdot \mathcal{J} = -\frac{d\rho}{dt} \tag{2.10a}$$

and

$$\nabla \cdot \mathbf{J} = -j\omega\rho \tag{2.10b}$$

for time-harmonic fields.

2.4 LOSS IN MEDIUM

Dielectrics used in electronic circuits are always nonperfect. Consequently, there is always loss present in any practical nonmagnetic dielectrics, known as *dielectric loss*, due to a nonzero conductivity of the medium. We can rewrite Maxwell's Eq. (2.5b), making use of the constitutive relations (2.7) and (2.9), as

$$\nabla \times \mathbf{H} = j\omega\varepsilon\left(1 - j\frac{\sigma}{\omega\varepsilon}\right)\mathbf{E} \tag{2.11}$$

or

$$\nabla \times \mathbf{H} = j\omega\varepsilon(1 - j\tan\delta)\mathbf{E} \tag{2.12}$$

where

$$\tan\delta = \frac{\sigma}{\omega\varepsilon} \tag{2.13}$$

is known as the *loss tangent* of the medium, which is normally used in practice to characterize the medium's loss. Compared to the ideal case of a lossless medium, we can then define a *complex dielectric constant* of a lossy medium as

$$\hat{\varepsilon} = \varepsilon' - j\varepsilon'' \tag{2.14a}$$

where

$$\varepsilon' = \varepsilon = \varepsilon_o\varepsilon_r \tag{2.14b}$$

and

$$\varepsilon'' = \frac{\sigma}{\omega} = \varepsilon\tan\delta \tag{2.14c}$$

Note that the real part ε' of the complex dielectric constant is the dielectric property that contributes to the stored electric energy in the medium. The imaginary part ε'' contains the finite conductivity and results in loss in the medium. As for ε, ε'' is also dependent on frequency. Figure 2.1 shows ε' and ε'' versus frequency for polystyrene [1]. It is apparent, from Eq. (2.14), that the loss tangent is equal to the ratio between the imaginary and real parts of the complex dielectric constant. The complex dielectric constant of a dielectric and, hence, its relative dielectric constant ε_r and loss tangent $\tan\delta$ can be measured. The loss tangents and relative dielectric constants of substrates are supplied by manufacturers at

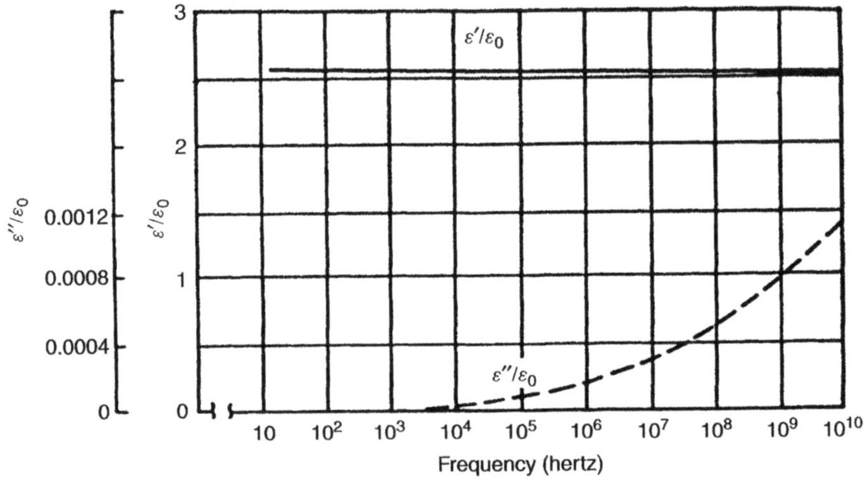

Figure 2.1 Real (ε') and imaginary (ε'') parts of the complex dielectric constant of polystyrene versus frequency at 25°C.

TABLE 2.1 Relative Dielectric Constant (ε_r) and Loss Tangent (tan δ) of Typical Microwave Substrates

Material	Frequency (GHz)	ε_r	tan δ@25°C
Styrofoam-103.7	3	1.03	0.0001
Rexolite-1422	3	2.54	0.0005
GaAs	10	12.9	0.006
Sapphire	10	9.4–11.5	0.0001
Alumina (96%)	10	8.9	0.0006
Alumina (99.5%)	10	9.8	0.0003
Quartz (fused)	10	3.78	0.0001
Teflon	10	2.1	0.0004
Silicon	10	11.9	0.004
RT/Duroid 5880	10	2.2	0.0009
RT Duroid 6010	10	10.2, 10.5, 10.8	0.0028 max.

particular frequencies and used by RF and microwave engineers. Table 2.1 shows the parameters of substrates commonly used at microwave frequencies.

2.5 BOUNDARY CONDITIONS

Maxwell's equations and constitutive relations may be used to obtain general solutions for electromagnetic fields existing in any microwave structures. To obtain unique solutions for the fields in a particular structure, such as *coplanar waveguide*, we must, however, enforce the structure's boundary conditions. This is in fact similar to using *Kirchhoff's voltage* and *current laws* in a lumped-element circuit to obtain unique solutions for the voltages and currents in that circuit. For time-harmonic fields, the boundary conditions between two different media, shown in Fig. 2.2, are given as

$$\mathbf{n} \times (\mathbf{E}_1 - \mathbf{E}_2) = 0 \tag{2.15a}$$

$$\mathbf{n} \times (\mathbf{H}_1 - \mathbf{H}_2) = \mathbf{J}_s \tag{2.15b}$$

$$\mathbf{n} \cdot (\mathbf{D}_1 - \mathbf{D}_2) = \rho_s \tag{2.15c}$$

$$\mathbf{n} \cdot (\mathbf{B}_1 - \mathbf{B}_2) = 0 \tag{2.15d}$$

where the subscripts 1 and 2 indicate media 1 and 2, respectively. $\mathbf{n}$ is the unit vector normal to the surface and pointing into medium 1. $\mathbf{J}_s$ and ρ_s are the (linear) surface current density (in A/m) and surface charge density (in C/m^2) existing at the boundary, respectively. General time-varying fields also follow these boundary conditions. These boundary conditions become simpler for special cases, such as between perfect dielectrics ($\rho_s = 0$ and $\mathbf{J}_s = 0$), between nonperfect dielectrics ($\mathbf{J}_s = 0$), and between a perfect dielectric and a perfect conductor.

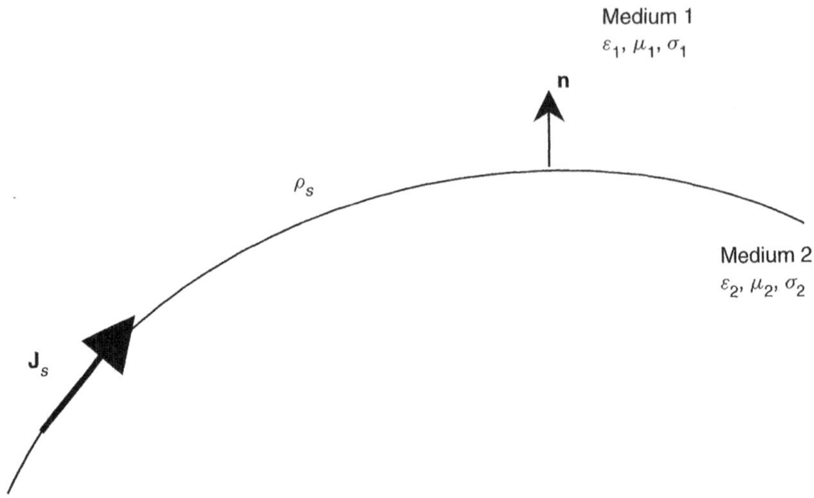

Figure 2.2 Boundary between two different media.

For instance, when media 1 and 2 are assumed to be a perfect dielectric and a perfect conductor, respectively, the boundary conditions become

$$\mathbf{n} \times \mathbf{E} = 0 \tag{2.16a}$$

$$\mathbf{n} \times \mathbf{H} = \mathbf{J}_s \tag{2.16b}$$

$$\mathbf{n} \cdot \mathbf{D} = \rho_s \tag{2.16c}$$

$$\mathbf{n} \cdot \mathbf{B} = 0 \tag{2.16d}$$

where the normal unit vector **n** points outward from the conductor surface. The tangential electric field along a perfect conductor is, therefore, always zero. For many practical problems, especially at low frequencies, good results can be obtained assuming good dielectrics and conductors are perfect. The boundary conditions (2.16b) and (2.16c) provide simple means for determining the current and charge induced on a conductor when fields are present. It should also be noted that the boundary conditions for the normal and tangential components of the fields between any two media are not independent of each other.

2.6 SKIN DEPTH

One of the most important parameters of a medium is its *skin depth* or *depth of penetration*. The skin depth is defined as the distance from the medium surface, over which the magnitudes of the fields of a wave traveling in the medium are reduced to $1/e$, or approximately 37%, of those at the medium's surface. The skin depth δ of a good conductor is approximately given as

$$\delta = \sqrt{\frac{2}{\omega\mu\sigma}} \tag{2.17}$$

The skin depths of good conductors are very small, especially at high frequencies, causing currents to reside near the conductors' surfaces. This subsequently results in a low conduction loss.

2.7 POWER FLOW

When a wave propagates in a medium, it carries the electric and magnetic fields and power. The power density at any location in the medium is given by the *Poynting vector*

$$\mathcal{S} = \mathcal{E} \times \mathcal{H} \tag{2.18}$$

Note that $\mathcal{S}$ is instantaneous power with a unit of watt per square meter (W/m^2). For time-harmonic fields, we define a *phasor Poynting vector*

$$\mathbf{S} = \mathbf{E} \times \mathbf{H}^* \tag{2.19}$$

where $\mathbf{H}^*$ is the complex conjugate of $\mathbf{H}$. The time average of the instantaneous power density $\mathcal{S}$ or the *average power density* can be derived as

$$\mathbf{S}_{\text{av}} = \tfrac{1}{2}\,\text{Re}(\mathbf{E} \times \mathbf{H}^*) \tag{2.20}$$

where Re$(\cdot)$ stands for the real part of a complex quantity. This power vector not only gives the magnitude of the power flow but also its direction. The direction of power flow or wave propagation is determined by the right-hand rule of the cross product and is always perpendicular to both $\mathbf{E}$ and $\mathbf{H}$. The total average power crossing a surface S is then given as

$$\mathbf{P}_{\text{av}} = \tfrac{1}{2}\,\text{Re} \iint_S \mathbf{E} \times \mathbf{H} \cdot d\mathbf{S} \tag{2.21}$$

2.8 POISSON'S AND LAPLACE'S EQUATIONS

Under the static assumption, the voltage $V(x, y, z)$ at any location of a structure having an electric charge density $\rho(x, y, z)$ is governed by *Poisson's equation,*

$$\nabla^2 V - \frac{\rho}{\varepsilon} \tag{2.22}$$

Note that ρ acts as a source producing the fields. When there is no charge, Poisson's equation reduces to

$$\nabla^2 V = 0 \tag{2.23}$$

known as *Laplace's equation*. Laplace's equation is frequently employed to determine the *static* or *quasi-static characteristic impedance* and *effective relative dielectric constant* of a transmission line. These static parameters are easier to obtain than their dynamic counterparts but are only valid at dc. In practice, however, many RF and microwave engineers use the static parameters for the analysis and design of microwave circuits even when the operating frequencies are high, perhaps more than 18 GHz or so, and still obtain good results.

2.9 WAVE EQUATIONS

Electromagnetic fields may be determined by using Maxwell's equations and constitutive relations directly. However, the most convenient way of obtaining these fields is solving a special class of equations known as the *wave equations*. We shall derive these equations as follows.

We consider a medium that is source free ($\rho = J = 0$) and simple (homogeneous, isotropic, and linear) and assume that the fields are time harmonic. The medium is characterized by a dielectric constant ε and permeability μ. Taking the curl of Maxwell's Eq. (2.5a) and making use of Maxwell's Eq. (2.5b) and constitutive relations (2.7) and (2.8) yields

$$\nabla \times \nabla \times \mathbf{E} - k^2\mathbf{E} = 0 \tag{2.24}$$

where $k = \omega\sqrt{\varepsilon\mu}$ is the *wave number*. Using the vector identity

$$\nabla \times \nabla \times \mathbf{A} = \nabla(\nabla \cdot \mathbf{A}) - \nabla^2\mathbf{A} \tag{2.25}$$

where $\mathbf{A}$ is an arbitrary vector, we can then rewrite Eq. (2.24) as

$$\nabla^2\mathbf{E} + k^2\mathbf{E} = 0 \tag{2.26}$$

where ∇^2 denotes the *Laplacian operator*. This equation is called the *wave equation for the electric field*. Similarly, the *wave equation for the magnetic field* can be derived as

$$\nabla^2\mathbf{H} + k^2\mathbf{H} = 0 \tag{2.27}$$

Both of these wave equations are also known as *Helmholtz equations*.

Other commonly used wave equations are those in the plane transverse to the direction of wave propagation. Let's assume that the direction of propagation is z. We separate the operator ∇ into the transverse, ∇_t, and longitudinal, ∇_z, components as

$$\nabla = \nabla_t + \nabla_z \tag{2.28}$$

where

$$\nabla_t = \mathbf{a}_x \frac{\partial}{\partial_x} + \mathbf{a}_y \frac{\partial}{\partial_y} \tag{2.29}$$

for the rectangular coordinates, and

$$\nabla_z = \mathbf{a}_z \frac{\partial}{\partial_z} = \pm\gamma \mathbf{a}_z \tag{2.30}$$

γ is the propagation constant with the $\pm$ signs denoting the $\mp z$ propagating directions, respectively. The Laplacian operator ∇^2 can be written as

$$\nabla^2 = \nabla_t^2 + \gamma^2 \tag{2.31}$$

where ∇_t^2 represents the *transverse* (to the z-axis) *Laplacian operator*. Substituting Eq. (2.31) into (2.26) and (2.27) we obtain

$$\nabla_t^2 \mathbf{E}(x, y) + k_c^2 \mathbf{E}(x, y) = 0 \tag{2.32a}$$

$$\nabla_t^2 \mathbf{H}(x, y) + k_c^2 \mathbf{H}(x, y) = 0 \tag{2.32b}$$

where

$$k_c^2 = \omega_c^2 \varepsilon\mu = k^2 + \gamma^2 \tag{2.33}$$

k_c and ω_c are referred to as the *cutoff wave number* and *cutoff frequency*, respectively, due to the fact that they reduce to the corresponding parameters at the cutoff ($\gamma = 0$). Equation (2.32) is known as the *wave* or *Helmholtz equation in the transverse plane*, with z as the direction of propagation.

2.10 ELECTRIC AND MAGNETIC POTENTIALS

The fields in a microwave structure, in general, and in a transmission line, in particular, may be determined by directly solving the wave equations subject to appropriate boundary conditions. In practice, however, these fields are normally obtained via intermediate terms known as *electric* and *magnetic vector* or *scalar potentials* to simplify the mathematical analysis. These potentials are also solutions of the wave equations. In this section, we will derive these parameters for a source-free medium and their corresponding wave equations under the assumption of time-harmonic fields.

From Maxwell's Eq. (2.5c), we can describe the electric field as

$$\mathbf{E} = -j\omega\mu\nabla \times \boldsymbol{\psi}^h \tag{2.34}$$

where $\psi^h(x, y, z)$ is a vector defined as the *magnetic vector potential*. Substituting Eq. (2.34) into (2.5b) yields

$$\nabla \times \mathbf{H} = k^2 \nabla \times \psi^h \tag{2.35}$$

from which we can obtain

$$\mathbf{H} = k^2 \psi^h + \nabla \Phi \tag{2.36}$$

where Φ is an arbitrary scalar function. Now substituting Eq. (2.34) into the left-hand side of (2.5a), we get

$$\nabla \times \mathbf{E} = -j\omega\mu \left[\nabla(\nabla \cdot \psi^h) - \nabla^2 \psi^h\right] \tag{2.37}$$

The right-hand side of Eq. (2.5a) becomes, after replacing $\mathbf{H}$ by Eq. (2.36),

$$-j\omega\mu\mathbf{H} = -j\omega\mu(k^2 \psi^h + \nabla \Phi) \tag{2.38}$$

Equating Eqs. (2.37) and (2.38) gives

$$\nabla(\nabla \cdot \psi^h) - \nabla^2 \psi^h = k^2 \psi^h + \nabla \Phi \tag{2.39}$$

Choosing

$$\nabla \cdot \psi^h = \Phi \tag{2.40}$$

according to the *Lorentz* condition then yields

$$\nabla^2 \psi^h + k^2 \psi^h = 0 \tag{2.41}$$

which is also known as the *wave* or *Helmholtz equation for the magnetic vector potential* ψ^h. Once solving for ψ^h from Eq. (2.41), we can determine the electric and magnetic fields from Eqs. (2.34), (2.36), and (2.40) as

$$\mathbf{E} = -j\omega\mu \nabla \times \psi^h \tag{2.42a}$$

$$\mathbf{H} = k^2 \psi^h + \nabla(\nabla \cdot \psi^h) \tag{2.42b}$$

Following the same approach, we can also derive the following *wave* or *Helmholtz equation for the electric vector potential* $\psi^e(x, y, z)$ as

$$\nabla^2 \psi^e + k^2 \psi^e = 0 \tag{2.43}$$

whose solution can be used to determine the magnetic and electric fields as

$$\mathbf{E} = k^2 \psi^e + \nabla(\nabla \cdot \psi^e) \tag{2.44a}$$

$$\mathbf{H} = j\omega\varepsilon \nabla \times \psi^e \tag{2.44b}$$

A remark needs to be made at this point. By setting the frequency to zero, we reduce the wave Eqs. (2.41) and (2.43) to the familiar Laplace's Eq. (2.23) used under the static condition. Note that both $\psi^e(x, y, z)$ and $\psi^h(x, y, z)$ are now identical to $V(x, y, z)$.

We can also write the wave equations for the field vector potentials in the plane transverse to the propagating direction. For instance, let z be the direction of wave propagation, we separate the vector potentials $\boldsymbol{\psi}^h(x, y, z)$ and $\boldsymbol{\psi}^e(x, y, z)$ into the transverse and longitudinal components as

$$\begin{aligned} \boldsymbol{\psi}^h &= \boldsymbol{\psi}_t^h + \boldsymbol{\psi}_z^h \\ &= \boldsymbol{\psi}_t^h(x, y)e^{\pm\gamma_h z} + \mathbf{a}_z \psi^h(x, y)e^{\pm\gamma_h z} \end{aligned} \tag{2.45a}$$

$$\begin{aligned} \boldsymbol{\psi}^e &= \boldsymbol{\psi}_t^e + \boldsymbol{\psi}_z^e \\ &= \boldsymbol{\psi}_t^e(x, y)e^{\pm\gamma_e z} + \mathbf{a}_z \psi^e(x, y)e^{\pm\gamma_e z} \end{aligned} \tag{2.45b}$$

where γ_h and γ_e are the corresponding propagation constants, with the $\pm$ signs indicating the $\mp z$-directions of propagation, respectively. Using these equations in (2.41) and (2.43), we obtain, after decomposing the Laplacian operator ∇^2 into the transverse and longitudinal components,

$$\nabla_t^2 \boldsymbol{\psi}_t^h(x, y) + k_{c,h}^2 \boldsymbol{\psi}_t^h(x, y) = 0 \tag{2.46a}$$

$$\nabla_t^2 \psi^h(x, y) + k_{c,h}^2 \psi^h(x, y) = 0 \tag{2.46b}$$

$$\nabla_t^2 \boldsymbol{\psi}_t^e(x, y) + k_{c,e}^2 \boldsymbol{\psi}_t^e(x, y) = 0 \tag{2.46c}$$

$$\nabla_t^2 \psi^e(x, y) + k_{c,e}^2 \psi^e(x, y) = 0 \tag{2.46d}$$

where

$$k_{c,h}^2 = k^2 + \gamma_h^2 \tag{2.47a}$$

$$k_{c,e}^2 = k^2 + \gamma_e^2 \tag{2.47b}$$

$\boldsymbol{\psi}_t^h(x, y)$ and $\boldsymbol{\psi}_t^e(x, y)$ are the *transverse magnetic* and *electric vector potentials*, respectively. $\psi^h(x, y)$ and $\psi^e(x, y)$ are the longitudinal components of the magnetic and electric vector potentials, respectively, and are referred to as the *magnetic* and *electric scalar potentials*. Note that, as for the case of the magnetic and electric fields in the transverse wave Eqs. (2.32), all these potentials are a function of x and y only.

2.11 WAVE TYPES AND SOLUTIONS

2.11.1 Wave Types

The two most commonly used waveguides† are the *rectangular waveguide* and the *transmission line*. Waves propagating along different waveguides possess

† We define a waveguide as any structure that guides waves.

different electromagnetic field distributions. It is these field distributions that dictate the nature of the waveguides. With reference to a particular wave-propagating direction, we can classify different *wave* (or *mode*) types based on possible combinations of the electric and magnetic fields in that direction. These are, assuming z-direction of wave propagation:

Transverse Electric (TE) Wave or Mode: This wave has the electric field only in the plane transverse to the direction of propagation. That is, the longitudinal components $E_z = 0$ and $H_z \neq 0$.

Transverse Magnetic (TM) Wave or Mode: This wave has only the magnetic field in the transverse plane. That is, $H_z = 0$ and $E_z \neq 0$.

Hybrid Wave or Mode: This wave is characterized as having both $E_z \neq 0$ and $H_z \neq 0$ and, therefore, is a combination of both TE and TM waves.

Transverse Electromagnetic (TEM) Wave or Mode‡: The electric and magnetic fields of this wave have only transverse components. Both E_z and H_z are equal to zero.

Note that, in general, different modes have different cutoff frequencies or cutoff wave numbers and hence different propagation constants. A rectangular waveguide is a special waveguide that has the same cutoff frequency for the corresponding TE_{mn} and TM_{mn} modes. When the modes have the same cutoff frequency, they are classified as *degenerate modes*. Otherwise, they are said to be *nondegenerate modes*.

2.11.2 Wave Solutions

Assuming $\boldsymbol{\psi}^h(x, y, z)$ and $\boldsymbol{\psi}^e(x, y, z)$ have only longitudinal z components such as

$$\boldsymbol{\psi}^h = \mathbf{a}_z \psi^h(x, y) e^{\pm\gamma_h z} \tag{2.48a}$$

$$\boldsymbol{\psi}^e = \mathbf{a}_z \psi^e(x, y) e^{\pm\gamma_e z} \tag{2.48b}$$

we can prove easily that the corresponding longitudinal electric and magnetic fields are equal to zero. This implies that the magnetic, $\psi^h(x, y)$, and electric, $\psi^e(x, y)$, scalar potentials may be used to determine the fields for TE and TM modes, respectively. This result will be used to derive fields for the four principal classes of modes discussed earlier.

Using the principle of superposition, we can express the fields in any waveguide as a summation of those of TE and TM modes. These fields are given, making use of Eqs. (2.42) and (2.44), as

$$\begin{aligned}\mathbf{E} &= \mathbf{E}_{\text{TM}} + \mathbf{E}_{\text{TE}} \\ &= k^2\boldsymbol{\psi}^e + \nabla(\nabla \cdot \boldsymbol{\psi}^e) - j\omega\mu\nabla \times \boldsymbol{\psi}^h \end{aligned} \tag{2.49a}$$

‡ For TEM mode to exist exactly on a transmission line, all conductors must be perfect.

$$\begin{aligned}\mathbf{H} &= \mathbf{H}_{\text{TM}} + \mathbf{H}_{\text{TE}} \\ &= j\omega\varepsilon\nabla \times \boldsymbol{\psi}^e + k^2\boldsymbol{\psi}^h + \nabla(\nabla \cdot \boldsymbol{\psi}^h)\end{aligned} \tag{2.49b}$$

where the subscripts TE and TM indicate the TE and TM modes, respectively. Substituting Eq. (2.48) into (2.49a) and replacing the ∇ operator by its transverse and longitudinal components yields

$$\mathbf{E} = \mathbf{a}_z k^2 \psi^e e^{\pm\gamma_e z} + \left(\nabla_t + \mathbf{a}_z \frac{\partial}{\partial z}\right)\left(\pm\gamma_e \psi^e e^{\pm\gamma_e z}\right) - j\omega\mu\left(\nabla_t + \mathbf{a}_z \frac{\partial}{\partial z}\right) \times \left(\psi^h e^{\pm\gamma_h z}\mathbf{a}_z\right) \tag{2.50}$$

from which, we obtain the z and transverse components of the electric field as

$$E_z = k_{c,e}^2 \psi^e e^{\pm\gamma_e z} \tag{2.51a}$$

$$\mathbf{E}_t = \pm\gamma_e e^{\pm\gamma_e z}\nabla_t \psi^e + j\omega\mu e^{\pm\gamma_h z}\mathbf{a}_z \times \nabla_t \psi^h \tag{2.51b}$$

respectively. Note that, for degenerate modes, $\gamma_e = \gamma_h$. Equation (2.51b) can then be used to derive individual x and y components of the electric field as

$$E_x = \pm\gamma_e e^{\pm\gamma_e z}\frac{\partial\psi^e}{\partial x} - j\omega\mu e^{\pm\gamma_h z}\frac{\partial\psi^h}{\partial y} \tag{2.52a}$$

$$E_y = \pm\gamma_e e^{\pm\gamma_e z}\frac{\partial\psi^e}{\partial y} + j\omega\mu e^{\pm\gamma_h z}\frac{\partial\psi^h}{\partial x} \tag{2.52b}$$

Similarly, by expanding Eq. (2.49b), we can express the longitudinal and transverse magnetic-field components as

$$H_z = k_{c,h}^2 \psi^h e^{\pm\gamma_h z} \tag{2.53a}$$

$$\mathbf{H}_t = \pm\gamma_h e^{\pm\gamma_h z}\nabla_t \psi^h - j\omega\varepsilon e^{\pm\gamma_e z}\mathbf{a}_z \times \nabla_t \psi^e \tag{2.53b}$$

Further expanding Eq. (2.53b) then yields

$$H_x = \pm\gamma_h e^{\pm\gamma_h z}\frac{\partial\psi^h}{\partial x} + j\omega\varepsilon e^{\pm\gamma_e z}\frac{\partial\psi^e}{\partial y} \tag{2.54a}$$

$$H_y = \pm\gamma_h e^{\pm\gamma_h z}\frac{\partial\psi^h}{\partial y} - j\omega\varepsilon e^{\pm\gamma_e z}\frac{\partial\psi^e}{\partial x} \tag{2.54b}$$

We can also determine the x and y components from the z components of the fields by using Maxwell's equations directly. We can easily prove that both E_z and H_z cannot be even or odd with respect to x and y simultaneously. The results derived so far are very general and so are applicable to any possible modes existing in any waveguide. We will now use them to obtain solutions for TE, TM, hybrid, and TEM modes.

TE Modes The TE modes, as indicated earlier, correspond only to the magnetic scalar potential $\psi^h(x, y)$. Therefore, to determine the TE fields, we simply let $\psi^e(x, y)$ equal zero in the foregoing general equations. From Eqs. (2.51) and (2.53), we then have

$$E_z = 0 \tag{2.55a}$$

$$\mathbf{E}_t = j\omega\mu e^{\pm\gamma_h z}\mathbf{a}_z \times \nabla_t \psi^h \tag{2.55b}$$

$$H_z = k_{c,h}^2 \psi^h e^{\pm\gamma_h z} \tag{2.55c}$$

$$\mathbf{H}_t = \pm\gamma_h e^{\pm\gamma_h z}\nabla_t \psi^h \tag{2.55d}$$

E_t and H_t are related by

$$\mathbf{E}_t = \pm Z_h \mathbf{a}_z \times \mathbf{H}_t \tag{2.56}$$

where $Z_h = j\omega\mu/\gamma_h$ is known as the *characteristic wave impedance of the TE mode*. Z_h is real and inductive for propagating and evanescent TE modes, respectively. This impedance is defined as a ratio between the transverse electric and magnetic components,

$$Z_h = \frac{E_x}{H_y} = -\frac{E_y}{H_x} \tag{2.57}$$

for the $+z$ direction, or

$$Z_h = -\frac{E_x}{H_y} = \frac{E_y}{H_x} \tag{2.58}$$

for the $-z$ direction. The individual transverse components of the TE fields may be obtained from Eqs. (2.52) and (2.54) or Eqs. (2.55b) and (2.55d), as

$$E_x = -j\omega\mu e^{\pm\gamma_h z}\frac{\partial\psi^h}{\partial y} \tag{2.59a}$$

$$E_y = j\omega\mu e^{\pm\gamma_h z}\frac{\partial\psi^h}{\partial x} \tag{2.59b}$$

$$H_x = \pm\gamma_h e^{\pm\gamma_h z}\frac{\partial\psi^h}{\partial x} \tag{2.59c}$$

$$H_y = \pm\gamma_h e^{\pm\gamma_h z}\frac{\partial\psi^h}{\partial y} \tag{2.59d}$$

Note that ψ^h is obtained by solving Eq. (2.46b), which is repeated here for completeness:

$$\nabla_t^2 \psi^h(x, y) + k_{c,h}^2 \psi^h(x, y) = 0 \tag{2.60}$$

TM Modes The TM modes are determined solely from the electric scalar potential $\psi^e(x, y)$. Their longitudinal and transverse fields can thus be determined by letting $\psi^h(x, y)$ equal zero in Eqs. (2.51) and (2.53). This gives

$$\mathbf{H}_z = 0 \tag{2.61a}$$

$$\mathbf{H}_t = -j\omega\varepsilon e^{\pm\gamma_e z}\mathbf{a}_z \times \nabla_t \psi^e \tag{2.61b}$$

$$E_z = k_{c,e}^2 \psi^e e^{\pm\gamma_e z} \tag{2.61c}$$

$$\mathbf{E}_t = \pm\gamma_e e^{\pm\gamma_e z}\nabla_t \psi^e \tag{2.61d}$$

E_t and H_t are related by

$$\mathbf{H}_t = \mp\frac{\mathbf{a}_z \times \mathbf{E}_t}{Z_e} \tag{2.62}$$

where $Z_e = j\omega\varepsilon/\gamma_e$ is called the *characteristic wave impedance of the TM mode.* This impedance is real and capacitive for propagating and evanescent TM modes, respectively. The transverse components of the TM fields can be expressed, from Eqs. (2.52) and (2.54) or Eqs. (2.61b) and (2.61d), as

$$E_x = \pm\gamma_e e^{\pm\gamma_e z}\frac{\partial\psi^e}{\partial x} \tag{2.63a}$$

$$E_y = \pm\gamma_e e^{\pm\gamma_e z}\frac{\partial\psi^e}{\partial y} \tag{2.63b}$$

$$H_x = j\omega\varepsilon e^{\pm\gamma_e z}\frac{\partial\psi^e}{\partial y} \tag{2.63c}$$

$$H_y = -j\omega\varepsilon e^{\pm\gamma_e z}\frac{\partial\psi^e}{\partial x} \tag{2.63d}$$

ψ^e is the solution of Eq. (2.46d) and is given again below:

$$\nabla_t^2 \psi^e(x, y) + k_{c,e}^2 \psi^e(x, y) = 0 \tag{2.64}$$

Hybrid Modes A hybrid mode is a combination of both TE and TM modes. The general results Eqs. (2.51)–(2.54), derived earlier can therefore be used directly to determine the fields of the hybrid modes.

TEM Modes Solution for the TEM mode can be viewed as a special solution of either the TE or TM mode when H_z or E_z is set to zero, respectively. For instance, we consider the TE mode and let H_z in Eq. (2.55c) equal zero. This leads to $k_{c,h} = 0$ and, consequently, Eq. (2.46b) becomes

$$\nabla_t^2 \psi^h(x, y) = 0 \tag{2.65}$$

which is basically Laplace's equation in the transverse plane. The transverse fields can be obtained from Eqs. (2.55c) and (2.55d) as

$$\mathbf{E}_t = j\omega\mu e^{\pm\gamma z}\mathbf{a}_z \times \nabla_t \psi^h \tag{2.66a}$$

$$\mathbf{H}_t = \pm\gamma e^{\pm\gamma z}\nabla_t \psi^h \tag{2.66b}$$

where γ is the TEM mode's propagation constant. Proceeding with the TM mode also gives identical results. It should be noted that for the TEM case, both $\psi^e(x, y)$ and $\psi^h(x, y)$ are equal to the two-dimensional voltage or potential $V(x, y)$.

2.12 ORTHOGONALITY RELATIONS

Consider a general waveguide as shown in Fig. 2.3 and assume that it has perfect conducting walls. In general, there exists an infinite number of modes in the structure. To signify different modes, we will use the subscripts m and n here, with their possible values from 0 to infinity. As indicated in Section 2.11.2, the TE_{mn}, TM_{mn}, and hybrid modes correspond to $\psi^h_{mn}(x, y)$, $\psi^e_{mn}(x, y)$, and both of these potentials, respectively. Following the approach described in Collin [2], we can derive the *orthogonality relations* between $\psi^h_{mn}(x, y)$ and $\psi^e_{mn}(x, y)$, between $E_{mn}(x, y)$, between $H_{mn}(x, y)$, and between $E_{mn}(x, y)$ and $H_{mn}(x, y)$.

2.12.1 Orthogonality Relations Between $\psi^h_{mn}(x, y)$ and Between $\psi^e_{mn}(x, y)$

Let us consider two different TE or TM modes characterized by (m, n) and (k, l). The corresponding scalar potentials $\psi^i_{mn}(x, y)$ or $\psi^i_{kl}(x, y)$, with i being h or e,

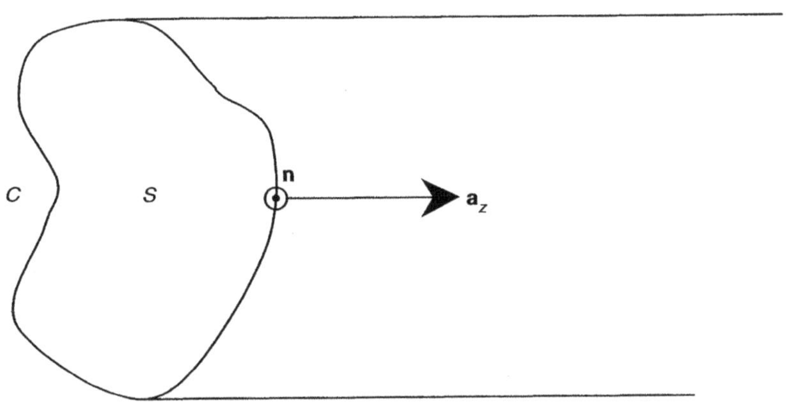

Figure 2.3 A waveguide of arbitrary shape. **n** is a unit vector perpendicular to the wall and pointing outward; **t** is a unit vector tangential to the wall; and $\mathbf{a}_z$ is a unit vector along the waveguide. **n**, **t**, and $\mathbf{a}_z$ form an orthogonal coordinate system.

satisfy the following wave equations:

$$\nabla_t^2 \psi_{mn}^i + k_{c,i,mn}^2 \psi_{mn}^i = 0 \tag{2.67a}$$

$$\nabla_t^2 \psi_{kl}^i + k_{c,i,kl}^2 \psi_{kl}^i = 0 \tag{2.67b}$$

where $k_{c,i,mn}$ and $k_{c,i,kl}$ are the corresponding cutoff wave numbers. We multiply Eqs. (2.67a) and (2.67b) with ψ_{kl}^i and ψ_{mn}^i, respectively, and subtract the resulting equations to obtain

$$\psi_{kl}^i \nabla_t^2 \psi_{mn}^i - \psi_{mn}^i \nabla_t^2 \psi_{kl}^i = \left(k_{c,i,kl}^2 - k_{c,i,mn}^2\right) \psi_{mn}^i \psi_{kl}^i \tag{2.68}$$

Taking the surface integral and using Green's second identity in two dimensions,

$$\iint_S \left(u \nabla_t^2 v - v \nabla_t^2 u\right) dS = \oint_C \left(u \frac{\partial u}{\partial n} - v \frac{\partial v}{\partial n}\right) dl \tag{2.69}$$

where u and v are arbitrary scalar functions, S is the surface, and C is the closed contour bounding that surface, we obtain

$$\left(k_{c,i,kl}^2 - k_{c,i,mn}^2\right) \iint_S \psi_{kl}^i \psi_{mn}^i \, dS = \oint_C \left(\psi_{kl}^i \frac{\partial \psi_{mn}^i}{\partial n} - \psi_{mn}^i \frac{\partial \psi_{kl}^i}{\partial n}\right) dl \tag{2.70}$$

Note that $\partial/\partial n$ denotes the derivative with respect to the normal direction $\mathbf{n}$. Along the perfectly conducting walls of the waveguide, the scalar potentials must satisfy the following *Neumann's* and *Dirichlet's conditions*:

Neumann's Condition: $$\frac{\partial \psi_j^h}{\partial n} = 0 \tag{2.71a}$$

Dirichlet's Condition: $$\psi_j^e = 0 \tag{2.71b}$$

where $j = mn$ or kl. Imposing these conditions on Eq. (2.70) gives

$$\left(k_{c,i,kl}^2 - k_{c,i,mn}^2\right) \iint_S \psi_{kl}^i \psi_{mn}^i \, dS = 0 \tag{2.72}$$

which implies that

$$\iint_S \psi_{mn}^e \psi_{kl}^e \, dS = 0, \qquad m \neq k \quad \text{or} \quad n \neq l \tag{2.73a}$$

$$\iint_S \psi_{mn}^h \psi_{kl}^h \, dS = 0, \qquad m \neq k \quad \text{or} \quad n \neq l \tag{2.73b}$$

for two different TM and TE modes, respectively, provided that they have different cutoff wave numbers; that is, they are nondegenerate modes. The relationship described by Eq. (2.73) is known as the *orthogonality condition between the scalar potentials*. It states that the scalar electric or magnetic potentials of two nondegenerate modes are always orthogonal to each other. It should be noted that, due to Neumann's and Dirichlet's conditions, this orthogonality only holds for waveguides with perfectly conducting walls.

When the modes are degenerate, Eq. (2.73) may not be satisfied, and so the scalar electric or magnetic potentials of two degenerate modes may not be orthogonal. Using a procedure analogous to the *Gram–Schmidt process* [3], however, we can construct a new set of mutually orthogonal modes, each of which is a linear combination of certain modes of the nonorthogonal degenerate modes. This process is described as follows. Let Ψ^i_{mn} and Ψ^i_{kl} be the two new modes defined by

$$\Psi^i_{mn} = \psi^i_{mn} \tag{2.74a}$$

$$\Psi^i_{kl} = \psi^i_{kl} - C\psi^i_{mn} \tag{2.74b}$$

where C is a constant. These new functions are required to be mutually orthogonal; that is,

$$\iint_S \Psi^i_{mn}\Psi^i_{kl} = 0 \tag{2.75}$$

which then leads to

$$C = \frac{\displaystyle\iint_S \psi^i_{mn}\psi^i_{kl}\,dS}{\displaystyle\iint_S \left(\psi^i_{mn}\right)^2\,dS} \tag{2.76}$$

provided that the integral $\iint (\psi^i_{mn})^2\,dS$ exists and is nonzero. This process can be applied to more than two nonorthogonal modes to determine a new set of modes that are mutually orthogonal.

From Eq. (2.73) and the fact that the norm $\|\psi^i_{mn}\| \neq 0$ for any combination of (m, n), the scalar potentials ψ^i_{mn} represent an orthogonal set and thus are linearly independent. Therefore, a field function at any location in a waveguide can be expressed as a summation of the scalar potentials of all possible modes as

$$\prod(x, y) = \sum_{i=e,h}\sum_m\sum_n C_{mn}\psi^i_{mn}(x, y) \tag{2.77}$$

where C_{mn} are called the *orthogonal coefficients* and may be computed by

$$C_{mn} = \frac{\displaystyle\iint_S \prod \psi^i_{mn}\,dS}{\displaystyle\iint_S \left(\psi^i_{mn}\right)^2\,dS} \tag{2.78}$$

Equation (2.77) implies that there is always a unique solution for the scalar potentials and, hence, electromagnetic fields for waveguides having perfect conductors.

2.12.2 Orthogonality Relations Between Electric Fields and Between Magnetic Fields

Let's consider two different TM_{mn} and TM_{kl} modes. The surface integral of the dot product of the transverse electric fields is given, using Eq. (2.61d), as

$$\iint_S \mathbf{E}_{t,mn}^e \cdot \mathbf{E}_{t,kl}^e \, dS = \pm \gamma_{e,mn} \gamma_{e,kl} e^{(\gamma_{e,mn} \pm \gamma_{e,kl})z} \iint_S \nabla_t \psi_{mn}^e \cdot \nabla_t \psi_{kl}^e \, dS \tag{2.79}$$

Applying the two-dimensional Green's first identity,

$$\iint_S \left(\nabla_t u \cdot \nabla_t v + u \nabla_t^2 v \right) dS = \oint_C u \frac{\partial v}{\partial n} dl \tag{2.80}$$

where u and v are arbitrary functions, and the Dirichlet condition (2.71b), we can rewrite Eq. (2.79) as

$$\iint_S \mathbf{E}_{t,mn}^e \cdot \mathbf{E}_{t,kl}^e \, dS = \mp \gamma_{e,mn} \gamma_{e,kl} e^{(\pm \gamma_{e,mn} \pm \gamma_{e,kl})z} \iint_S \psi_{mn}^e \nabla_t^2 \psi_{kl}^e \, dS \tag{2.81}$$

Making use of the wave equation for ψ_{kl}^e,

$$\nabla_t^2 \psi_{kl}^e + k_{c,kl}^2 \psi_{kl}^e = 0 \tag{2.82}$$

and the fact that ψ_{mn}^e and ψ_{kl}^e, with $m \neq n$ and $k \neq l$, are mutually orthogonal, we finally obtain

$$\iint_S \mathbf{E}_{t,mn}^e \cdot \mathbf{E}_{t,kl}^e \, dS = 0 \tag{2.83}$$

This result indicates that the transverse electric fields of two different TM modes are always orthogonal to each other. Following the same approach, we can also derive the other orthogonality relationships for the transverse electromagnetic fields of TM, TE, and hybrid modes. All the orthogonality relationships are given as follows.

TM Modes

$$\iint_S \mathbf{E}_{t,mn}^e \cdot \mathbf{E}_{t,kl}^e \, dS = 0 \tag{2.84a}$$

$$\iint_S \mathbf{H}_{t,mn}^e \cdot \mathbf{H}_{t,kl}^e \, dS = 0 \tag{2.84b}$$

TE Modes

$$\iint_S \mathbf{E}^h_{t,mn} \cdot \mathbf{E}^h_{t,kl}\, dS = 0 \tag{2.85a}$$

$$\iint_S \mathbf{H}^h_{t,mn} \cdot \mathbf{H}^h_{t,kl}\, dS = 0 \tag{2.85b}$$

Hybrid Modes

$$\iint_S \mathbf{E}^e_{t,mn} \cdot \mathbf{E}^h_{t,kl}\, dS = 0 \tag{2.86a}$$

$$\iint_S \mathbf{H}^e_{t,mn} \cdot \mathbf{H}^h_{t,kl}\, dS = 0 \tag{2.86b}$$

Note that Eq. (2.86) is valid even when $m = k$ and $n = l$. Making use of the characteristics of an orthogonal system, we can then express the transverse fields at any location in a waveguide as a summation of the transverse fields of all possible modes as

$$\mathbf{E}_t = \sum_m \sum_n (C^e_{mn}\mathbf{E}^e_{t,mn} + C^h_{mn}\mathbf{E}^h_{t,mn}) \tag{2.87a}$$

$$\mathbf{H}_t = \sum_m \sum_n (D^e_{mn}\mathbf{H}^e_{t,mn} + D^h_{mn}\mathbf{H}^h_{t,mn}) \tag{2.87b}$$

where C^e_{mn}, C^h_{mn}, D^e_{mn}, and D^h_{mn} are the orthogonal coefficients. As for the case of the scalar potentials, when the modes are degenerate and nonorthogonal, we can construct a new set of transverse fields, each of which is a linear combination of certain fields of the nonorthogonal degenerate modes, such that the new fields are mutually orthogonal.

2.12.3 Orthogonality Relations Between Electric and Magnetic Fields

Let's consider again two different TM_{mn} and TM_{kl} modes. Assuming z is the direction of propagation, the fields of these modes can be expressed as a sum of the transverse and longitudinal fields. For instance, the fields of the TM_{mn} mode are given as

$$\begin{aligned} \mathbf{E}^e_{mn}(x, y, z) &= \mathbf{E}^e_{t,mn}(x, y, z) + \mathbf{E}^e_{z,mn}(x, y, z) \\ &= \mathbf{e}^e_{mn}(x, y)e^{-\gamma_{e,mn}z} + \mathbf{e}^e_{z,mn}(x, y)e^{-\gamma_{e,mn}z} \end{aligned} \tag{2.88a}$$

$$\begin{aligned} \mathbf{H}^e_{mn}(x, y, z) &= \mathbf{H}^e_{t,mn}(x, y, z) + \mathbf{H}^e_{z,mn}(x, y, z) \\ &= \mathbf{h}^e_{mn}(x, y)e^{-\gamma_{e,mn}z} + \mathbf{h}^e_{z,mn}(x, y)e^{-\gamma_{e,mn}z} \end{aligned} \tag{2.88b}$$

Note that we have introduced the notation **e**, **h** and $\mathbf{e}_z$, $\mathbf{h}_z$ to signify the two-dimensional transverse and longitudinal components, respectively. The TM fields satisfy the Maxwell equations,

$$\nabla \times \mathbf{E}_j^e = -j\omega\mu\mathbf{H}_j^e \tag{2.89a}$$

$$\nabla \times \mathbf{H}_j^e = j\omega\varepsilon\mathbf{E}_j^e \tag{2.89b}$$

where $j = mn, kl$. Making use of Eq. (2.89), we can write the following equation:

$$\begin{aligned}\nabla \cdot \left(\mathbf{E}_{mn}^e \times \mathbf{H}_{kl}^e - \mathbf{E}_{kl}^e \times \mathbf{H}_{mn}^e\right) &= j\omega\mu\left(\mathbf{H}_{kl}^e \cdot \mathbf{H}_{mn}^e - \mathbf{H}_{kl}^e \cdot \mathbf{H}_{mn}^e\right) \\ &+ j\omega\varepsilon\left(\mathbf{E}_{kl}^e \cdot \mathbf{E}_{mn}^e - \mathbf{E}_{kl}^e \cdot \mathbf{E}_{mn}^e\right) = 0\end{aligned} \tag{2.90}$$

Separating the ∇ operator in Eq. (2.90) into the transverse and longitudinal parts gives

$$\begin{aligned}\nabla \cdot \left(\mathbf{E}_{mn}^e \times \mathbf{H}_{kl}^e - \mathbf{E}_{kl}^e \times \mathbf{H}_{mn}^e\right) &= \nabla_t \cdot \left(\mathbf{E}_{mn}^e \times \mathbf{H}_{kl}^e - \mathbf{E}_{kl}^e \times \mathbf{H}_{mn}^e\right) \\ &+ \mathbf{a}_z \frac{\partial}{\partial z} \cdot \left(\mathbf{E}_{mn}^e \times \mathbf{H}_{kl}^e - \mathbf{E}_{kl}^e \times \mathbf{H}_{mn}^e\right) = 0\end{aligned} \tag{2.91}$$

Now taking the surface integral of Eq. (2.91) and applying the divergence theorem in two dimensions,

$$\iint_S \nabla_t \cdot \mathbf{A}\, dS = \oint_C \mathbf{n} \cdot \mathbf{A}\, dl \tag{2.92}$$

where **A** is an arbitrary vector, we obtain

$$\begin{aligned}\iint_S \nabla_t \cdot \left(\mathbf{E}_{mn}^e \times \mathbf{H}_{kl}^e - \mathbf{E}_{kl}^e \times \mathbf{H}_{mn}^e\right) dS &= \oint_C (\mathbf{n} \times \mathbf{E}_{mn} \cdot \mathbf{H}_{kl} - \mathbf{n} \times \mathbf{E}_{kl} \cdot \mathbf{H}_{mn})\, dl \\ &= 0\end{aligned} \tag{2.93}$$

since

$$\mathbf{n} \times \mathbf{E}_j^e = 0; \qquad j = mn, kl \tag{2.94}$$

on perfectly conducting walls. Substituting Eq. (2.93) into (2.91), we have

$$\iint_S \mathbf{a}_z \cdot \frac{\partial}{\partial z}\left(\mathbf{E}_{mn}^e \times \mathbf{H}_{kl}^e - \mathbf{E}_{kl}^e \times \mathbf{H}_{mn}^e\right) dS = 0 \tag{2.95}$$

Substituting Eq. (2.88) into (2.95) and taking the derivatives leads to

$$(\gamma_{e,mn} + \gamma_{e,kl}) \iint_S \mathbf{a}_z \cdot \left(\mathbf{e}_{mn}^e \mathbf{h}_{kl}^e - \mathbf{e}_{kl}^e \times \mathbf{h}_{mn}^e\right) dS = 0 \tag{2.96}$$

Note that this equation is obtained assuming both TM_{mn} and TM_{kl} modes correspond to the $+z$ direction of propagation. If we now assume that the TM_{mn} mode still corresponds to the $+z$ propagation direction, but the TM_{kl} mode corresponds to the $-z$ direction, then we can write E^e_{kl} and H^e_{kl} as

$$\mathbf{E}^e_{kl} = \mathbf{e}^e_{kl} e^{\gamma_{e,kl} z} - \mathbf{e}^e_{z,kl} e^{\gamma_{e,kl} z} \tag{2.97a}$$

$$\mathbf{H}^e_{kl} = -\mathbf{h}^e_{kl} e^{\gamma_{e,kl} z} - \mathbf{h}^e_{z,kl} e^{\gamma_{e,kl} z} \tag{2.97b}$$

which implies that we can derive the following relation by simply changing h_{kl} to $-h_{kl}$ in Eq. (2.96):

$$(\gamma_{e,mn} - \gamma_{e,kl}) \iint_S \mathbf{a}_z \cdot \left(-\mathbf{e}^e_{mn} \times \mathbf{h}^e_{kl} - \mathbf{e}^e_{kl} \times \mathbf{h}^e_{mn}\right) dS = 0 \tag{2.98}$$

Adding and subtracting Eqs. (2.96) and (2.98) gives

$$\iint_S \mathbf{a}_z \cdot \gamma_{e,mn} \mathbf{e}^e_{kl} \times \mathbf{h}^e_{mn}\, dS = \iint_S \mathbf{a}_z \cdot \gamma_{e,kl} \mathbf{e}^e_{mn} \times \mathbf{h}^e_{kl}\, dS \tag{2.99a}$$

$$\iint_S \mathbf{a}_z \cdot \gamma_{e,kl} \mathbf{e}^e_{kl} \times \mathbf{h}^e_{mn}\, dS = \iint_S \mathbf{a}_z \cdot \gamma_{e,mn} \mathbf{e}^e_{mn} \times \mathbf{h}^e_{kl}\, dS \tag{2.99b}$$

Assuming nondegeneracy between these modes, that is, $\gamma_{kl} \neq \gamma_{mn}$, we can then obtain directly from Eq. (2.99)

$$\iint_S \mathbf{e}^e_{mn} \times \mathbf{h}^e_{kl} \cdot \mathbf{a}_z\, dS = 0 \tag{2.100a}$$

$$\iint_S \mathbf{e}^e_{kl} \times \mathbf{h}^e_{mn} \cdot \mathbf{a}_z\, dS = 0 \tag{2.100b}$$

which is the *orthogonality relationship between the transverse electric and magnetic fields*. When the structure is lossless, we can prove that

$$\iint_S \mathbf{e}^e_{mn} \times \mathbf{h}^{e*}_{kl} \cdot \mathbf{a}_z\, dS = 0 \tag{2.101a}$$

$$\iint_S \mathbf{e}^e_{mn} \times \mathbf{h}^{h*}_{kl} \cdot \mathbf{a}_z\, dS = 0 \tag{2.101b}$$

where $\mathbf{h}^{e*}_{kl}$ is the complex conjugate of $\mathbf{h}^e_{kl}$, assuming the modes are nondegenerate. Similarly, for two nondegenerate TE_{mn} and TE_{kl} modes, we can derive

$$\iint_S \mathbf{e}_{mn}^h \times \mathbf{h}_{kl}^h \cdot \mathbf{a}_z \, dS = 0 \tag{2.102a}$$

$$\iint_S \mathbf{e}_{kl}^h \times \mathbf{h}_{mn}^h \cdot \mathbf{a}_z \, dS = 0 \tag{2.102b}$$

$$\iint_S \mathbf{e}_{mn}^h \times \mathbf{h}_{kl}^{h*} \cdot \mathbf{a}_z \, dS = 0 \tag{2.103a}$$

$$\iint_S \mathbf{e}_{mn}^h \times \mathbf{h}_{kl}^{e*} \cdot \mathbf{a}_z \, dS = 0 \tag{2.103b}$$

for lossless waveguides.

When the modes are degenerate, Eqs. (2.100)–(2.103) still hold when one mode is TE and the other is TM. In general, however, these equations may not be satisfied and so the fields are not mutually orthogonal. In that case, as for the case of the scalar potentials discussed earlier, we may define new fields, which are linear combinations of the old fields such that the orthogonality holds.

2.12.4 Power Orthogonality for Lossless Structures

Let's assume that the waveguide is lossless and there are only two hybrid modes, characterized by (m, n) and (k, l), exist in the structure. Effectively, there will be four different modes propagating in the waveguide, namely, TE_{mn}, TM_{mn}, TE_{kl}, and TM_{kl}. The fields of these hybrid modes can therefore be expressed as a sum of those of the corresponding TE and TM modes. For instance, the fields of the hybrid (m, n) mode is given as

$$\begin{aligned}\mathbf{E}_{mn} &= \mathbf{E}_{mn}^e + \mathbf{E}_{mn}^h \\ &= \mathbf{e}_{mn}^e e^{-j\beta_{e,mn}z} + \mathbf{e}_{z,mn}^e e^{-j\beta_{e,mn}z} + \mathbf{e}_{mn}^h e^{-j\beta_{h,mn}z} + \mathbf{e}_{z,mn}^h e^{-j\beta_{h,mn}z}\end{aligned} \tag{2.104a}$$

$$\begin{aligned}\mathbf{H}_{mn} &= \mathbf{H}_{mn}^e + \mathbf{H}_{mn}^h \\ &= \mathbf{h}_{mn}^e e^{-j\beta_{e,mn}z} + \mathbf{h}_{z,mn}^e e^{-j\beta_{e,mn}z} + \mathbf{h}_{mn}^h e^{-j\beta_{h,mn}z} + \mathbf{h}_{z,mn}^h e^{-j\beta_{h,mn}z}\end{aligned} \tag{2.104b}$$

The total average power flow along the waveguide is given by

$$\begin{aligned}\mathbf{P}_{\text{av}} &= \tfrac{1}{2}\,\text{Re} \iint_S \mathbf{E} \times \mathbf{H}^* \cdot \mathbf{a}_z \, dS \\ &= \tfrac{1}{2}\,\text{Re} \iint_S (\mathbf{E}_{mn} + \mathbf{E}_{kl}) \times \left(\mathbf{H}_{mn}^* + \mathbf{H}_{kl}^*\right) \cdot \mathbf{a}_z \, dS\end{aligned} \tag{2.105}$$

Substituting Eq. (2.104) into (2.105) and expanding the cross products gives

$$\mathbf{P}_{\text{av}} = \tfrac{1}{2}\,\text{Re} \iint_S \left(\mathbf{e}^e_{mn} \times \mathbf{h}^{e*}_{mn} + \mathbf{e}^h_{mn} \times \mathbf{h}^{h*}_{mn} + \mathbf{e}^e_{kl} \times \mathbf{h}^{e*}_{kl} + \mathbf{e}^h_{kl} \times \mathbf{h}^{h*}_{kl}\right) \cdot \mathbf{a}_z\, dS$$

$$+ \tfrac{1}{2}\,\text{Re} \iint_S \Big[\mathbf{e}^e_{mn} \times \mathbf{h}^{h*}_{mn} e^{-j(\beta_{e,mn}-\beta_{h,mn})z} + \mathbf{e}^e_{kl} \times \mathbf{h}^{h*}_{kl} e^{-j(\beta_{e,kl}-\beta_{h,kl})z}$$

$$+ \mathbf{e}^e_{mn} \times \mathbf{h}^{e*}_{kl} e^{-j(\beta_{e,mn}-\beta_{e,kl})z} + \mathbf{e}^h_{mn} \times \mathbf{h}^{h*}_{kl} e^{-j(\beta_{h,mn}-\beta_{h,kl})z}$$

$$+ \mathbf{e}^e_{kl} \times \mathbf{h}^{e*}_{mn} e^{-j(\beta_{e,kl}-\beta_{e,mn})z} + \mathbf{e}^h_{kl} \times \mathbf{h}^{h*}_{mn} e^{-j(\beta_{h,kl}-\beta_{h,mn})z}$$

$$+ \mathbf{e}^e_{mn} \times \mathbf{h}^{h*}_{kl} e^{-j(\beta_{e,mn}-\beta_{h,kl})z} + \mathbf{e}^h_{mn} \times \mathbf{h}^{e*}_{kl} e^{-j(\beta_{h,mn}-\beta_{e,kl})z}$$

$$+ \mathbf{e}^e_{kl} \times \mathbf{h}^{h*}_{mn} e^{-j(\beta_{e,kl}-\beta_{h,mn})z} + \mathbf{e}^h_{kl} \times \mathbf{h}^{e*}_{mn} e^{-j(\beta_{h,kl}-\beta_{e,mn})z}\Big] \cdot \mathbf{a}_z\, dS \qquad (2.106)$$

The second integral of the right-hand side represents the power resulting from the interaction between the TE and TM modes. This power term is zero by virtue of Eqs. (2.101) and (2.103). The total average power is therefore given as

$$\mathbf{P}_{\text{av}} = \tfrac{1}{2}\,\text{Re} \iint_S \left(\mathbf{e}^e_{mn} \times \mathbf{h}^{e*}_{mn} + \mathbf{e}^h_{mn} \times \mathbf{h}^{h*}_{mn} + \mathbf{e}^e_{kl} \times \mathbf{h}^{e*}_{kl} + \mathbf{e}^h_{kl} \times \mathbf{h}^{h*}_{kl}\right) \cdot \mathbf{a}_z\, dS$$

$$= \mathbf{P}^e_{mn} + \mathbf{P}^h_{mn} + \mathbf{P}^e_{kl} + \mathbf{P}^h_{kl} \qquad (2.107)$$

where P^e_{mn}, P^h_{mn}, P^e_{kl}, and P^h_{kl} are the powers carried by the TM_{mn}, TE_{mn}, TM_{kl}, and TE_{kl} modes, respectively. Generalizing this result to multiple hybrid modes we obtain

$$\mathbf{P}_{\text{av}} = \sum_m \sum_n \mathbf{P}^e_{mn} + \mathbf{P}^h_{mn} \qquad (2.108)$$

Equation (2.108) suggests that the total power flow in a lossless waveguide is equal to the summation of the powers carried by individual modes. This further implies that each mode carries power independent of the other modes. This condition is known as the *power orthogonality*. Equation (2.108) is only valid for hybrid, TE, and TM modes that are nondegenerate. For degenerate modes, it only holds when the modes are of the same kind (TE or TM). For degenerate modes not satisfying Eq. (2.108), we can, however, choose new modes related to these degenerate modes such that the new modes follow the relation (2.101). Possible choices for the fields of the new modes are

$$\mathbf{E}'_{mn} = \mathbf{E}_{mn} \qquad (2.109a)$$

$$\mathbf{H}'_{mn} = \mathbf{H}_{mn} \qquad (2.109b)$$

$$\mathbf{E}'_{kl} = \mathbf{E}_{kl} - C\mathbf{E}_{mn} \qquad (2.109c)$$

$$\mathbf{H}'_{kl} = \mathbf{H}_{kl} - C\mathbf{H}_{mn} \qquad (2.109d)$$

where the constant C is chosen to satisfy the power orthogonality. It should be noted that the power orthogonality is only approximately held for low-loss waveguides propagating nondegenerate modes. For degenerate modes, strong couplings between these modes occur.

One remark needs to be made at this point. All the derived orthogonality relations are completely satisfied only if the modes were calculated exactly. Normally, the eigenmodes in planar transmission lines can only be determined approximately. Under this condition, it is easily proved that the orthogonality relations are not satisfied. The satisfaction of these orthogonality relations, such as Eq. (2.100) or (2.101), may therefore serve as a check of the accuracy of the computed modes.

REFERENCES

1. R. F. Harrington, *Time-Harmonic Electromagnetic Fields*, McGraw-Hill, New York, 1961, p. 24.
2. R. E. Collin, *Field Theory of Guided Waves*, IEEE Press, New York, 1991, pp. 329–337.
3. A. E. Taylor and W. R. Mann, *Advanced Calculus*, John Wiley & Sons, New York, 1983, pp. 277–279.

PROBLEMS

2.1 Derive the boundary conditions (2.15) between two different media as shown in Fig. 2.2.

2.2 Using the Poynting vector, prove that the average power density of a signal propagating in a waveguide is given by Eq. (2.20).

2.3 Show that TE modes can be characterized only by the magnetic scalar potential $\psi^h(x, y)$.

2.4 Derive Eqs. (2.51)–(2.54).

2.5 Show that TM modes can be characterized only by the electric scalar potential $\psi^e(x, y)$.

2.6 Prove that, in any waveguides, both E_z and H_z cannot be even or odd simultaneously.

2.7 Consider a general waveguide with perfectly conducting walls as shown in Fig. 2.3. Derive the following boundary conditions along the surface of the conductor for both TE and TM modes:

TE Modes: $\dfrac{\partial \psi^h}{\partial n} = 0$ (Neumann's Condition)

TM Modes: $\psi^e = 0$ (Dirichlet's Condition)

2.8 Verify the power orthogonality relationship for nondegenerate modes in a lossless circular waveguide.

2.9 Verify the power orthogonality relationship for nondegenerate modes in a lossless rectangular waveguide.

2.10 The electric and magnetic fields of the eigenmodes existing in planar transmission lines are normally determined approximately. Show that, under this condition, the orthogonality relationship (2.100) does not hold. Show also that, if these fields were calculated with a good accuracy, the orthogonality relation is well satisfied.

2.11 Derive the constant C in Eq. (2.109) that describes the fields of a new set of modes in terms of the fields of nonorthogonal degenerate modes, such that the new modes satisfy the power orthogonality (2.108).

2.12 Derive the orthogonality relations (2.84b) and (2.85) for TM and TE modes, respectively.

2.13 Derive the orthogonality relations (2.86) for hybrid modes.

2.14 Derive the orthogonality relation (2.101) and (2.103) for a lossless waveguide.

2.15 In general, there exist many degenerate modes in a waveguide. Some are coupled together while others are not. Prove that the mode coupling does not take place between the two degenerate TE_{mn} and TM_{mn} modes in a rectangular waveguide: that is, prove that the power-interaction terms

$$P_{\mathrm{TE}}^{\mathrm{TM}} = \frac{1}{2} \iint_S \mathbf{E}_{\mathrm{TE}_{mnn}} \times \mathbf{H}^*_{\mathrm{TM}_{mnn}} \cdot dS = 0$$

and

$$P_{\mathrm{TM}}^{\mathrm{TE}} = \frac{1}{2} \iint_S \mathbf{E}_{\mathrm{TM}_{mnn}} \times \mathbf{H}^*_{\mathrm{TE}_{mnn}} \cdot dS = 0.$$

2.16 Prove that, for a lossless waveguide, the orthogonality relationship (2.101) holds for nondegenerate modes.

CHAPTER THREE

Green's Function

Green's function is one of the most commonly used functions in solving microwave boundary-value problems. It represents a response (e.g., an electric field) due to a source of unit amplitude (e.g., a unit current). Green's function has been used in finding solutions for many microwave problems such as scattering and transmission line analysis. Its particular use in analyzing transmission lines is described in Chapters 6 (Variational Methods) and 7 (Spectral-Domain Method).

Green's function is described in details in [1] and [2]. In this chapter, we will present essential information on Green's function in the space domain. Its treatment in the spectral domain can be found in Chapters 6 and 7.

3.1 DESCRIPTIONS OF GREEN'S FUNCTION

Solution to a microwave boundary-value problem would involve finding the response due to a source in the microwave structure directly or indirectly. In essence, the main task of analyzing a microwave structure can thus revolve around finding the response caused by a source of unit amplitude (i.e., a Green's function) in that particular structure. Once Green's function is found, the total response can easily be determined by taking a summation or integral. To illustrate this principle, we will describe and use Green's function in obtaining the solutions of the two basic but most commonly used equations in microwave boundary-value problems: Poisson's equation and the wave equation.

3.1.1 Solution of Poisson's Equation Using Green's Function

We consider a medium characterized by a permittivity ε and permeability μ as shown in Fig. 3.1. We assume that the medium contains a charge density $\rho(x', y', z')$. This charge distribution represents a source in the structure and

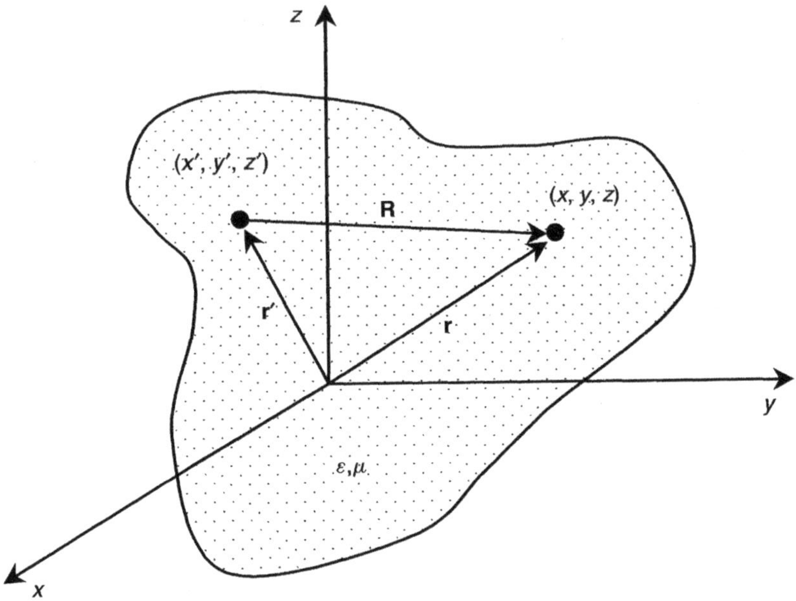

Figure 3.1 A medium characterized by ε and μ.

therefore would produce a potential V at every location (x, y, z), which is governed by Poisson's equation,

$$\nabla^2 V(x, y, z) = \frac{\rho(x', y', z')}{\varepsilon} \tag{3.1}$$

and the boundary conditions of the considered structure. Let $G(x, y, z; x', y', z')$ represent the Green's function of the structure, which is the potential at point (x, y, z) due to a unit charge located at point (x', y', z') in the medium. This Green's function must also satisfy Poisson's equation and corresponding boundary conditions of the structure. That is,

$$\nabla^2 G(x, y, z; x', y', z') = -\frac{\delta(x, y, z; x', y', z')}{\varepsilon} \tag{3.2}$$

where $\delta(x, y, z; x', y', z')$ is the *Dirac delta function*, defined as

$$\begin{aligned} \delta(x, y, z; x', y', z') &= \delta(x - x')\,\delta(y - y')\,\delta(z - z') \\ &= \begin{cases} 1, & x = x'; y = y', z = z' \\ 0, & \text{otherwise} \end{cases} \end{aligned} \tag{3.3}$$

The solution of Eq. (3.2) is

$$G(x, y, z; x', y', z') = -\frac{1}{4\pi\varepsilon R} \tag{3.4}$$

The potential due to the total charge distributed over the entire medium contained in a volume V' is obtained from the Green's function as

$$V(x, y, z) = \iiint\limits_{V'} G(x, y, z; x', y', z')\rho(x', y', z')\,dV' \tag{3.5}$$

or, upon using Eq. (3.4),

$$V(x, y, z) = \iiint\limits_{V'} \frac{\rho(x', y', z')}{4\pi\varepsilon R}\,dV' \tag{3.6}$$

which is exactly the same as the solution obtained by solving Poisson's equation directly.

3.1.2 Solution of the Wave Equation Using Green's Function

We consider again the medium in Fig. 3.1 and assume that a (vector) current distribution $\mathbf{J}(x', y', z')$ exists in the region. This current distribution produces a magnetic vector potential $\mathbf{A}(x, y, z)$, which can be obtained from the wave equation

$$\nabla^2\mathbf{A}(x, y, z) + k^2\mathbf{A}(x, y, z) = -\mu\,\mathbf{J}(x', y', z') \tag{3.7}$$

and the structure's boundary conditions. $k = \omega\sqrt{\varepsilon\mu}$ is the wave number. As for the case of Poisson's equation, we let $\mathbf{G}(x, y, z; x', y', z')$ be the Green's function of the structure, which now represents the magnetic vector potential at point (x, y, z) due to a unit current located at point (x', y', z') in the medium. This Green's function must also be the solution of the wave equation

$$\nabla^2\mathbf{G}(x, y, z; x', y', z') + k^2\mathbf{G}(x, y, z; x', y', z') = -\mu\delta(x, y, z; x', y', z')\,\mathbf{a}_J \tag{3.8}$$

subject to the structure's appropriate boundary conditions. $\mathbf{a}_J$ is the unit vector of the current at (x', y', z'). $\mathbf{G}$ can be derived as

$$\mathbf{G}(x, y, z; x', y', z') = \frac{\mu}{4\pi}\frac{e^{-jkR}}{R}\mathbf{a}_J \tag{3.9}$$

The magnetic vector potential due to the total current distributed over the entire medium can be obtained from the Green's function as

$$\mathbf{A}(x, y, z) = \frac{\mu}{4\pi}\iiint\limits_{V'} \mathbf{G}(x, y, z; x', y', z')\,\mathbf{J}(x', y', z')\,dV' \tag{3.10}$$

which becomes, after using Eq. (3.9),

$$\mathbf{A}(x, y, z) = \frac{\mu}{4\pi} \iiint_{V'} \mathbf{J}(x', y', z') \frac{e^{-jkR}}{R} \, dV' \tag{3.11}$$

This magnetic vector potential is identical to the direct solution of the wave equation. The magnetic and electric fields produced by the current can readily be obtained from the magnetic vector potential as

$$\mathbf{H}(x, y, z) = \frac{1}{\mu} \nabla \times \mathbf{A}(x, y, z) \tag{3.12}$$

$$\mathbf{E}(x, y, z) = \frac{1}{j\omega\varepsilon} \nabla \times \mathbf{H}(x, y, z) \tag{3.13}$$

These simple analyses demonstrate that, instead of solving directly Poisson's equation, wave equations, or other equations, resulting from a microwave boundary-value problem, for a desired field quantity, we can first determine the Green's function of the considered problem and then use it to obtain the field quantity. As will be seen, the Green's function satisfies the so-called *Sturm–Liouville* equation, which can uniquely be obtained for a microwave boundary-value problem.

3.2 STURM–LIOUVILLE EQUATION

The Sturm–Liouville equation is a differential equation of the following form:

$$Ly = f(x) \tag{3.14}$$

where

$$L = \left[\frac{d}{dx} \left(P_1(x) \frac{d}{dx} \right) + \frac{P}{2}(x) \right] \tag{3.15}$$

is called the Sturm–Liouville operator. The function $f(x)$ represents a source. This function is given as a direct function of x but may also depend on another unknown function of x (e.g., $y(x)$).

Many microwave boundary-value problems produce differential equations that can be converted to the Sturm–Liouville equation. To show this, we consider the following one-dimensional, second-order differential equation:

$$Fy = S(x) \tag{3.16}$$

where $S(x)$ represents a source existing in the microwave structure and F is the operator defined as

$$F = \left(C_1(x)\frac{d^2}{dx^2} + C_2(x)\frac{d}{dy} + C_3(x) \right) \tag{3.17}$$

Equation (3.16) is a generalization of one-dimensional differential equations resulting from a microwave boundary-value problem (e.g., Poisson's and wave equations). Equation (3.16) can easily be converted to the Sturm–Liouville Eq. (3.14) as follows:

Expanding Eq. (3.14) and dividing it by $P_1(x)$ gives

$$\frac{d^2y}{dx^2} + \frac{1}{P_1}\frac{dP_1}{dx}\frac{dy}{dx} + \frac{P_2}{P_1}y = \frac{f(x)}{P_1} \tag{3.18}$$

Dividing Eq. (3.16) by $C_1(x)$ produces

$$\frac{d^2y}{dx^2} + \frac{C_2}{C_1}\frac{dy}{dx} + \frac{C_3}{C_1}y = \frac{S(x)}{C_1} \tag{3.19}$$

In order for Eqs. (3.18) and (3.19) to be equivalent, we set their coefficients equal as

$$\frac{C_2(x)}{C_1(x)} = \frac{1}{P_1(x)}\frac{dP_1(x)}{dx} \tag{3.20}$$

$$\frac{C_3(x)}{C_1(x)} = \frac{P_2(x)}{P_1(x)} \tag{3.21}$$

$$\frac{S(x)}{C_1(x)} = \frac{f(x)}{P_1(x)} \tag{3.22}$$

which are then solved to obtain

$$P_1(x) = \exp\left(\int^x \frac{C_2(t)}{C_1(t)}\,dt \right) \tag{3.23}$$

$$P_2(x) = P_1(x)\frac{C_3(x)}{C_1(x)} \tag{3.24}$$

$$f(x) = P_1(x)\frac{S(x)}{C_1(x)} \tag{3.25}$$

These equations facilitate the conversion of the one-dimensional, second-order differential Eq. (3.16) to the Sturm–Liouville Eq. (3.14).

As an example, we consider the following *Bessel differential equation*, normally obtained in microwave problems involving cylindrical coordinates:

$$x^2\frac{d^2y}{dx^2} + x\frac{dy}{dx} + (x^2 - \lambda^2)y = 0 \tag{3.26}$$

Using Eqs. (3.23)–(3.25), we can derive

$$P_1(x) = \exp\left(\int^X \frac{t}{t^2}\,dt\right) = \exp(\ln x) = x \tag{3.27}$$

$$P_2(x) = \frac{x^2 - \lambda^2}{x} \tag{3.28}$$

and

$$f(x) = 0 \tag{3.29}$$

which, after substituting into Eq. (3.14), gives the equivalent Sturm–Liouville form of the Bessel differential equation,

$$\frac{d}{dx}\left(x\frac{dy}{dx}\right) + \frac{x^2 - \lambda^2}{x}y = 0 \tag{3.30}$$

It is now apparent that the Green's function for a microwave structure can be obtained as the solution of the Sturm–Liouville equation when the source has unit amplitude, subject to appropriate boundary conditions.

3.3 SOLUTIONS OF GREEN'S FUNCTION

In general, Green's function can be described in three forms: closed form, series form, and integral form [2]. The formulation of these functions is given as follows.

3.3.1 Closed-Form Green's Function

For the sake of generality, we now consider the more general Sturm–Liouville equation

$$[L + \lambda P_3(x)]y = f(x) \tag{3.31}$$

where $a \leq x \leq b$; L is again the Sturm–Liouville operator; and λ is a constant. The Green's function $G(x; x')$ is the solution of the Sturm–Liouville equation corresponding to a unit source and thus must satisfy the equation

$$[L + \lambda P_3(x)]G(x; x') = \delta(x; x') \tag{3.32}$$

where $\delta(x; x')$ is the one-dimensional Dirac delta function. The Green's function has the following properties:

1. For any $x \neq x'$, $G(x; x')$ satisfies the equation

$$[L + \lambda P_3(x)]G = 0 \tag{3.33}$$

2. $G(x;x')$ satisfies appropriate boundary conditions at $x = a$ and $x = b$.
3. $G(x;x')$ is symmetrical with respect to x and x'.
4. $G(x;x')$ is continuous at $x = x'$.
5. The derivative of $G(x;x')$ has a discontinuity of magnitude $-1/P_1(x')$ at $x = x'$. That is,

$$\left.\frac{dG}{dx}\right|_{x=x'^-}^{x'^+} = G'(x'^+) - G'(x'^-) = -\frac{1}{P_1(x')} \tag{3.34}$$

Properties 1 and 2 are readily seen, while Properties 3, 4, and 5 can easily be proved. We assume that the function $P_1(x)$ is continuous and nonzero at any point within the interval $[a, b]$. The discontinuity of the derivative of $G(x;x')$ in Property 5 thus has a finite value. We also assume that $P_1(x)$ and $P_2(x)$ are continuous in the interval $[a, b]$. Once $G(x;x')$ is found, the solution to Eq. (3.31) can be obtained as

$$y(x) = \int_a^b G(x;x')f(x)\,dx' \tag{3.35}$$

We now find the Green's function by dividing the interval $[a, b]$ into two separate regions: $[a, x']$ and $[x', b]$.

Region $a \leq x < x'$ Let $y_1(x)$ be a nontrivial solution of the associated homogeneous differential equation of (3.31),

$$[L + \lambda P_3(x)]y = 0 \tag{3.36}$$

which satisfies the boundary condition at $x = a$. At $x \neq x'$, $G(x;x')$ must satisfy

$$[L + \lambda P_3(x)]G = 0 \tag{3.37}$$

and the boundary condition at $x = a$, according to Properties 1 and 2 of the Green's function, respectively. From Eqs. (3.36) and (3.37), we can then write

$$G(x;x') = \alpha_1 y_1(x) \tag{3.38}$$

where α_1 is an unknown constant.

Region $x' < x \leq b$ Similarly, we let $y_2(x)$ be a nontrivial solution of the homogeneous differential Eq. (3.36), which satisfies the boundary condition at $x = b$, and obtain

$$G(x;x') = \alpha_2 y_2(x) \tag{3.39}$$

where α_2 is an unknown constant.

Using Property 4, we obtain from Eqs. (3.38) and (3.39)

$$\alpha_2 y_2(x') - \alpha_1 y_1(x') = 0 \tag{3.40}$$

Applying Property 5, we take the derivatives of Eqs. (3.38) and (3.39) and substitute into Eq. (3.34) to obtain

$$\alpha_2 y_2'(x') - \alpha_1 y_1'(x') = -\frac{1}{P_1(x)} \tag{3.41}$$

Solving Eqs. (3.40) and (3.41) gives

$$\alpha_1 = \frac{y_2(x')}{P_1(x')W(x')} \tag{3.42}$$

$$\alpha_2 = \frac{y_1(x')}{P_1(x')W(x')} \tag{3.43}$$

where

$$W(x') = y_1(x')y_2'(x') - y_2(x')y_1'(x') \tag{3.44}$$

is called the *Wronskian* of y_1 and y_2 at $x = x'$. The solutions α_1 and α_2 in Eqs. (3.42) and (3.43) exist and are unique when the Wronskian differs from zero, which is always valid unless y_1 and y_2 are linearly dependent.

The closed-form solution of the Green's function is now obtained from Eqs. (3.38), (3.39), (3.42), and (3.43) as

$$G(x;x') = \begin{cases} y_1(x) = \dfrac{y_2(x')}{P_1(x')W(x')}, & a \le x \le x' \\ y_2(x) = \dfrac{y_1(x')}{P_1(x')W(x')}, & x' \le x \le b \end{cases} \tag{3.45}$$

It is apparent that this closed-form Green's function is only useful if the solution to the homogeneous differential Eq. (3.36) can be found within the interval $[a, b]$.

As a demonstration of the procedure for finding the Green's function in closed form, we consider a shielded microstrip line as shown in Fig. 3.2. The shield and strip are assumed to be perfect conductors, and the strip thickness is negligible. We assume there exists a current source $\mathbf{J}$ on the strip. We wish to find the closed-form Green's function and the z component of the electric field, E_z.

The electric field component E_{zi} in region i $(i = 0, 1)$ is produced by the current component J_z. It satisfies the two-dimensional Helmholtz wave equation

$$(\nabla_t^2 + k_{c,i}^2)E_{zi}(x, y) = j\omega\mu_i J_z(x', y') \tag{3.46}$$

and the boundary conditions

$$E_{zi}(0, y) = E_{zi}(a, y) = 0, \quad y \in [0, b] \tag{3.47a}$$

$$E_{z1}(x, 0) = E_{z0}(x, b) = 0, \quad x \in [0, a] \tag{3.47b}$$

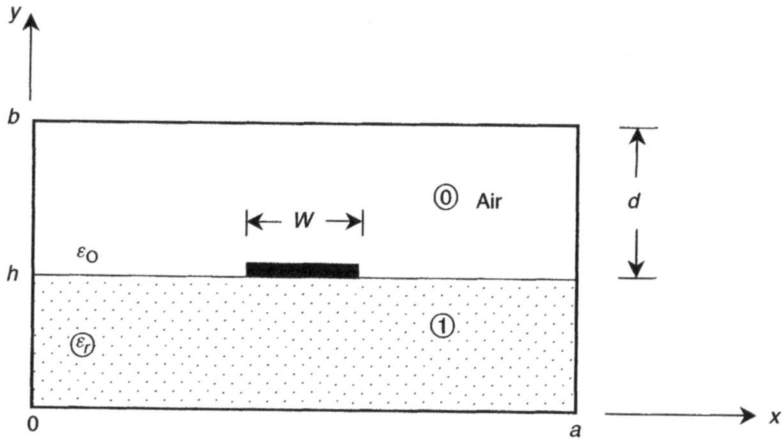

Figure 3.2 Cross section of shielded microstrip line.

where $k_{c,i} = \sqrt{k_i^2 + \gamma^2}$ is the cutoff wave number in region i, with $k_i = \omega\sqrt{\varepsilon_i \mu_i}$ and γ being the corresponding wave number and propagation constant, respectively. $\nabla_t^2 = \partial^2/\partial x^2 + \partial^2/\partial y^2$ represents the transverse Laplacian operator.

The Green's function $G(x, y; x', y')$ then satisfies

$$\nabla_t^2 G(x, y; x', y') + k_{c,i}^2 G(x, y; x', y') = \delta(x; x)\,\delta(y; y') \tag{3.48}$$

and the boundary conditions

$$G(x = 0, y; x', y' = h) = G(a, y; x', h) = 0 \tag{3.49a}$$

$$G(x, y = 0; x', y' = h) = G(x, b; x', h) = 0 \tag{3.49b}$$

We now express $G(x, y; x', y')$ as a Fourier series of sine functions that satisfy the boundary conditions along the x axis (i.e., at $x = 0, a$):

$$G(x, y; x', y') = \sum_{m=1}^{\infty} g_m(y; x', y') \sin\left(\frac{m\pi x}{a}\right) \tag{3.50}$$

Substituting Eq. (3.50) into (3.48) yields

$$\sum_{m=1}^{\infty} \left\{ \left[k_{c,i}^2 - \left(\frac{m\pi}{a}\right)^2 \right] g_m(y; x', h) \sin\left(\frac{m\pi x}{a}\right) + \sin\left(\frac{m\pi x}{a}\right) \frac{d^2 g_m(y; x', h)}{dy^2} \right\} = \delta(x; x')\,\delta(y; y') \tag{3.51}$$

Multiplying both sides by $\sin(m\pi x/a)$, integrating from 0 to a with respect to x, and applying the orthogonality relation, we find

$$\int_0^\ell \sin\left(\frac{m\pi x}{\ell}\right)\sin\left(\frac{n\pi x}{\ell}\right)dx = \begin{cases} \ell/2, & m = n \\ 0, & m \neq n \end{cases} \tag{3.52}$$

leads to

$$\frac{d^2 g_m(y;x',h)}{dy^2} - \lambda_{m,i}^2 g_m(y;x',h) = \frac{2}{a}\sin\left(\frac{m\pi x'}{a}\right)\delta(y;h) \tag{3.53}$$

where

$$\lambda_{m,i}^2 = \left(\frac{m\pi}{a}\right) - k_{c,i}^2 \tag{3.54}$$

Equation (3.53) is a one-dimensional differential equation, which can be solved using the procedure discussed earlier as follows.

We break the interval $[0, b]$ along the y axis into two separate regions, $[0, h]$ and $[h, b]$, and find the solution of the corresponding homogeneous differential equation

$$\frac{d^2 g_m(y;x',h)}{dy^2} - \lambda_{m,i}^2\, g_m(y;x',h) = 0 \tag{3.55}$$

This solution, which satisfies the boundary conditions at $y = 0$ and b, is given as

$$g_m^1 = C_m^1(x',h)\sinh(\lambda_{m,1}y), \qquad 0 \leq y < h \tag{3.56}$$

$$g_m^0 = C_m^0(x',h)\sinh[\lambda_{m,0}(b-y)], \qquad h < y \leq b \tag{3.57}$$

The Wronskian of g_m^0 and g_m^1 at $y = y' = h$ becomes, after substituting $y_1 = g_m^0$ and $y_2 = g_m^1$ in Eq. (3.44),

$$W(x',h) = C_m^0\, C_m^1\, Q_m \tag{3.58}$$

where

$$\begin{aligned} Q_m = {} & \lambda_{m,0}\sinh(\lambda_{m,1}h)\cosh(\lambda_{m,0}d) \\ & + \lambda_{m,1}\sinh(\lambda_{m,0}d)\cosh(\lambda_{m,1}h) \end{aligned} \tag{3.59}$$

Comparing Eq. (3.53) with the Sturm–Liouville equation

$$\frac{d}{dy}\left(P_1(y)\frac{du}{dy} - P_2(y)u\right) + \lambda P_3(y)u = f(y) \tag{3.60}$$

we obtain

$$\begin{aligned} P_1(y) &= P_3(y) = 1 \\ P_2(y) &= 0 \\ \lambda &= -\lambda_{m,i}^2 \end{aligned} \tag{3.61}$$

The closed-form solution of Eq. (3.55) is

$$g_m(y; x', h) = \begin{cases} \dfrac{y_1(y) y_2(h)}{P_1(h) W(x', h)}, & 0 \leq y \leq h \\ \dfrac{y_1(h) y_2(y)}{P_1(h) W(x', h)}, & h \leq y \leq b \end{cases} \tag{3.62}$$

from which we can obtain the solution of Eq. (3.53) as

$$g_m(y; x', h) = \begin{cases} \dfrac{2}{a} \sin\left(\dfrac{m\pi x'}{a}\right) \dfrac{\sinh(\lambda_{m,1} h) \sinh[\lambda_{m,0}(b-y)]}{Q_m}, & 0 \leq y \leq h \\ \dfrac{2}{a} \sin\left(\dfrac{m\pi x'}{a}\right) \dfrac{\sinh[\lambda_{m,0}(b-h)] \sinh(\lambda_{m,1} y)}{Q_m}, & h \leq y \leq b \end{cases} \tag{3.63}$$

The Green's function is then derived, from Eqs. (3.50) and (3.63), as

$$G(y : x', h) = \begin{cases} \dfrac{2}{a} \displaystyle\sum_{m=1}^{\infty} \sin\left(\dfrac{m\pi x'}{a}\right) \dfrac{\sinh(\lambda_{m,1} h) \sinh[\lambda_{m,0}(b-y)]}{Q_m}, \\ \qquad 0 \leq x \leq a, 0 \leq y \leq h \\ \dfrac{2}{a} \displaystyle\sum_{m=1}^{\infty} \sin\left(\dfrac{m\pi x'}{a}\right) \dfrac{\sinh[\lambda_{m,0}(b-h)] \sinh(\lambda_{m,1} y)}{Q_m}, \\ \qquad 0 \leq x \leq a, h \leq y \leq b \end{cases} \tag{3.64}$$

which can be used to find the electric field component E_z as

$$E_{zi} = j\omega\mu_i \int_0^a J_z(x', h) G(x, y; x', h)\, dx' \tag{3.65}$$

We can also derive another closed-form expression for the Green's function by expanding it as a series of functions that satisfy the boundary conditions along the y axis (i.e., at $y = 0$ and b).

3.3.2 Series-Form Green's Function

We consider again the Sturm–Liouville Eq. (3.31) over the interval $[a, b]$ along with the following (general) mixed boundary conditions:

$$\alpha_1 y(x)|_{x=a} + \alpha_2 \frac{dy(x)}{dx}|_{x=a} = \alpha_1 y(a) + \alpha_2 \frac{dy(a)}{dx} = 0 \tag{3.66}$$

$$\beta_1 y(x)|_{x=b} + \beta_2 \frac{dy(x)}{dx}|_{x=b} = \beta_1 y(b) + \beta_2 \frac{dy(b)}{dx} = 0 \tag{3.67}$$

where α_i and β_i $(i = 1, 2)$ are constants. Accordingly, the Green's function $G(x; x')$ must satisfy Eq. (3.32) and the boundary conditions

$$\alpha_1 G(a; x') + \alpha_2 \frac{dG(a; x')}{dx} = 0 \tag{3.68}$$

$$\beta_1 G(b; x') + \beta_2 \frac{dG(b; x')}{dx} = 0 \tag{3.69}$$

Let $\Phi_n(x)$ be the eigenfunctions with the eigenvalues λ_n. $\Phi_n(x)$ satisfy the homogeneous differential equation

$$[L + \lambda_n P_3(x)]\Phi_n(x) = 0 \tag{3.70}$$

and the same boundary conditions for $G(x; x')$ in Eqs. (3.68) and (3.69). Note that $\Phi_n(x)$ form a complete set of orthonormal eigenfunctions, and so, for $a \leq x \leq b$, these functions follow the orthogonality condition

$$\int_a^b \Phi_m(x)\Phi_n(x)P_3(x)\, dx = \delta_{mn} \tag{3.71}$$

where

$$\delta_{mn} = \begin{cases} 1, & m = n \\ 0, & m \neq n \end{cases} \tag{3.72}$$

is the *Kronecker delta function*. We can therefore let

$$G(x; x') = \sum_{n=1}^{\infty} a_n(x')\Phi_n(x) \tag{3.73}$$

where a_n is the amplitude coefficient. Multiplying both sides of Eq. (3.73) by $\Phi_n(x)P_3(x)$, integrating from a to b, and applying the orthogonality condition (3.71), we get

$$a_n(x') = \int_a^b G(x; x')\Phi_n(x)P_3(x)\, dx \tag{3.74}$$

Now multiplying Eq. (3.32) by $\Phi_n(x)$ and Eq. (3.70) by $G(x; x')$, subtracting from each other, and taking the integral from a to b gives

$$\int_a^b [\Phi_n(x)\, LG\,(x; x') - G(x; x')L\Phi_n(x)]\, dx$$

$$= -(\lambda - \lambda_n) \int_a^b G(x; x')\Phi_n(x)P_3(x)\, dx + \int_a^b \delta(x; x')\Phi_n(x)\, dx \tag{3.75}$$

whose left-hand side vanishes upon using Eqs. (3.73) and (3.70). Equation (3.75) hence becomes

$$-(\lambda - \lambda_n) \int_a^b G(x;x')\Phi_n(x)P_3(x)\,dx + \int_a^b \delta(x;x')\Phi_n(x)\,dx = 0 \tag{3.76}$$

Substituting Eq. (3.74) into (3.76) and solving for a_n produces

$$a_n(x') = \frac{\Phi_n(x)}{\lambda - \lambda_n} \tag{3.77}$$

where

$$\Phi_n(x') = \int_a^b \delta(x;x')\Phi_n(x)\,dx \tag{3.78}$$

The series-form Green's function of Eq. (3.73) now becomes, upon substitution of Eq. (3.77),

$$G(x;x') = \sum_{n=1}^{\infty} \frac{\Phi_n(x')}{\lambda - \lambda_n}\Phi_n(x) \tag{3.79}$$

The Green's function in series form is useful for bounded microwave structures, in which the orthogonal functions $\Phi_n(x)$ used to describe the Green's function represent standing waves, and when the eigenvalue spectrum formed by λ_n is discrete.

As an illustration for the formulation of the series-form Green's function, we consider a rectangular-waveguide resonator whose dimensions along the x, y, and z axes are a, b, and d, respectively. We assume that a current source, represented by the current density $\mathbf{J}(x, y, z)$, exists in the resonator and TE modes are excited. The wave equation governing the magnetic field component H_z produced by the current can be derived as

$$\nabla^2 H_z(x, y, z) + \omega^2\varepsilon\mu H_z(x, y, x) = \frac{\partial J_x(x, y, z)}{\partial y} - \frac{\partial J_y(x, y, z)}{\partial x} \tag{3.80}$$

H_z satisfies the following boundary conditions:

$$H_z(x, y, 0) = H_z(x, y, d) = 0 \tag{3.81a}$$

$$\frac{\partial H_z}{\partial x}\Big|_{x=0,a} = \frac{\partial H_z}{\partial y}\Big|_{y=0,b} = 0 \tag{3.81b}$$

The Green's function $G(x, y, z; x', y', z')$ is the solution of

$$\nabla^2 G(x, y, z; x', y', z') + k^2 G(x, y, z; x', y', z') = \delta(x;x')\,\delta(y;y')\delta(z;z') \tag{3.82}$$

Let $\Phi_{mnp}(x, y, z)$ represent the orthogonal eigenfunctions, which satisfy the homogeneous differential equation

$$\nabla^2 \Phi_{mnp}(x, y, z) + k^2 \Phi_{mnp}(x, y, z) = 0 \tag{3.83}$$

and the same boundary conditions as in Eq. (3.81). Using the method of separation of variables, we can write

$$\Phi_{mnp}(x, y, z) = C_{mnp} \Phi_m(x) \Phi_n(y) \Phi_p(y) \tag{3.84}$$

where $\Phi_m(x)$, $\Phi_n(y)$, and $\Phi_p(z)$ are functions of only x, y, and z, respectively, and C_{mnp} is a constant. Substituting Eq. (3.84) into (3.83) and solving the resulting equation, subject to the boundary conditions, yields

$$\Phi_{mnp}(x, y, z) = C_{mnp} \cos\left(\frac{m\pi x}{a}\right) \cos\left(\frac{n\pi y}{b}\right) \sin\left(\frac{p\pi z}{d}\right) \tag{3.85}$$

The eigenvalues are derived as

$$\lambda_{mnp} = \left(\frac{m\pi}{a}\right)^2 + \left(\frac{n\pi}{b}\right)^2 + \left(\frac{p\pi}{d}\right)^2 \tag{3.86}$$

Multiplying Eq. (3.85) by Φ_{qrs}, taking the integral from 0 to a, b, and d, and applying the orthogonality relation

$$\int_0^a \int_0^b \int_0^d \Phi_{mnp}(x, y, z) \Phi_{qrs}(x, y, z)\, dz\, dy\, dx = \begin{cases} 1; & m = q, n = r, p = s \\ 0; & m \neq q, n \neq r, p \neq s \end{cases} \tag{3.87}$$

leads to

$$C_{mnp} = \frac{2\sqrt{2}}{\sqrt{abd}} \tag{3.88}$$

Substituting Eq. (3.88) into (3.85) gives

$$\Phi_{mnp}(x, y, z) = \frac{2\sqrt{2}}{\sqrt{abd}} \cos\left(\frac{m\pi x}{a}\right) \cos\left(\frac{n\pi y}{b}\right) \sin\left(\frac{p\pi z}{d}\right) \tag{3.89}$$

The Green's function is now determined from Eq. (3.79), noting that for TE modes, $m = 0, 1, 2, \ldots$; $n = 0, 1, 2, \ldots$; $p = 1, 2 \ldots$; and m and n cannot be zero simultaneously, as

$$G(x, y; x', y') = \sum_{m=0}^{\infty} \sum_{n=0}^{\infty} \sum_{p=1}^{\infty} \frac{\Phi_{mnp}(x', y', z') \Phi_{mnp}(x, y, z)}{k^2 - \lambda_{mnp}^2} \tag{3.90}$$

or, upon using Eqs. (3.89) and (3.86),

$$G(x, y; x', y') = \frac{2\sqrt{2}}{\sqrt{abd}} \sum_{m=0}^{\infty} \sum_{n=0}^{\infty} \sum_{p=1}^{\infty}$$
$$\times \frac{\cos\left(\frac{m\pi x'}{a}\right) \cos\left(\frac{n\pi y'}{b}\right) \sin\left(\frac{p\pi z'}{d}\right) \cos\left(\frac{m\pi x}{a}\right) \cos\left(\frac{n\pi y}{b}\right) \sin\left(\frac{p\pi z}{d}\right)}{\omega^2 \varepsilon\mu - \left[\left(\frac{m\pi}{a}\right)^2 + \left(\frac{n\pi}{b}\right)^2 + \left(\frac{p\pi}{d}\right)^2\right]} \tag{3.91}$$

The magnetic field component H_z can now be found, using $G(x, y; x', y')$ in Eq. (3.91), as

$$H_z(x, y, z) = \frac{2\sqrt{2}}{\sqrt{abd}} \sum_{m=0}^{\infty} \sum_{n=0}^{\infty} \sum_{p=1}^{\infty} \frac{\cos\left(\frac{m\pi x}{a}\right) \cos\left(\frac{n\pi y}{b}\right) \sin\left(\frac{p\pi z}{d}\right)}{\omega^2 \varepsilon\mu - \left[\left(\frac{m\pi}{a}\right)^2 + \left(\frac{n\pi}{b}\right)^2 + \left(\frac{p\pi}{d}\right)^2\right]}$$
$$\times \int_0^a \int_0^b \int_0^d \left(\frac{\partial J_x}{\partial y} - \frac{\partial J_y}{\partial x}\right) \cos\left(\frac{m\pi x'}{a}\right)$$
$$\times \cos\left(\frac{n\pi y'}{b}\right) \sin\left(\frac{p\pi z'}{d}\right) dz'\, dy'\, dx' \tag{3.92}$$

Note that H_z has a singularity at

$$f_r = \frac{1}{2\pi\sqrt{\varepsilon\mu}} \left[\left(\frac{m\pi}{a}\right)^2 + \left(\frac{n\pi}{b}\right)^2 + \left(\frac{p\pi}{d}\right)^2\right]^{1/2} \tag{3.93}$$

which is known as the resonant frequency of the rectangular-waveguide resonator.

3.3.3 Integral-Form Green's Function

We now discuss another form of the Green's function, the integral form. This form is useful when the eigenvalue spectrum is continuous. Under this condition, the summation in Eq. (3.79) becomes an integral. The integral-form Green's function is especially desirable when at least one of the boundary conditions is at infinity. For instance, a source (e.g., a biased transistor) integrated with an antenna is radiating in an unbounded medium. The orthogonal functions $\Phi_n(x)$ can also be used to describe the integral-form Green's function, but here they represent traveling waves.

We consider the one-dimensional Helmholtz equation

$$\frac{d^2\psi(x)}{dx} + \beta_0^2 \psi(x) = f(x) \tag{3.94}$$

that describes the field $\psi(x)$ produced by the source represented by $f(x)$. Here, $\psi(x)$ is assumed to vanish at infinity; that is, the source is radiating in an unbounded medium. β_0 is a constant. The Green's function $G(x;x')$ then satisfies

$$\frac{d^2G(x;x')}{dx^2} + \beta_0^2 G(x;x') = \delta(x;x') \tag{3.95}$$

subject to the same boundary conditions of $\psi(x)$. Making use of the Fourier transform, we can write

$$G(x;x') = \frac{1}{\sqrt{2\pi}} \int_{-\infty}^{\infty} \tilde{G}(\beta;x') e^{-j\beta x}\, d\beta \tag{3.96}$$

and

$$\tilde{G}(\beta;x') = \frac{1}{\sqrt{2\pi}} \int_{-\infty}^{\infty} G(x;x') e^{j\beta x}\, dx \tag{3.97}$$

$G(x;x')$ represented by Eq. (3.96) can be interpreted as consisting of a continuous spectrum of waves having amplitudes $\tilde{G}(\beta;x')$ and traveling in a lossless medium in the $+x$ direction, which is in essence similar to the series form of Eq. (3.79). Similarly, the function $\delta(x;x')$ and its Fourier transform, $\tilde{\delta}(\beta;x')$, are related by

$$\begin{aligned}\tilde{\delta}(\beta;x') &= \frac{1}{\sqrt{2\pi}} \int_{-\infty}^{\infty} \delta(x;x') e^{j\beta x}\, dx \\ &= \frac{1}{\sqrt{2\pi}} e^{j\beta x'}\end{aligned} \tag{3.98}$$

and

$$\begin{aligned}\delta(x;x') &= \frac{1}{\sqrt{2\pi}} \int_{-\infty}^{\infty} \tilde{\delta}(\beta;x') e^{-j\beta x}\, d\beta \\ &= \frac{1}{\sqrt{2\pi}} \int_{-\infty}^{\infty} \frac{1}{\sqrt{2\pi}} e^{j\beta x'} e^{-j\beta x}\, d\beta\end{aligned} \tag{3.99}$$

Substituting Eqs. (3.96) and (3.99) into Eq. (3.95) yields

$$\frac{1}{\sqrt{2\pi}} \int_{-\infty}^{\infty} \left((\beta_0^2 - \beta^2)\tilde{G}(\beta;x') - \frac{1}{\sqrt{2\pi}} e^{j\beta x'} \right) e^{-j\beta x}\, d\beta = 0 \tag{3.100}$$

from which we obtain

$$\tilde{G}(\beta;x') = \frac{1}{\sqrt{2\pi}} \frac{e^{j\beta x'}}{\beta_0^2 - \beta^2} \tag{3.101}$$

The Green's function then becomes, upon substituting Eq. (3.101) into (3.96),

$$G(x;x') = \frac{1}{\sqrt{2\pi}} \int_{-\infty}^{\infty} \frac{e^{-j\beta(x-x')}}{\beta_0^2 - \beta^2}\, d\beta \tag{3.102}$$

To illustrate the derivation of this Green's function, we consider an open, linear, isotropic and homogeneous charge-free medium characterized by ε and μ. The medium contains a current source $\mathbf{J}(x', y', z'; t)$. The time-domain wave equation for the electric field is derived as

$$\nabla^2 \mathbf{E}(\mathbf{r};t) - \mu\varepsilon \frac{\partial^2 \mathbf{E}(\mathbf{r};t)}{\partial t^2} = \mu \frac{\partial \mathbf{J}(\mathbf{r}';t)}{\partial t} \tag{3.103}$$

where $\mathbf{r} = (x, y, z)$ and $\mathbf{r}' = (x', y', z')$. We now define the Green's functions in the time domain, $g(\mathbf{r} - \mathbf{r}'; t)$, and frequency domain, $\tilde{G}(r - r'; \omega)$, which are related by

$$g(\mathbf{r} - \mathbf{r}'; t) = \frac{1}{\sqrt{2\pi}} \int_{-\infty}^{\infty} \tilde{G}(\mathbf{r} - \mathbf{r}'; \omega) e^{j\omega t}\, d\omega \tag{3.104}$$

$$\tilde{G}(\mathbf{r} - \mathbf{r}'; \omega) = \frac{1}{\sqrt{2\pi}} \int_{-\infty}^{\infty} g(\mathbf{r} - \mathbf{r}'; t) e^{-j\omega t}\, dt \tag{3.105}$$

$g(\mathbf{r} - \mathbf{r}'; t)$ satisfies

$$\left(\nabla^2 - \mu\varepsilon \frac{\partial^2}{\partial t^2}\right) g(r - r'; t) = -\delta(\mathbf{r} - \mathbf{r}')\,\delta\,(t; t') \tag{3.106}$$

where $\delta(t;t')$ is the unit impulse function, which is derived in the frequency domain as

$$\begin{aligned} \tilde{\delta}(\omega) &= \frac{1}{\sqrt{2\pi}} \int_{-\infty}^{\infty} \delta\left(t; t'\right) e^{-j\omega t}\, dt \\ &= \frac{1}{\sqrt{2\pi}} e^{-j\omega t'} \end{aligned} \tag{3.107}$$

$\delta(t;t')$ is then given by, upon using Eq. (3.107),

$$\begin{aligned} \delta(t;t') &= \frac{1}{\sqrt{2\pi}} \int_{-\infty}^{\infty} \tilde{\delta}(\omega) e^{j\omega t}\, d\omega \\ &= \frac{1}{2\pi} \int_{-\infty}^{\infty} e^{j\omega(t-t')}\, d\omega \end{aligned} \tag{3.108}$$

The Green's function is due to a point source, and so we can use the spherical coordinates and assume spherical symmetry. Equation (3.106) then becomes, letting $R = |\mathbf{r} - \mathbf{r}'|$ and making use of Eqs. (3.104) and (3.108),

$$\frac{1}{\sqrt{2\pi}} \int_{-\infty}^{\infty} \left\{ \left(\frac{1}{R^2} \frac{\partial}{\partial R} \left(R^2 \frac{\partial}{\partial R} \right) \right) - \mu\varepsilon \frac{\partial^2}{\partial t^2} \right\} \tilde{G} \left(\mathbf{r} - \mathbf{r}'; \omega \right) e^{j\omega t} \, d\omega$$
$$= -\frac{1}{2\pi} \int_{-\infty}^{\infty} \delta \left(\mathbf{r} - \mathbf{r}' \right) e^{j\omega(t-t')} \, d\omega \tag{3.109}$$

This leads to

$$\frac{1}{R^2} \frac{\partial}{\partial R} \left(R^2 \frac{\partial}{\partial R} \tilde{G}(\mathbf{r} - \mathbf{r}'; \omega) \right) + \omega^2 \varepsilon\mu \tilde{G}(\mathbf{r} - \mathbf{r}'; \omega) = \frac{\delta(\mathbf{r} - \mathbf{r}')}{\sqrt{2\pi}} \tag{3.110}$$

We have chosen $t' = 0$. The solution to Eq. (3.110) is

$$\tilde{G}(\mathbf{r} - \mathbf{r}'; \omega) = \frac{1}{\sqrt{2\pi}} \frac{e^{-jkR}}{R} \tag{3.111}$$

Using Eq. (3.111) in (3.104) gives

$$g(\mathbf{r} - \mathbf{r}'; t) = \frac{1}{R} \frac{1}{\sqrt{2\pi}} \int_{-\infty}^{\infty} e^{j\omega(t - R\sqrt{\varepsilon\mu})} \, d\omega \tag{3.112}$$

which becomes, after applying Eq. (3.108),

$$g(\mathbf{r} - \mathbf{r}'; t) = \frac{1}{R} \delta(t - R\sqrt{\varepsilon\mu}) \tag{3.113}$$

REFERENCES

1. R. E. Collin, *Field Theory of Guided Waves*, 2nd edition, IEEE Press, New York, 1991, Chap. 2.
2. C. A. Balanis, *Advanced Engineering Electromagnetics*, John Wiley & Sons, New York, 1989, Chap. 14.

PROBLEMS

3.1 Show that the two-dimensional Laplace equation,

$$\frac{\partial^2 V}{\partial x^2} + \frac{\partial^2 V}{\partial y^2} = 0 \tag{P3.1}$$

can be converted to two one-dimensional Sturm–Liouville equations.

3.2 In solving microwave boundary-value problems involving spherical coordinates, one often encounters the following partial differential equation:

$$\frac{1}{R}\frac{d}{dr}\left(r^2\frac{dR}{dr}\right) - n(n+1) + k^2r^2 = 0 \qquad \text{(P3.2)}$$

where $R = f(r)$, n is an integer, and k is a constant. Convert the above equation to the Sturm–Liouville equation.

3.3 Prove the following properties of the Green's function $G(x; x')$:

(a) $G(x; x')$ is symmetrical with respect to x and x'.

(b) $G(x; x')$ is continuous at $x = x'$.

(c) The derivative of $G(x; x')$ has a discontinuity of $-1/P_1(x')$ at $x = x'$:

$$\left.\frac{dG}{dx}\right|_{x=x'^-}^{x'^+} = G'(x'^+) - G'(x'^-) = -\frac{1}{P_1(x')} \qquad \text{(P3.3)}$$

3.4 Show that the Green's function associated with the Bessel differential equation

$$\frac{d}{dx}\left(x\frac{dy}{dx}\right) - \frac{n^2}{x}y = f(x) \qquad \text{(P3.4)}$$

and the boundary conditions $y(0) = y(1) = 0$ is given as

$$G(x;x') = \begin{cases} \dfrac{1}{2n}\left(\dfrac{x}{x'}\right)^n\left[1-(x')^{2n}\right], & x < x' \\ \dfrac{1}{2n}\left(\dfrac{x'}{x}\right)^n\left(1-x^{2n}\right), & x > x' \end{cases} \qquad \text{(P3.5)}$$

when $n \neq 0$.

3.5 Consider the Sturm–Liouville equation

$$Ly = f(x) \qquad \text{(P3.6)}$$

whose domain is $[a, b]$, and the boundary conditions

$$\alpha_1 y(x)|_{x=a} + \alpha_2\frac{dy(x)}{dx}|_{x=a} = \alpha_1 y(a) + \alpha_2\frac{dy(a)}{dx} = 0 \qquad \text{(P3.7a)}$$

$$\beta_1 y(x)|_{x=b} + \beta_2\frac{dy(x)}{dx}|_{x=b} = \beta_1 y(b) + \beta_2\frac{dy(b)}{dx} = 0 \qquad \text{(P3.7b)}$$

where α_i and β_i $(i = 1, 2)$ are constants. The solution to Eq. (P3.6) subject to the boundary conditions (P3.7) is given as

$$y(x) = \int_a^b G(x;x')f(x')\,dx' \tag{P3.8}$$

where $G(x; x')$ is the Green's function. Derive this equation by performing the following:

(a) Show that

$$\int_a^b G(x;x')\,Ly(x)\,dx = \int_a^b G(x;x')f(x)\,dx \tag{P3.9}$$

(b) Show that the left-hand side of Eq. (P3.9), upon division into two separate regions $[a, x']$ and $[x', b]$, can be written as

$$\begin{aligned}\int_a^b G(x;x')\,Ly(x)\,dx &= \int_a^{x'} G_1(x;x')\,Ly(x)\,dx + \int_{x'}^b G_2(x;x')\,Ly(x)\,dx\\ &= -P_1(x')\left(G_1(x';x')y'(x') - \frac{\partial G_1(x;x')}{\partial x}\Big|_{x=x'}\,y(x')\right)\\ &\quad + P_1(x')\left(G_2(x';x')y'(x') - \frac{\partial G_2(x;x')}{\partial x}\Big|_{x=x'}\,y(x')\right) = y(x')\end{aligned} \tag{P3.10}$$

where G_1 and G_2 are the Green's functions in the intervals $[a, x']$ and $[x', b]$, respectively.

(c) Deduce Eq. (P3.8) by appropriate changes of variables.

3.6 Consider a rectangular waveguide whose dimensions along the x and y directions are a and b, respectively. Assume that the waveguide is infinitely long and its dielectric is air. Let the source in the waveguide be represented by a current density $\mathbf{J}(x, y, z)$. Find the series-form expressions for the Green's function and the z component of the electric field for TM modes.

3.7 Consider an infinitely long and thin conducting wire enclosed by an infinitely long rectangular conducting pipe (i.e., a rectangular waveguide) as shown in Fig. P3.1. The medium inside the pipe has a permittivity ε. The pipe is held at zero potential. The charge distribution $\rho(x, y)$ along the wire is assumed to be uniform. Find the Green's function in closed form and the corresponding potential distribution in the medium. The Green's function may be expressed as a Fourier series of sine functions that satisfy the boundary conditions at $x = 0$ and a.

3.8 Repeat Problem 3.7 using the series-form Green's function.

3.9 Consider the shielded microstrip line of Fig. 3.2 having a current distribution $\mathbf{J}$ on the strip. Derive the closed-form Green's function and the corresponding electric field component E_z by expressing the Green's function

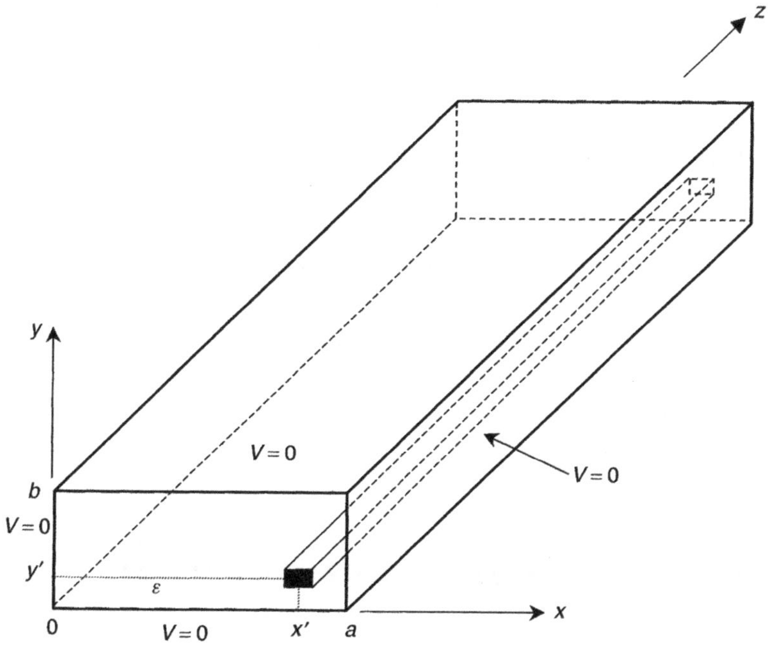

Figure P3.1 A conducting wire inside a rectangular waveguide.

as a Fourier series of sine functions that satisfy the boundary conditions along the y direction.

3.10 Repeat Problem 3.9 using the Green's function in series form.

3.11 Consider again the shielded microstrip line of Fig. 3.2 having a current distribution $\mathbf{J}$ on the strip. Derive the Green's function and the magnetic field component H_z in closed form. The Green's function may be expressed as a Fourier series of cosine functions that satisfy the boundary conditions along the x axis.

3.12 Repeat Problem 3.11 using the Green's function described as a Fourier series of cosine functions that satisfy the boundary conditions along the y axis.

3.13 Derive Eq. (A3.1) and (A3.2) in the Appendix describing Green's first identity.

3.14 Derive Eq. (A3.3) and (A3.4) of Green's second identity.

3.15 Consider a rectangular-waveguide resonator with dimensions a, b, and d along the x, y, and z directions, respectively. Let $\mathbf{J}(x, y, z)$ represent the current source in the resonator and assume that only TM modes are excited. Derive the Green's function in series form and the associated electric field component E_z.

3.16 Consider a circular-waveguide resonator whose radius is a and length along the z direction is d. The resonator is excited by a current source $\mathbf{J}(r, \phi, z)$. For TE modes, derive the Green's function and the magnetic field component H_z in series form.

3.17 Repeat problem 3.16 for TM modes and electric field.

3.18 A strip line is shown in Fig. P3.2. It is assumed that the strip is a perfect conductor and has zero thickness. Also, the ground plane is assumed to be infinitely wide and a perfect conductor. The dielectric is assumed to be perfect. Derive the Green's function and the potential distribution in closed form.

3.19 A shielded asymmetrical three-layer microstrip line is shown in Fig. P3.3. The shield and strip are perfect conductors and the strip thickness is infinitesimally thin. The dielectrics are perfect. Derive the closed-form expressions for the Green's function and the potential distribution.

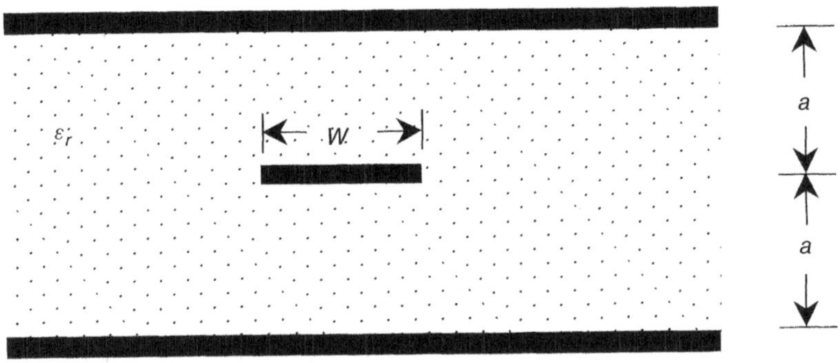

Figure P3.2 Cross section of strip line.

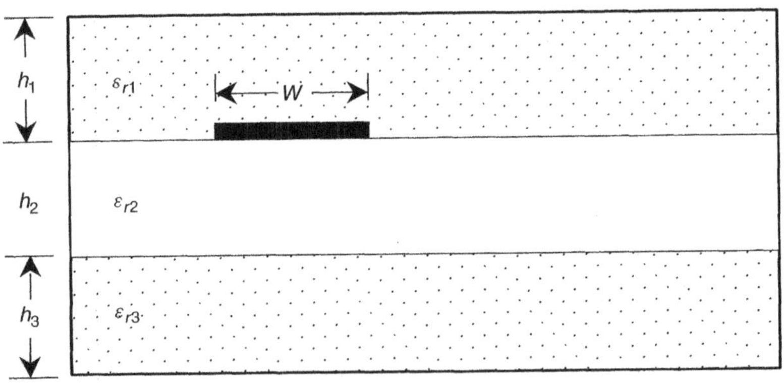

Figure P3.3 Cross section of shielded asymmetrical three-layer microstrip line.

3.20 Consider the following differential equation, which appears in problems involving transmission lines, waveguides, and so on:

$$\frac{d^2\varphi(x)}{dx^2} + \beta_0^2\varphi(x) = f(x) \tag{P3.11}$$

where $\beta_0^2 = \omega^2\mu\varepsilon$. The boundary conditions are $\varphi(0) = \varphi(L) = 0$.

(a) Derive the Green's function in series form.

(b) Derive the Green's function in closed form at dc.

3.21 A shielded stripline is shown in Fig. P3.4. It is assumed that the shield and strip are perfect conductors, the strip has negligible thickness, and the dielectric is perfect. The charge distribution $\rho(x, y)$ along the strip is uniform. Derive closed-form expressions for the Green's function and the potential distribution.

3.22 Consider an infinitely long coaxial transmission line, whose inner and outer radii are a and b, respectively. The dielectric between these conductors has a permittivity of ε. The outer conductor is grounded and the inner conductor is held at potential V_0. The charge distribution along the inner conductor is assumed to be uniform. Find the Green's function and potential distribution in closed form.

3.23 Repeat Problem 3.22 using series form.

3.24 Consider a microstrip line shown in Fig. P3.5 having a current distribution **J** on the strip. It is assumed that the ground plane and strip are perfect conductors, the strip's thickness is infinitely thin, and the substrate is

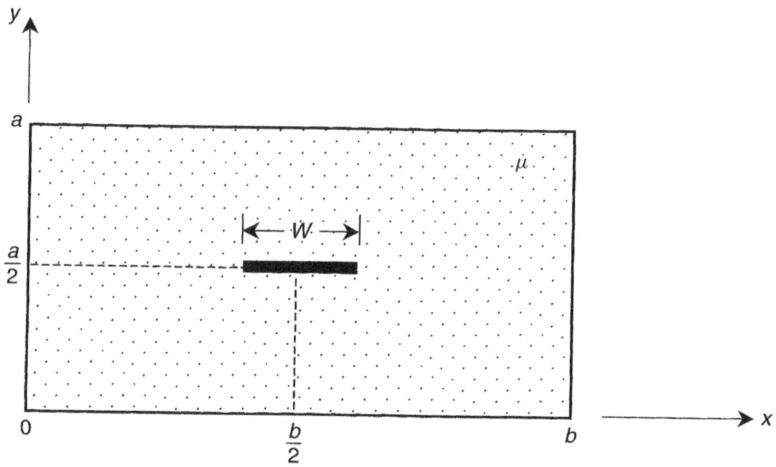

Figure P3.4 Cross section of shielded strip line.

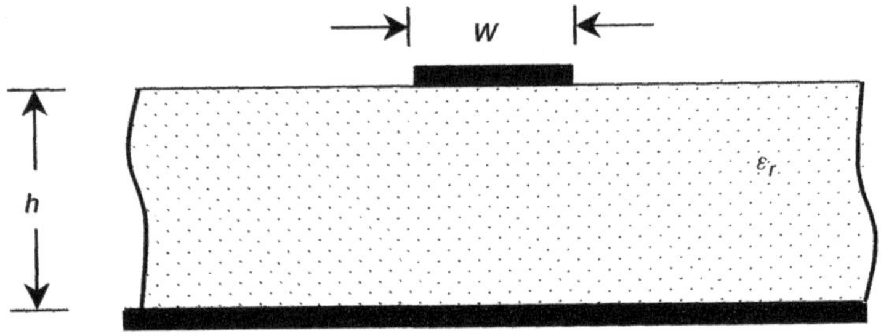

Figure P3.5 Cross section of microstrip line.

perfect. Derive the integral-form Green's function and the corresponding electric field.

APPENDIX: GREEN'S IDENTITIES

Green's First Identity

$$\oiint_S u\nabla v \cdot d\mathbf{S} = \iiint_V (u\nabla^2 v + \nabla u \cdot \nabla v)\,dV \tag{A3.1}$$

or

$$\oiint_S u\frac{\partial v}{\partial n}\,dS = \iiint_V (u\nabla^2 v + \nabla u \cdot \nabla v)\,dV \tag{A3.2}$$

where V is a volume and S is the surface enclosing that volume. u and v are two scalar functions. $\partial v/\partial n$ is the directional derivative of v along the direction of the normal vector $\mathbf{n}$.

Green's Second Identity

$$\oiint_S (u\nabla v - v\nabla u) \cdot d\mathbf{S} = \iiint_V (u\nabla^2 v - v\nabla^2 u)\,dV \tag{A3.3}$$

or

$$\oiint_S \left(u\frac{\partial v}{\partial n} - v\frac{\partial u}{\partial n}\right) dS = \iiint_V (u\nabla^2 v = v\nabla^2 u)\,dV \tag{A3.4}$$

Green's second identity is also called *Green's theorem*.

CHAPTER FOUR

Planar Transmission Lines

Virtually all electronic circuits, from low-frequency digital to high-frequency RF and microwave circuits, employ transmission lines. Among transmission lines, printed circuit structures, also known as planar transmission lines, are most useful in modern electronics. Planar transmission lines have evolved from the advances of electronic integrated circuits (ICs) and, in turn, they have helped the progress of IC technology, such as making ICs that are more compact and more versatile and that have better interconnections and improved performance. Not only have planar transmission lines fulfilled their most fundamental objective of delivering signals, but they can also be exploited to create various RF and microwave components, such as wideband hybrid junctions, by appropriately combining them. The most well-known and commonly used planar transmission line is perhaps the microstrip line proposed in 1952 [1]. It, together with the strip line, forms the first planar version of the coaxial transmission line. Since then, various other planar transmission lines have been developed for use in RF and microwave ICs. Figure 1.1 and Table 1.1 describe commonly used planar transmission lines. Currently, the most commonly used structures for RF and microwave ICs are the *microstrip line*, the *coplanar waveguide* (CPW), the *coplanar strip* (CPS), the *strip line*, and the *slot line*.

In this chapter, we will first discuss the transmission line's static and dynamic parameters. We will then present a brief discussion of the most commonly used planar transmission lines, including the microstrip line, the coplanar waveguide (CPW), the coplanar strip (CPS), the strip line, and the slot line, along with the closed-form formulas for computing their characteristic impedances, effective dielectric constants, and losses. Although these parameters can accurately be determined using the methods presented in Chapters 5, 6, 7, and 8, closed-form expressions allow convenient and fast computer-aided design and analysis of RF and microwave circuits.

4.1 TRANSMISSION LINE PARAMETERS

Transmission lines with perfect conductors embedded in a homogeneous medium, such as the strip line, support a pure *transverse electromagnetic (TEM) mode* or *wave*. It is, however, also well known that transmission lines with more than two dielectrics, such as the microstrip line, can only support a *quasi-TEM mode*, which resembles a TEM mode. The TEM or quasi-TEM mode† is the dominant mode in a transmission line, having no cutoff frequency. The most important parameters of a transmission line for circuit design purposes are perhaps the characteristic impedance and effective dielectric constant for the TEM mode. As the operating frequency increases, an infinite number of modes may exist in a transmission line, including the TEM and hybrid modes. Hybrid modes are higher order modes with cutoff frequencies different from dc (0 Hz) and are undesirable. As discussed in Chapter 2, these modes are a combination of both the transverse electric (TE) and transverse magnetic (TM) modes and thus have the longitudinal components of both the electric and the magnetic fields.

There are two approaches in analyzing a transmission line: *static* or *quasi-static* and *dynamic* or *full-wave* approaches. The static or quasi-static† approach only produces transmission line parameters for the TEM mode that are valid only at dc. On the other hand, the dynamic approach can produce the transmission line parameters not only for the TEM mode but also for the hybrid modes, whose parameters are functions of frequency. In Chapters 5, 6, and 7, we present the conformal-mapping, variational, and spectral-domain methods for the static analysis. Dynamic approaches based on the spectral-domain and mode-matching methods are given in Chapters 7 and 8.

Transmission line parameters for the TEM mode obtained by the static approach are only valid theoretically at dc. In practice, however, these results can also be used at higher frequencies. Most engineers want to know up to what frequency they can use static results. There is no clear-cut answer for this. Some tend to believe that as long as the frequency is different from zero, static results cannot be used. On the other hand, most engineers use static results up to frequencies as high as 18 GHz. In fact, a number of millimeter-wave circuits operating up to W-band (75–110 GHz) have successfully been designed using only static results. Nevertheless, at high frequencies, especially those in the millimeter-wave region, a dynamic approach should be employed for more accurate determination of the transmission line parameters. The main trade-off between using static versus dynamic results is that the former is simpler to calculate but less accurate.

4.1.1 Static Analysis

Static analysis produces transmission line parameters, which are frequency independent. We now derive simple equations for the static characteristic impedance

† In general, we will use TEM to denote both TEM and quasi-TEM, and static for both static and quasi-static.

Figure 4.1 Cross section of a microstrip line with an overlay dielectric.

and effective dielectric constant of a transmission line. Without lost of generality, we consider a lossless microstrip line with an overlay dielectric, as shown in Fig. 4.1, as a representative transmission line.

The characteristic impedance of this transmission line is given as

$$Z_o = \sqrt{\frac{L}{C}} \tag{4.1}$$

where L and C are the inductance and capacitance per unit length of the transmission line, respectively. Let L_a and C_a represent the inductance and capacitance per unit length of the transmission line when the dielectrics are replaced by air. We can then rewrite Eq. (4.1), using the fact that the inductance per unit length does not depend on the surrounding substrates, as

$$\begin{aligned} Z_o &= \sqrt{\frac{L_a C_a}{C_a C}} \\ &= \frac{1}{c\sqrt{C C_a}} \end{aligned} \tag{4.2}$$

where $c = 3 \times 10^8$ m/s is the speed of light in free space.

The phase velocity of the quasi-TEM wave propagating along the transmission line is given as

$$v_p = \frac{c}{\sqrt{\varepsilon_{\text{eff}}}} \tag{4.3}$$

where ε_{eff} is the effective dielectric constant. Squaring both sides of Eq. (4.3) and making use of $L = L_a$, we obtain

$$\varepsilon_{\text{eff}} = \frac{C}{C_a} \tag{4.4}$$

The wavelength is given as

$$\lambda = \frac{\lambda_0}{\sqrt{\varepsilon_{\text{eff}}}} \tag{4.5}$$

where $\lambda_0 = c/f$ is the free-space wavelength.

All these equations are applicable to the quasi-TEM mode in any transmission line. It is now apparent that in order to determine the static characteristic impedance and effective dielectric constant of any transmission line, we only need to calculate the capacitances per unit length of the transmission line with and without the substrates. To do so, we would need to solve Laplace's equation as discussed in Chapters 5, 6, and 7.

4.1.2 Dynamic Analysis

Dynamic analysis produces a frequency-dependent characteristic impedance and propagation constant and, hence, attenuation and effective dielectric constants of a transmission line for TEM as well as hybrid modes. This approach begins by solving the wave equations discussed in Chapter 2 for the scalar electric and magnetic potentials subject to appropriate boundary conditions, and then using them to obtain solutions for the electric and magnetic fields. The process also yields an eigenvalue equation whose eigenvalue is the propagation constant $\gamma = \alpha + j\beta$, where α and β are the attenuation and phase constants, respectively. Illustrations of this approach using the spectral-domain and mode-matching methods are given in Chapters 6 and 7, respectively.

The phase constant β can be used to calculate the effective dielectric constant as

$$\varepsilon_{\text{eff}} = \left(\frac{\beta}{k_0}\right)^2 \tag{4.6}$$

where $k_0 = \omega\sqrt{\varepsilon_0\mu_0}$ is the free-space wave number. It should be noted that the phase constant must follow the condition

$$\omega\sqrt{\varepsilon_0\mu_0} \le \beta \le \omega\sqrt{\varepsilon_0\mu_0\varepsilon_r} \tag{4.7}$$

where ε_r has the highest value of relative dielectric constants used for the transmission line.

Using the electric and magnetic fields, we can calculate the (dynamic) characteristic impedance as a function of frequency. It is well known that the characteristic impedance for the TEM mode is unique since the mode's voltage and current are uniquely defined. For a hybrid mode, however, the characteristic impedance is not unique. There exists various definitions for the characteristic impedance, and the common ones are based on the voltage and current, power and current, and power and voltage as given respectively below:

$$Z_o^{VI} = \frac{V_o}{I_o} \tag{4.8}$$

$$Z_o^{PI} = \frac{2P_{\text{avg}}}{|I_o|^2} \tag{4.9}$$

$$Z_o^{PV} = \frac{|V_o|^2}{2P_{\text{avg}}} \tag{4.10}$$

where V_o is the voltage across the conductors, I_o is the longitudinal current, and P_{avg} represents the time-average power transmitted across the transmission line's cross section. These three characteristic impedances are related by

$$Z_o^{VI} = \sqrt{Z_o^{PI} Z_o^{PV}} \tag{4.11}$$

It should be noted that different definitions of the characteristic impedance give different numerical results, except at dc. For TEM mode, however, all three definitions give identical results. These definitions are also identical if $P_{\text{avg}} = \frac{1}{2}VI^*$. Generally, this condition is not valid for hybrid modes in a transmission line. The definitions based on the power and current and power and voltage are deduced from a two-conductor transmission line. The choice of a particular definition is not very clear. A possible choice is to use a definition according to a specific use of the transmission line in a circuit. A common definition for transmission lines containing slots, such as the slot line and CPW, is based on the power and voltage. For instance, the characteristic impedance of a CPW, shown in Fig. 4.3(a), can be obtained from Eq. (4.10) with the voltage and power given as

$$V_o = \int_a^b E_x(x, h)\, dx \tag{4.12}$$

$$P_{\text{avg}} = \tfrac{1}{2} \operatorname{Re} \int_{-\infty}^{\infty} \int_{-\infty}^{\infty} (E_x H_y^* - E_y H_x^*)\, dx\, dy \tag{4.13}$$

The characteristic impedance of the slot line, shown in Fig. 4.6, can also be defined by Eq. (4.10), where the power is again given by Eq. (4.13) and the voltage is obtained as

$$V_o = \int_{-W/2}^{W/2} E_x(x, h)\, dx \tag{4.14}$$

For other transmission lines without slots, the definitions based on the current and power and voltage and current are normally employed. For the case of microstrip line, shown in Fig. 4.2, the current, voltage, and power are given as

$$I_o = \int_{-W/2}^{W/2} J_z(x, h)\, dx \tag{4.15}$$

$$V_o = -\int_0^h E_y\, dy \tag{4.16}$$

$$P_{\text{avg}} = \tfrac{1}{2}\,\text{Re}\int_0^{\infty}\int_{-\infty}^{\infty}(E_x H_y^* - E_y H_x^*)\,dx\,dy \tag{4.17}$$

where J_z is the current density on the strip along the z direction.

4.2 MICROSTRIP LINE

The microstrip line was first proposed in 1952 [1], and since then it has become perhaps the best known and most widely used planar transmission line for RF and microwave ICs. This popularity and widespread use are due to its planar nature, ease of fabrication using photolithographic processes, easy integration with solid-state devices, good heat sinking, good mechanical support, and vast design information.

The closed-form expressions for the effective dielectric constant, ε_{eff}, and characteristic impedance, Z_o, of a microstrip line as shown in Fig. 4.2, assuming zero strip thickness ($t = 0$), are given as [2]

$$\varepsilon_{\text{eff}} = \frac{\varepsilon_{r+1}}{2} + \frac{\varepsilon_{r-1}}{2}\left(1 + 10\frac{h}{W}\right)^{-B} \tag{4.18}$$

and

$$Z_o = \frac{60}{\sqrt{\varepsilon_{\text{eff}}}}\ln\left[\frac{h}{W}A + \sqrt{1 + \left(\frac{2h}{W}\right)^2}\right] \tag{4.19}$$

where

$$A = 6 + (2\pi - 6)\exp\left[-\left(\frac{30.666h}{W}\right)^{0.7528}\right] \tag{4.20}$$

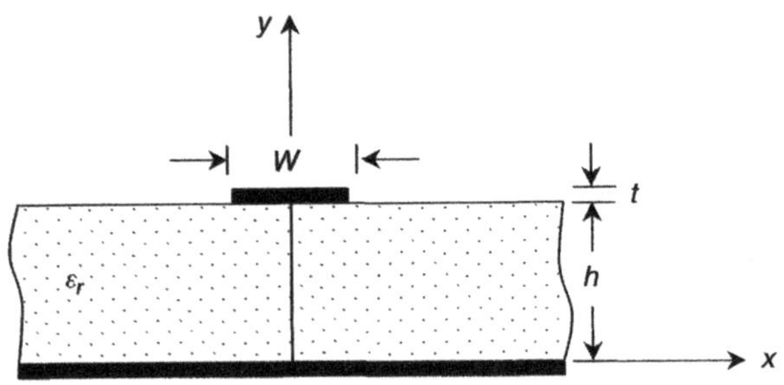

Figure 4.2 Cross section of a microstrip line.

$$B = 0.564\left\{1 + \frac{1}{49}\ln\left(\frac{(W/h)^4 + (W/52h)^2}{(W/h)^4 + 0.432}\right) + \frac{1}{18.7}\ln\left[1 + \left(\frac{W}{18.1h}\right)^3\right]\right\}\left(\frac{\varepsilon_r - 0.9}{\varepsilon_r + 3}\right)^{0.053} \tag{4.21}$$

Accuracy of the effective dielectric constants obtained from Eq. (4.18) is better than 0.2% for $\varepsilon_r \leq 128$ and $0.01 \leq W/h \leq 100$. For the characteristic impedance, maximum errors are 0.01% and 0.03% for $W/h \leq 1$ and 1000, respectively.

The normalized strip width, W/h, can also be determined from the characteristic impedance and relative dielectric constant as [3]

$$\frac{W}{h} = \begin{cases} \dfrac{8\exp(C)}{\exp(2C) - 2}, & \dfrac{W}{h} \leq 2 \\ \dfrac{2}{\pi}\left[D - 1 - \ln(2D - 1) + \dfrac{\varepsilon_r - 1}{2\varepsilon_r} \times \left(\ln(D - 1) + 0.39 - \dfrac{0.61}{\varepsilon_r}\right)\right], & \dfrac{W}{h} \geq 2 \end{cases} \tag{4.22}$$

where

$$C = \frac{Z_0}{60}\sqrt{\frac{\varepsilon_r + 1}{2}} + \frac{\varepsilon_r - 1}{\varepsilon_r + 1}\left(0.23 + \frac{0.11}{\varepsilon_r}\right) \tag{4.23}$$

$$D = \frac{60\pi^2}{Z_o\sqrt{\varepsilon_r}} \tag{4.24}$$

A practical microstrip line has a finite strip thickness† t, and its effect is to increase the strip width. Equations (4.18) and (4.19) can be modified to take into account t for more accurate results as [2]

$$\varepsilon_{\mathrm{eff}}(t) = \left[\frac{\varepsilon_r + 1}{2} + \frac{\varepsilon_r - 1}{2}\left(1 + \frac{10h}{W_e}\right)^{-B_t}\right]\frac{Z_o(t', \varepsilon_r = 1)}{Z_o(t, \varepsilon_r = 1)} \tag{4.25}$$

and

$$Z_o(t) = \frac{60}{\sqrt{\varepsilon_{\mathrm{eff}}(t)}}\ln\left[\frac{h}{W_e}A_t + \sqrt{1 + \left(\frac{2h}{W_e}\right)^2}\right] \tag{4.26}$$

where

$$A_t = 6 + (2\pi - 6)\exp\left[-\left(\frac{30.666h}{W_e}\right)^{0.7528}\right] \tag{4.27}$$

† Typical microwave substrates have copper as the metallization in 0.5, 1, and 2 oz. These correspond to thicknesses of 0.0007, 0.0014, and 0.0028 inch, respectively.

$$B_t = 0.564\left\{1+\frac{1}{49}\ln\left(\frac{(W_e/h)^4+(W_e/52h)^2}{(W_e/h)^4+0.432}\right)\right.$$

$$\left.+\frac{1}{18.7}\ln\left[1+\left(\frac{W_e}{18.1h}\right)^3\right]\right\}\left(\frac{\varepsilon_r-0.9}{\varepsilon_r+3}\right)^{0.053} \tag{4.28}$$

$$W_e = W + \Delta W \tag{4.29}$$

$$\Delta W = \frac{\Delta W'}{2}\left(1+\frac{1}{\cos h\sqrt{\varepsilon_r-1}}\right) \tag{4.30}$$

$$\Delta W' = \frac{t}{\pi}\ln\left(1+\frac{4\exp(1)}{(t/h)\coth^2\sqrt{6.517W/h}}\right) \tag{4.31}$$

$$W' = W + \Delta W' \tag{4.32}$$

$Z_o(t, \varepsilon_r = 1)$ and $Z_o(t', \varepsilon_r = 1)$ are the characteristic impedances obtained from Eq. (4.26) with $\varepsilon_{\text{eff}}(t) = 1$ and W_e given by Eqs. (4.29) and (4.32), respectively.

The frequency-dependent effective dielectric constant and characteristic impedance can be found [2] as

$$\varepsilon_{\text{eff}}(f) = \varepsilon_r - \frac{\varepsilon_r - \varepsilon_{\text{eff}}(0)}{1+G(f/f_p)^2} \tag{4.33}$$

and

$$Z_o(f) = Z_o(0)\frac{\varepsilon_{\text{eff}}(f)-1}{\varepsilon_{\text{eff}}(0)-1}\sqrt{\frac{\varepsilon_{\text{eff}}(0)}{\varepsilon_{\text{eff}}(f)}} \tag{4.34}$$

where

$$f_p = \frac{Z_o(0)}{2\mu_0 h} \tag{4.35}$$

$$G = \frac{\pi^2}{12}\frac{\varepsilon_r-1}{\varepsilon_{\text{eff}}(0)}\sqrt{\frac{Z_o(0)}{60}} \tag{4.36}$$

$\varepsilon_{\text{eff}}(0)$ and $Z_o(0)$ are the quasi-static effective dielectric constant and characteristic impedance, respectively, and $\mu_0 = 4\pi \times 10^{-7}$ H/m is the *permeability of free space.*

As for any transmission line, loss in the microstrip line is due to imperfect conductors and dielectrics and is characterized by the attenuation constant $\alpha = \alpha_{c+}\alpha_d$, where α_c and α_d represent the conductor and dielectric attenuation constants, respectively. α_c (in dB/cm) can be determined as follows [4, 5].

For $0 < W/h \le 1/2\pi$,

$$\alpha_c = \frac{8.68R_s}{2\pi Z_o h}\left[1-\left(\frac{W_e}{4h}\right)^2\right]\left\{1+\frac{h}{W_e}+\frac{h}{\pi W_e}\left[\ln\left(\frac{4\pi W}{t}\right)+\frac{t}{W}\right]\right\} \tag{4.37}$$

For $1/2\pi < W/h \leq 2$,

$$\alpha_c = \frac{8.68R_s}{2\pi Z_o h}\left[1-\left(\frac{W_e}{4h}\right)^2\right]\left\{1+\frac{h}{W_e}+\frac{h}{\pi W_e}\left[\ln\left(\frac{2h}{t}\right)-\frac{t}{h}\right]\right\} \tag{4.38}$$

For $2 \leq W/h$,

$$\alpha_c = \frac{8.68R_s}{Z_o h}\left\{1+\frac{h}{W_e}+\frac{h}{\pi W_e}\left[\ln\left(\frac{2h}{t}\right)-\frac{t}{h}\right]\right\} \times\left(\frac{W_e}{h}+\frac{W_e/\pi h}{W_e/2h+0.94}\right)\left[\frac{W_e}{h}+\frac{2}{\pi}\ln\left(\frac{W_e}{2h}+0.94\right)\right]^{-2} \tag{4.39}$$

where $R_s = \sqrt{\omega\mu_0/2\sigma}$ is the surface resistivity of the conductor, with σ being the conductivity. W_e is the effective strip width, taking into account the strip's finite metallization thickness, from Eq. (4.29). α_d can be found from [6]

$$\alpha_d = \frac{27.3\varepsilon_r(\varepsilon_{\text{eff}}-1)\tan\delta}{\sqrt{\varepsilon_{\text{eff}}}(\varepsilon_r-1)\lambda_0} \quad \text{(dB/m)} \tag{4.40}$$

where $\tan\delta$ is the loss tangent of the dielectric and λ_0 is the free-space wavelength. For planar transmission lines, dielectric loss is normally much lower than conductor loss, except when the dielectrics are low-resistivity semiconductor substrates such as silicon.

4.3 COPLANAR WAVEGUIDE

The coplanar waveguide (CPW) was first proposed in 1969 [7]. Since then it has rapidly gained widespread use for RF and microwave ICs due to its many attractive features, such as elimination of via holes in connecting circuit elements to ground, easy integration with solid-state devices, ease in realizing compact balanced circuits, and reduction of cross talk between lines. Figure 4.3 shows cross sections of a conventional and a conductor-backed CPW.

Closed-form expressions for the effective dielectric constant and characteristic impedance of these transmission lines with zero strip thickness were derived using conformal-mapping methods [8, 9]. For the conventional CPW, they are given below [and also in Chapter 5, Eqs. (5.71a)–(5.71d)] according to [8]

$$\varepsilon_{\text{eff}} = 1 + \frac{\varepsilon_r - 1}{2}\frac{K(k')}{K(k)}\frac{K(k_1)}{K(k_1')} \tag{4.41}$$

and

$$Z_o = \frac{30\pi}{\sqrt{\varepsilon_{\text{eff}}}}\frac{K(k')}{K(k)} \tag{4.42}$$

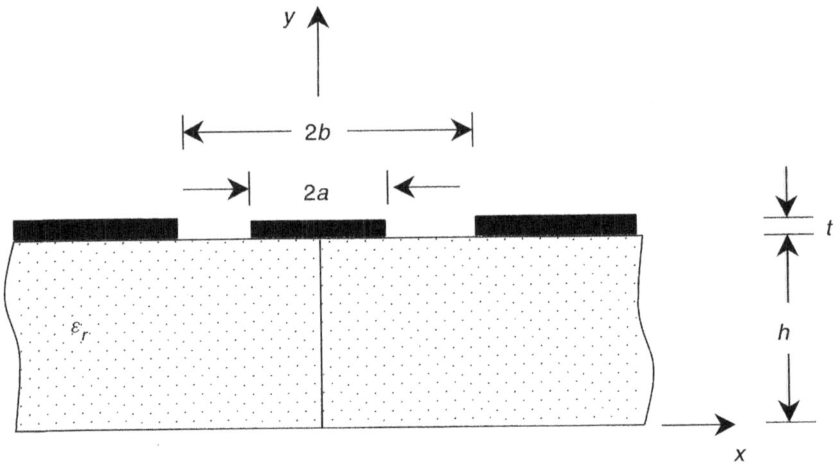

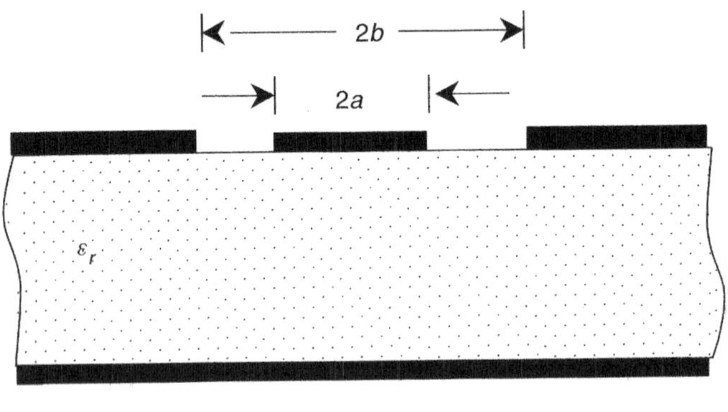

Figure 4.3 Cross sections of (a) convention and (b) conductor-backed CPWs.

where

$$k = \frac{a}{b} \tag{4.43}$$

$$k' = \sqrt{1 - k^2} \tag{4.44}$$

$$k_1 = \frac{\sinh(\pi a/2h)}{\sinh(\pi b/2h)} \tag{4.45}$$

$$k_1' = \sqrt{1 - k_1^2} \tag{4.46}$$

K represents the *complete integral of the first kind,* whose values can be determined from an integral or tabulated tables. The $K(k)/K(k')$ ratio can also be obtained approximately by [10]

$$\frac{K(k)}{K(k')} = \begin{cases} \dfrac{\pi}{\ln\left(2\dfrac{1+\sqrt{k'}}{1-\sqrt{k'}}\right)}, & 0 \le k \le 0.707 \\ \dfrac{1}{\pi}\ln\left(2\dfrac{1+\sqrt{k}}{1-\sqrt{k}}\right), & 0.707 \le k \le 1 \end{cases} \tag{4.47}$$

For the conductor-backed CPW, the effective dielectric constant and characteristic impedance can be evaluated from [9]

$$\varepsilon_{\text{eff}} = \frac{1 + \varepsilon_r \dfrac{K(k')}{K(k)} \dfrac{K(k_1)}{K(k_1')}}{1 + \dfrac{K(k')}{K(k)} \dfrac{K(k_1)}{K(k_1')}} \tag{4.48}$$

and

$$Z_o = \frac{60\pi}{\sqrt{\varepsilon_{\text{eff}}}} \frac{1}{\dfrac{K(k)}{K(k')} + \dfrac{K(k_1)}{K(k_1')}} \tag{4.49}$$

where

$$k = \frac{a}{b} \tag{4.50}$$

$$k' = \sqrt{1-k^2} \tag{4.51}$$

$$k_1 = \frac{\tanh(\pi a/2h)}{\tanh(\pi b/2h)} \tag{4.52}$$

$$k_1' = \sqrt{1-k_1^2} \tag{4.53}$$

These equations are also given in Chapter 5 [Eqs. (5.73a)–(5.73d)].

When the thickness t of the central strip and ground planes is considered, the widths of the strip and gaps effectively increase and decrease, respectively. The effective dielectric constant and characteristic impedance can be found, taking this effect into account, using the following formulas [11]:

$$\varepsilon_{\text{eff}}(t) = \varepsilon_{\text{eff}} - \frac{0.7(\varepsilon_{\text{eff}} - 1)\dfrac{t}{b-a}}{\dfrac{K(k)}{K(k')} + 0.7\dfrac{t}{b-a}} \tag{4.54}$$

and

$$Z_o = \frac{30\pi}{\sqrt{\varepsilon_{\text{eff}}(t)}} \frac{K(k_e')}{K(k_e)} \tag{4.55}$$

where

$$k_e = \frac{S_e}{S_e + 2W_e} \tag{4.56}$$

$$k'_e = \sqrt{1 - k_e^2} \tag{4.57}$$

$$S_e = 2a + \Delta \tag{4.58}$$

$$W_e = b - a - \Delta \tag{4.59}$$

$$\Delta = \frac{1.25t}{\pi}\left[1 + \ln\left(\frac{8\pi a}{t}\right)\right] \tag{4.60}$$

The conductor attenuation constant for the conventional CPW is given as [12]

$$\alpha_c = \frac{4.88 \times 10^4}{\pi} R_s \varepsilon_{\text{eff}}\, Z_o P \frac{b + a}{(b - a)^2} \times \left\{ \frac{\dfrac{1.25t}{\pi} \ln\left(\dfrac{8\pi a}{t}\right) + 1 + \dfrac{1.25t}{2\pi a}}{\left[2 + \dfrac{2a}{b - a} - \dfrac{1.25t}{\pi(b - a)}\left(1 + \ln\dfrac{8\pi a}{t}\right)\right]^2} \right\} \quad (\text{dB/m}) \tag{4.61}$$

where

$$P = \begin{cases} \dfrac{k}{(1 - k')(k')^{3/2}}\left[\dfrac{K(k)}{K(k')}\right]^2, & 0 \le k \le 0.707 \\ \dfrac{1}{(1 - k)\sqrt{k}}, & 0.707 \le k \le 1 \end{cases} \tag{4.62}$$

The expression for the dielectric attenuation constant is the same as that for the microstrip line, which is given in Eq. (4.40).

4.4 COPLANAR STRIPS

Coplanar strips (CPS), as shown in Fig. 4.4, employ two parallel strips on the same side of a dielectric substrate. This structure is also useful for RF and microwave ICs, especially balanced circuits due to its inherent balanced nature. The CPS structure allows easy connections for series and shunt solid-state devices. Its effective dielectric constant and characteristic impedance for zero strip thickness can be evaluated using the following closed-form equations derived from a conformal-mapping method [8]:

$$\varepsilon_{\text{eff}} = 1 + \frac{\varepsilon_r - 1}{2} \frac{K(k')}{K(k)} \frac{K(k_1)}{K(k'_1)} \tag{4.63}$$

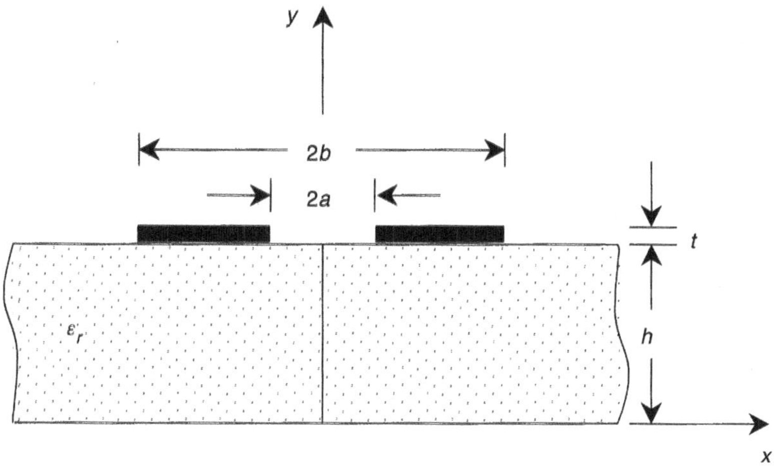

Figure 4.4 Cross section of coplanar strips.

and

$$Z_o = \frac{120\pi}{\sqrt{\varepsilon_{\text{eff}}}} \frac{K(k)}{K(k')} \tag{4.64}$$

where

$$k = \frac{a}{b} \tag{4.65}$$

$$k' = \sqrt{1 - k^2} \tag{4.66}$$

$$k_1 = \frac{\sinh(\pi a/2h)}{\sinh(\pi b/2h)} \tag{4.67}$$

$$k_1' = \sqrt{1 - k_1^2} \tag{4.68}$$

These equations are also listed in Chapter 5 [Eqs. (5.77a)–(5.77d)].

When the strip thickness is considered, the effective dielectric constant and characteristic impedance can be determined using [11]

$$\varepsilon_{\text{eff}}(t) = \varepsilon_{\text{eff}} - \frac{0.7(\varepsilon_{\text{eff}} - 1)\dfrac{t}{a}}{\dfrac{K(k')}{K(k)} + 0.7\dfrac{t}{a}} \tag{4.69}$$

and

$$Z_o(t) = \frac{120\pi}{\sqrt{\varepsilon_{\text{eff}}(t)}} \frac{K(k_e)}{K(k_e')} \tag{4.70}$$

where

$$k_e = \frac{S_e}{S_e + 2W_e} \tag{4.71}$$

$$k'_e = \sqrt{1 - k_e^2} \tag{4.72}$$

$$S_e = S - \Delta \tag{4.73}$$

$$W_e = W + \Delta \tag{4.74}$$

$$\Delta = \frac{1.25t}{\pi}\left[1 + \ln\left(4\pi\frac{b-a}{t}\right)\right] \tag{4.75}$$

The attenuation due to the conductor loss can be found from [12]

$$\alpha_c = \frac{4.34}{\pi}\frac{R_s}{Z_o}P\frac{a+b}{a^2}\left(\frac{\frac{1.25}{\pi}\ln\left(4\pi\frac{b-a}{t}\right) + 1 + \frac{1.25}{\pi}\frac{t}{b-a}}{\left\{1 + \frac{b-a}{a} + \frac{1.25}{2\pi}\frac{t}{a}\left[1 + \ln\left(4\pi\frac{b-a}{t}\right)\right]\right\}^2}\right) \tag{4.76}$$

where P is given in Eq. (4.62). The dielectric attenuation constant can be determined from Eq. (4.40).

4.5 STRIP LINE

The strip line, as shown in Fig. 4.5, is essentially a printed circuit version of the coaxial transmission line. Its dominant mode of propagation is pure TEM, assuming perfect conductors. Because of the possibility of exciting unwanted parallel-plate modes at high frequencies due to two parallel ground planes, the strip line is more suitable for use at low microwave frequencies. Its characteristic impedance for a zero-thickness strip can be determined using the following conformal-mapping formulas [13], which are also given in Chapter 5

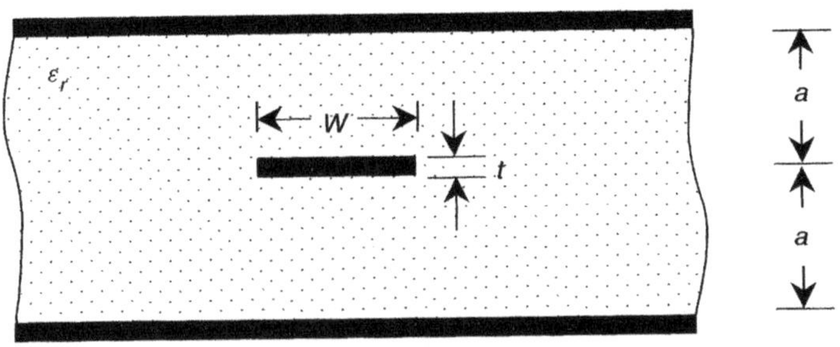

Figure 4.5 Cross section of a strip line.

[Eqs. (5.69a) and (5.69b)]:

$$Z_o = \frac{30\pi}{\sqrt{\varepsilon_r}} \frac{K(k')}{K(k)} \tag{4.77}$$

where

$$k = \tanh\left(\frac{\pi W}{4a}\right) \tag{4.78}$$

$$k' = \sqrt{1-k^2} \tag{4.79}$$

For finite strip thickness, the characteristic impedance can be found from [14]

$$Z_o = \frac{30}{\sqrt{\varepsilon_r}} \ln\left\{1 + \frac{4}{\pi}\frac{2a-t}{W_e}\left[\frac{8}{\pi}\frac{2a-t}{W_e} + \sqrt{\left(\frac{8}{\pi}\frac{2a-t}{W_e}\right)^2 + 6.27}\right]\right\} \tag{4.80}$$

where

$$\frac{W_e}{2a-t} = \frac{W}{2a-t} + \frac{\Delta W}{2a-t} \tag{4.81}$$

$$\frac{\Delta W}{2a-t} = \frac{x}{\pi(1-x)}\left\{1 - \frac{1}{2}\ln\left[\left(\frac{x}{2-x}\right)^2 + \left(\frac{0.0796x}{W/2a + 1.1x}\right)^m\right]\right\} \tag{4.82}$$

$$m = \frac{2}{1 + 2x/3(1-x)} \tag{4.83}$$

$$x = \frac{t}{2a} \tag{4.84}$$

The strip width can also be determined from the characteristic impedance and relative dielectric constant. For a zero-thickness strip, the expression for the strip width can be derived from Eqs. (4.77)–(4.79) as

$$\frac{W}{a} = \frac{4}{\pi}\tanh^{-1}\sqrt{p} \tag{4.85}$$

where

$$p = \begin{cases} \sqrt{1 - (e^{\pi q - 2}/e^{\pi q + 2})^4}, & q \geq 1 \\ [(e^{\pi/q} - 2)/(e^{\pi/q} + 2)]^2, & 0 \leq q \leq 1 \end{cases} \tag{4.86}$$

$$q = \frac{Z_o\sqrt{\varepsilon_r}}{30\pi} \tag{4.87}$$

Using Eqs. (4.80)–(4.84), we can derive the following expression for the strip width, taking into account the strip thickness:

$$\frac{W}{2a-t} = \frac{W_e}{2a-t} - \frac{\Delta W}{2a-t} \tag{4.88}$$

where

$$\frac{W_e}{2a - t} = \frac{8}{\pi A} \tag{4.89}$$

$$\frac{\Delta W}{2a - t} = \frac{x}{\pi(1 - x)} \left\{ 1 - \frac{1}{2} \ln \left[\left(\frac{x}{2 - x} \right)^2 + \left(\frac{0.0796x}{W_e/2a + 1.1x} \right)^m \right] \right\} \tag{4.90}$$

$$A = \frac{2B}{C} \tag{4.91}$$

$$B = \exp\left(\frac{Z_o \sqrt{\varepsilon_r}}{30} \right) - 1 \tag{4.92}$$

$$C = \sqrt{4B + 6.27} \tag{4.93}$$

$$m = \frac{2}{1 + 2x/3(1 - x)} \tag{4.94}$$

$$x = \frac{t}{2a} \tag{4.95}$$

The conductor attenuation constant (in dB/m) can be determined from

$$\alpha_c = \begin{cases} \dfrac{23.4 \times 10^{-3} R_s \varepsilon_r Z_o A}{30\pi(2a - t)}, & Z_o < \dfrac{120}{\sqrt{\varepsilon_r}} \\ \dfrac{1.4 R_s B}{2 Z_o a}, & Z_o > \dfrac{120}{\sqrt{\varepsilon_r}} \end{cases} \tag{4.96}$$

where

$$A = 1 + \frac{2W}{2a - t} + \frac{1}{\pi} \frac{2a + t}{2a - t} \ln\left(\frac{4a - t}{t} \right) \tag{4.97}$$

$$B = 1 + \frac{2a}{0.5W + 0.7t} \left(0.5 + \frac{0.414t}{W} + \frac{1}{2\pi} \ln \frac{4\pi W}{t} \right) \tag{4.98}$$

The dielectric attenuation constant is given by

$$\alpha_d = \frac{27.3 \sqrt{\varepsilon_r} \tan\delta}{\lambda_0} \quad \text{(dB/m)} \tag{4.99}$$

4.6 SLOT LINE

The slot line, as shown in Fig. 4.6, is also useful for RF and microwave ICs. Its balanced nature is especially attractive for circuits requiring a balanced topology. In contrast with conventional transmission lines, the slot line does not support

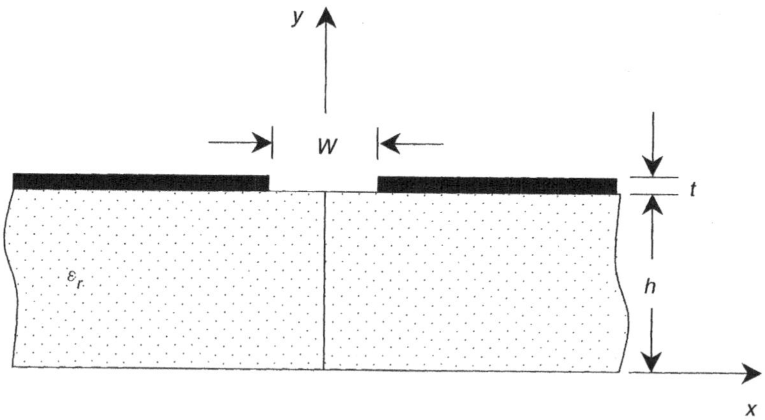

Figure 4.6 Cross section of a slot line.

a TEM propagation mode. Modes propagating on the slot the line are quasi-TE modes, which resemble the TE type. The dominant mode is quasi-TE_{10}, similar to the TE_{10} of the rectangular waveguide. The quasi-TE_{10} mode of the slot line, however, has no cutoff frequency. In order to be used as a good transmission line, the slot line must be fabricated using a substrate with a high dielectric constant to minimize radiation. On the other hand, an antenna using the slot line should be implemented using a low dielectric constant substrate.

Closed-form expressions for the characteristic impedance, Z_o, based on voltage and power and the wavelength, λ_g, for the slot line on high dielectric constant substrates ($9.7 \leq \varepsilon_r \leq 20$) were obtained [15] by curve fitting to Cohn's [16] numerical results. They are given as follows.

For $0.02 \leq W/h \leq 0.2$,

$$\begin{aligned} \lambda_g/\lambda_0 &= 0.923 - 0.195 \ln \varepsilon_r + 0.2W/h \\ &\quad - (0.126W/h + 0.02) \ln(h/\lambda_0 \times 10^2) \end{aligned} \tag{4.100}$$

$$\begin{aligned} Z_o &= 72.62 - 15.283 \ln \varepsilon_r + 50 \frac{(W/h - 0.02)(W/h - 0.1)}{W/h} \\ &\quad + \ln(W/h \times 10^2)(19.23 - 3.693 \ln \varepsilon_r) \\ &\quad - [0.139 \ln \varepsilon_r - 0.11 + W/h(0.465 \ln \varepsilon_r + 1.44)] \\ &\quad \times (11.4 - 2.636 \ln \varepsilon_r - h/\lambda_0 \times 10^2)^2 \end{aligned} \tag{4.101}$$

For $0.2 \leq W/h \leq 1.0$,

$$\begin{aligned} \lambda_g/\lambda_0 &= 0.987 - 0.21 \ln \varepsilon_r + W/h(0.111 - 0.0022\varepsilon_r) \\ &\quad - (0.053 + 0.041W/h - 0.0014\varepsilon_r) \ln(h/\lambda_0 \times 10^2) \end{aligned} \tag{4.102}$$

$$
\begin{aligned}
Z_o = {} & 113.19 - 23.257 \ln \varepsilon_r + 1.25W/h(114.59 - 22.531 \ln \varepsilon_r) \\
& + 20(W/h - 0.2)(1 - W/h) \\
& - [0.15 + 0.1 \ln \varepsilon_r + W/h(-0.79 + 0.899 \ln \varepsilon_r)] \\
& \times [10.25 - 2.171 \ln \varepsilon_r + W/h(2.1 - 0.617 \ln \varepsilon_r) - h/\lambda_0 \times 10^2]^2 \quad (4.103)
\end{aligned}
$$

These equations were obtained assuming infinitesimally thin conductors and are accurate to within 2% for the following ranges:

$$9.7 \le \varepsilon_r \le 20 \tag{4.104}$$

$$0.01 \le \frac{h}{\lambda_0} \le \left(\frac{h}{\lambda_0}\right)_c \tag{4.105}$$

where $(h/\lambda_0)_c$ is the cutoff value for the TE_{10} surface-wave mode on the slot line and is given as

$$\left(\frac{h}{\lambda_0}\right)_c = 0.25\sqrt{\varepsilon_r - 1} \tag{4.106}$$

Closed-form expressions for the characteristic impedance and wavelength for low dielectric constant substrates were also derived by curve fitting to results of the spectral-domain method [17].

REFERENCES

1. D. D. Grieg and H. F. Englemann, "Microstrip—A New Transmission Technique for the Kilomegacycle Range," *Proc. IRE*, Vol. 40, pp. 1644–1650, Dec. 1952.
2. E. O. Hammerstad and O. Jensen, "Accurate Models for Microstrip Computer-Aided Design," *1980 IEEE MTT-S Digest*, pp. 407–409.
3. E. O. Hammerstad, "Equation for Microstrip Circuit Design," *Proc. Eur. Microwave Conf.*, pp. 268–272, 1975.
4. R. A. Pucel, D. J. Masse, and C. P. Hartwig, "Losses in Microstrip," *IEEE Trans. Microwave Theory Tech.*, Vol. MTT-16, pp. 342–350, June 1968.
5. R. A. Pucel, D. J. Masse, and C. P. Hartwig, "Correction to Losses in Microstrip," *IEEE Trans. Microwave Theory Tech.*, Vol. MTT-16, p. 1064, Dec. 1968.
6. J. D. Welch and H. J. Pratt, "Losses in Microstrip Transmission Systems for Integrated Microwave Circuits," *NEREM Rec.*, Vol. 8, pp. 100–101, 1966.
7. C. P. Wen, "Coplanar Waveguide: A Surface Strip Transmission Line Suitable for Nonreciprocal Gyromagnetic Device Application," *IEEE Trans. Microwave Theory Tech.*, Vol. MTT-17, pp. 1087–1090, Dec. 1969.
8. G. Ghione and C. Naldi, "Analytical Formulas for Coplanar Lines in Hybrid and Monolithic MICs," *Electron. Lett.*, Vol. 20, No. 4, pp. 179–181, Feb. 1984.
9. G. Ghione and C. Naldi, "Parameters of Coplanar Waveguides with Lower Ground Plane," *Electron. Letters*, Vol. 19, No. 18, pp. 734–735, Sept. 1983.

10. W. Hilberg, "From Approximation to Exact Relations for Characteristic Impedances," *IEEE Trans. Microwave Theory Tech.*, Vol. MTT-13, pp. 29–38, Jan. 1965.
11. K. C. Gupta, R. Garg, and I. J. Bahl, *Microstrip Lines and Slotlines*, Artech House, Dedham, MA, 1979, pp. 277–280.
12. K. C. Gupta, R. Garg, and R. Chadha, *Computer-Aided Design of Microwave Circuits*, Artech House, Dedham, MA, 1981, p. 72.
13. S. B. Cohn, "Problems in Strip Transmission Lines," *IRE Trans. Microwave Theory Tech.*, Vol. MTT-2, pp. 52–55, July 1954.
14. H. A. Wheeler, "Transmission Line Properties of a Stripline Between Parallel Planes," *IEEE Trans. Microwave Theory Tech.*, Vol. MTT-26, pp. 866–876, Nov. 1978.
15. R. Garg and K. C. Gupta, "Expressions for Wavelength and Impedance of Slotline," *IEEE Trans. Microwave Theory Tech.*, Vol. MTT-24, p. 532, Aug. 1976.
16. S. B. Cohn, "Slot Line on a Dielectric Substrate," *IEEE Trans. Microwave Theory Tech.*, Vol. MTT-17, pp. 768–778, Oct. 1969.
17. R. Janaswamy and D. H. Schaubert, "Characteristic Impedance of a Wide Slotline on Low-Permittivity Substrates," *IEEE Trans. Microwave Theory Tech.*, Vol. MTT-34, pp. 900–902, Aug. 1986.

PROBLEMS

4.1 The quasi-TEM characteristic impedance of a lossless transmission line may be determined from

$$Z_o = \frac{1}{c\sqrt{CC_a}}$$

where C and C_a are the capacitances per unit length with and without the dielectrics, and c is the speed of light in air. Use this equation to derive Z_o of a lossless coaxial line having inner and outer radii of a and b, respectively. The coaxial line is filled with a dielectric having relative dielectric constant ε_r.

4.2 Prove that the three commonly used definitions of the dynamic characteristic impedance given in Eqs. (4.8)–(4.10) are identical if $P_{\text{avg}} = \frac{1}{2}\,\text{Re}(VI^*)$.

4.3 Show that the three definitions of characteristic impedance in Eqs. (4.8)–(4.10) are derived from the formula for the characteristic impedance of a TEM transmission line.

4.4 Derive the following characteristic impedances of a rectangular waveguide, having dimensions a and b along the x and y axes, respectively, for TE_{m0} modes:

(a) $Z_o^{VI} = \dfrac{V_o}{I_o} = \dfrac{m\pi b}{2a} Z_h; \quad m = 1, 3, 5, \ldots$

(b) $Z_o^{PI} = \dfrac{2P_{\text{avg}}}{|I_o|^2} = \dfrac{m^2\pi^2 b}{8a} Z_h; \quad m = 1, 3, 5, \ldots$

(c) $Z_o^{PV} = \dfrac{|V_o|^2}{2P_{\text{avg}}} = \dfrac{2b}{a} Z_h$

where $Z_h = j\omega\mu/\gamma$ is the wave impedance, and $\gamma = \sqrt{(m\pi/a)^2 - (2\pi/\lambda_g)^2}$ is the propagation constant.

4.5 Write a computer program to calculate the characteristic impedance, Z_o, and effective dielectric constant, ε_{eff}, of a microstrip line, with zero strip thickness, as a function of the normalized strip width, *W/h*. Compute and plot Z_o and ε_{eff} versus *W/h* from 0.05 to 5 for relative dielectric constant, ε_r, of 2.2 and 10.5.

4.6 Write a computer program to calculate the characteristic impedance, Z_o, and effective dielectric constant, ε_{eff}, of a microstrip line with zero strip thickness as a function of frequency, f. Compute and plot Z_o and ε_{eff} versus f from 0 to 40 GHz for normalized strip width, *W/h*, of 0.1, 0.5, 1, 2, and 5, and two different substrates: $\varepsilon_r = 10.5$, $h = 0.635$ mm (RT/Duroid 6010) and $\varepsilon_r = 2.2$, $h = 0.635$ mm (RT/Duroid 5880). Based on these results, compare the dispersions between high- and low-dielectric-constant substrates, and between thin and thick substrates.

4.7 Write a computer program to calculate the characteristic impedance, Z_o, and effective dielectric constant, ε_{eff}, of a microstrip line with finite strip thickness as a function of the normalized strip width, *W/h*. Compute and plot Z_o and ε_{eff} versus *W/h* from 0.05 to 5 for a relative dielectric constant, ε_r, of 2.2 and 10.5, and normalized strip thickness, *t/h*, of 0, 0.01, 0.02, 0.05, and 0.1.

4.8 Write a computer program to calculate the conductor, α_c, and dielectric, α_d, attenuation constants of a microstrip line as a function of the normalized strip width, *W/h*. Compute and plot α_c and α_d versus *W/h* from 0.05 to 5 for RT/Duroid 5880 substrate with relative dielectric constant, ε_r, of 2.2 and loss tangent of 0.0009, normalized strip thickness, *t/h*, of 0, 0.01, 0.02, 0.05, and 0.1, and a strip conductivity, σ, of 5.8×10^7 S/m (copper).

4.9 Write a computer program to calculate the normalized strip width, *W/h*, of a microstrip line with zero strip thickness as a function of the characteristic impedance, Z_o. Compute and plot *W/h* versus Z_o from 30 to 120 ohms for a relative dielectric constant, ε_r, of 2.2 and 10.5.

4.10 Write a computer program to calculate the characteristic impedance, Z_o, and effective dielectric constant, ε_{eff}, of a conventional CPW with zero strip thickness as a function of the dimension ratio, *a/b*. Compute and plot Z_o and ε_{eff} versus *a/b* from 0.1 to 0.9 for a relative dielectric constant, ε_r, of 2.2 and 10.5 and normalized substrate thickness, *h/b*, of 0.1, 0.5, 4, and 10.

4.11 Repeat Problem 4.10 for a conductor-backed CPW with zero strip thickness.

4.12 Write a computer program to calculate the characteristic impedance, Z_o, and effective dielectric constant, ε_{eff}, of a conventional CPW with finite strip thickness as a function of the dimension ratio, *a/b*. Compute and plot Z_o and ε_{eff} versus *a/b* from 0.1 to 0.9 for a relative dielectric constant, ε_r, of 10.5, normalized substrate thickness, *h/b*, of 0.1, 0.5, 4, and 10, and normalized strip thickness, *t/a*, of 0, 0.01, 0.02, 0.05, and 0.1.

4.13 Repeat Problem 4.12 for a conductor-backed CPW.

4.14 Write a computer program to calculate the conductor, α_c, and dielectric, α_d, attenuation constants of a conventional CPW as a function of the dimension ratio, *a/b*. Compute and plot α_c and α_d versus *a/b* from 0.1 to 0.9 for RT/Duroid 6010 substrate with a relative dielectric constant, ε_r, of 10.5 and loss tangent of 0.0028, normalized strip thickness, *t/a*, of 0, 0.01, 0.02, 0.05, and 0.1, and a strip conductivity, σ, of 5.8×10^7 S/m (copper).

4.15 Write a computer program to calculate the characteristic impedance, Z_o, and effective dielectric constant, ε_{eff}, of a CPS with zero strip thickness as a function of the dimension ratio, *a/b*. Compute and plot Z_o and ε_{eff} versus *a/b* from 0.1 to 0.9, for a relative dielectric constant, ε_r, of 2.2 and 10.5, and normalized substrate thickness, *h/b*, of 0.1, 0.5, 4, and 10.

4.16 Write a computer program to calculate the characteristic impedance, Z_o, and effective dielectric constant, ε_{eff}, of a CPS as a function of the dimension ratio, *a/b*. Compute and plot Z_o and ε_{eff} versus *a/b* from 0.1 to 0.9 for a relative dielectric constant, ε_r, of 10.5, normalized substrate thickness, *h/b*, of 0.1, 0.5, 4, and 10, and normalized strip thickness, *t/2a*, of 0, 0.01, 0.02, 0.05, and 0.1.

4.17 Write a computer program to calculate the conductor, α_c, and dielectric, α_d, attenuation constants of a CPS as a function of the dimension ratio *a/b*. Compute and plot α_c and α_d versus *a/b* from 0.1 to 0.9 for RT/Duroid 6010 substrate with a relative dielectric constant, ε_r, of 10.5 and loss tangent of 0.0028, normalized strip thickness, *t/2a*, of 0, 0.01, 0.02, 0.05, and 0.1, and a strip conductivity, σ, of 5.8×10^7 S/m (copper).

4.18 Write a computer program to calculate the characteristic impedance, Z_o, of a strip line with zero strip thickness as a function of the normalized strip width, *W/a*. Compute and plot Z_o versus *W/a* from 0.1 to 5 for a relative dielectric constant, ε_r, of 2.2 and 10.5.

4.19 Write a computer program to calculate the characteristic impedance, Z_o, of a strip line with finite strip thickness as a function of the normalized strip width, *W/a*. Compute and plot Z_o versus *W/a* from 0.1 to 5 for a relative dielectric constant, ε_r, of 2.2 and 10.5, and normalized strip thickness, *t/a*, of 0, 0.01, 0.02, 0.050, 0.1, and 0.2.

4.20 Write a computer program to calculate the normalized strip width, *W/a*, of a strip line with finite strip thickness as a function of the characteristic impedance, Z_o. Compute and plot *W/a* versus Z_o from 10 to 100 ohms for a relative dielectric constant, ε_r, of 2.2 and 10.5, and normalized strip thickness, t/a, of 0, 0.01, 0.02, 0.05, 0.1, and 0.2.

4.21 Write a computer program to calculate the conductor, α_c, and dielectric, α_d, attenuation constants of a strip line as a function of the normalized strip width, W/a. Compute and plot α_c and α_d versus W/a from 0.1 to 5 for RT/Duroid 5880 substrate with a relative dielectric constant, ε_r, of 2.2 and loss tangent of 0.0009, normalized strip thickness, t/a, of 0, 0.01, 0.02, 0.050, 0.1, and, 0.2, and a strip conductivity, σ, of 5.8×10^7 S/m (copper).

4.22 Write a computer program to calculate the characteristic impedance, Z_o, and effective dielectric constant, ε_{eff}, of a slot line as a function of the normalized slot width, W/h. Compute and plot Z_o and ε_{eff} versus W/h from 0.1 to 2 for a relative dielectric constant, ε_r, of 10.5, and frequency of 1, 10, and 20 GHz.

CHAPTER FIVE

Conformal Mapping

Conformal mapping is a useful mathematical analysis for solving many boundary problems in various engineering disciplines. Perhaps the most elegant features of conformal mapping are its simplicity and ability to produce closed-form formulas. Conformal mapping can be employed to obtain solutions for many electromagnetic problems for both static and dynamic situations. In particular, this analytical method has been used to analyze various transmission lines used in RF and microwave integrated circuits (e.g., [1]). In this chapter, we will present the fundamentals of this method [2] and apply it to determine the static or quasi-static parameters, such as characteristic impedances and effective dielectric constants, of planar transmission lines.

5.1 PRINCIPLES OF MAPPINGS

Mapping by itself is a very elementary geometric technique. In many mathematical problems, we have encountered a function f describing the relation between two complex variables $z = x + jy$ and $w = u + jv$ as $w = f(z)$. Geometrically, these complex variables are represented by points $z = (x, y)$ and $w = (u, v)$, respectively. Consequently, u and v are effectively a function of both x and y. In polar coordinates, z and w are given as $z = re^{j\theta}$ and $w = \rho e^{j\phi}$, where r, ρ and θ, ϕ are the corresponding magnitudes or moduli and the angles or arguments of the complex numbers. In this mathematical description, the function f performs a *transformation* or *mapping* between the two corresponding points z and w in general or, specifically, *transforms* or *maps* point z into the corresponding point w according to a rule described by the function f. With this concept, f is thus referred to as a transformation or mapping. Geometrically, the function f then maps or transforms a curve or a region containing points (x, y) into another curve or region consisting of points (u, v). For simplicity and convenience of mathematical analysis, separate planes are normally used for the coordinates of points z and w. We will refer to the planes containing points z and w as the (complex) z and w planes, respectively. Figure 5.1 illustrates the principle of the mapping of z into w, in which C and C' contain the corresponding points z and w, respectively.

As an example, we consider point z in the z plane described in polar form as $z = re^{j\theta}$. We assume that point z is related to point $w(u, v)$ in the w plane by the function or transformation

$$w = f(z) = \ln z \tag{5.1}$$

Assume that point z lies in a circle of radius a. We wish to map this circle into a curve in the w plane. From Eq. (5.1), we have $u = \ln r$ and $v = \theta$, which indicate that points z on the circle with $r = a$ and $0 \leq \theta \leq 2\pi$ in the z plane correspond to points w having $u = \ln a$ and $0 \leq v \leq 2\pi$ in the w plane. The circle of radius a in the z plane is thus mapped onto the line segment $u = a$, $0 \leq v \leq 2\pi$ of the w plane. This mapping of a circle onto a line segment is useful in analyses of some microwave structures. Figure 5.2 shows this mapping.

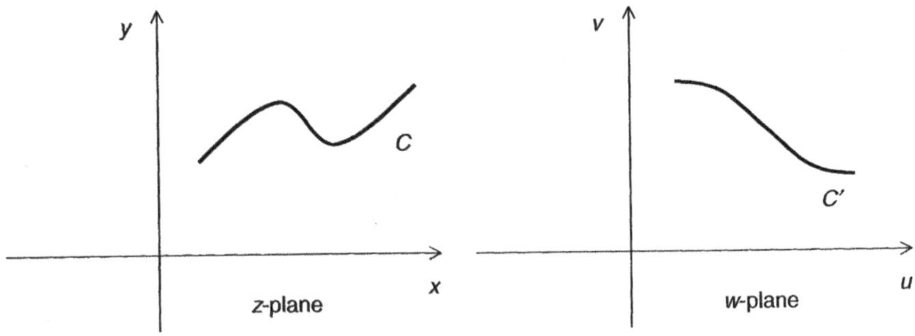

Figure 5.1 General mapping from $z = (x, y)$ into $w = (u, v) = f(z)$.

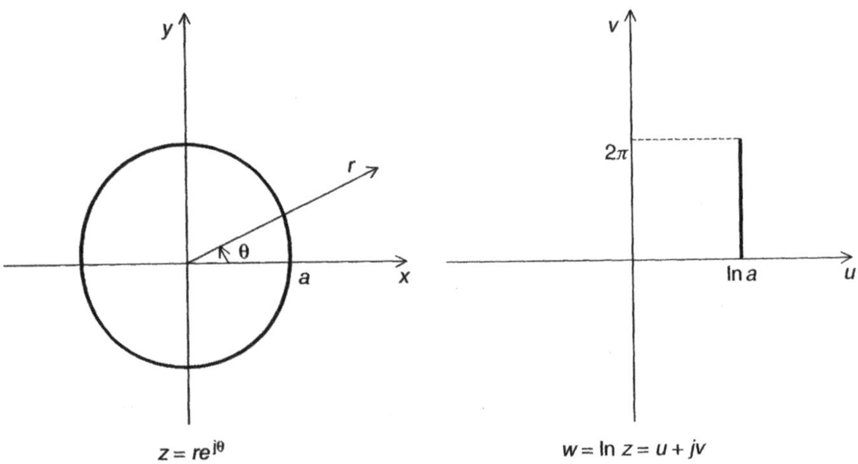

Figure 5.2 Mapping between a circle and a line segment.

This example demonstrates that we can map a rather complex geometry into a simpler one. This leads to the possibility that we may be able to transform mathematically a geometrically complicated transmission line into a much simpler structure whose solution is known or can easily be found. This is the principle behind applications of conformal mapping in microwave problems. Figure 5.3 shows some useful transformations of regions along with the mapping functions [2].

5.2 FUNDAMENTALS OF CONFORMAL MAPPING

We consider a transformation $w = f(z)$ that maps point $z(x, y)$ into point $w(u, v)$. If the function f is *analytic* at point z and $f'(z) = df/dz \neq 0$, then the mapping is said to be *conformal. Conformal mapping*, or *conformal transformation*, is therefore a mapping that employs an analytic function whose derivative is always nonzero on the z plane where the function is defined. As will be seen, this kind of mapping is especially useful for analyzing planar transmission lines.

A function f is analytic at a point z if its derivative exists and is unique at z. The function is then said to be analytic in a particular region if it is analytic at every point in that region. This requirement is met if f satisfies both the necessary and sufficient conditions for analyticity. A necessary, but not sufficient, condition for f to be analytic is that the first-order partial derivatives of its real part u and imaginary part v with respect to x and y must satisfy the *Cauchy–Riemann equations*:

$$\frac{\partial u}{\partial x} = \frac{\partial v}{\partial y} \tag{5.2a}$$

$$\frac{\partial u}{\partial y} = -\frac{\partial v}{\partial x} \tag{5.2b}$$

Any two functions that satisfy the Cauchy–Riemann equations are called *harmonic conjugate functions* or simply *conjugate functions*. A sufficient condition for the existence of analyticity is that the first-order partial derivatives of u and v must exist in the neighborhood of z, where f is defined, and are continuous and satisfy the Cauchy–Riemann equations at z.

Taking the partial derivatives of Eqs. (5.2a) and (5.2b) with respect to x and y, respectively, and adding the resulting equations, we obtain

$$\frac{\partial^2 u}{\partial x^2} + \frac{\partial^2 u}{\partial y^2} = 0 \tag{5.3}$$

Similarly, differentiating Eqs. (5.2a) and (5.2b) with respect to y and x, respectively, and subtracting the results gives

$$\frac{\partial^2 v}{\partial x^2} = \frac{\partial^2 v}{\partial y^2} = 0 \tag{5.4}$$

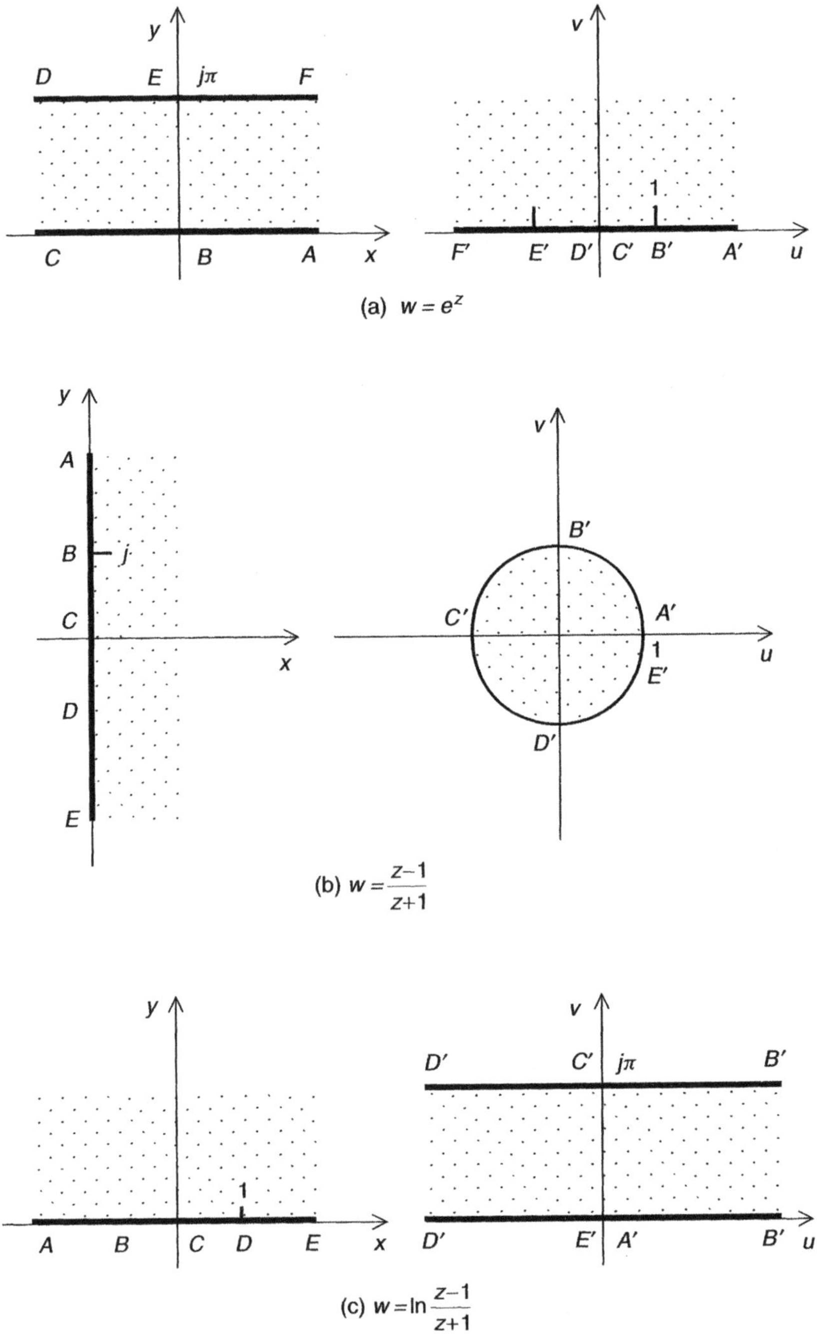

Figure 5.3 Some useful mappings.

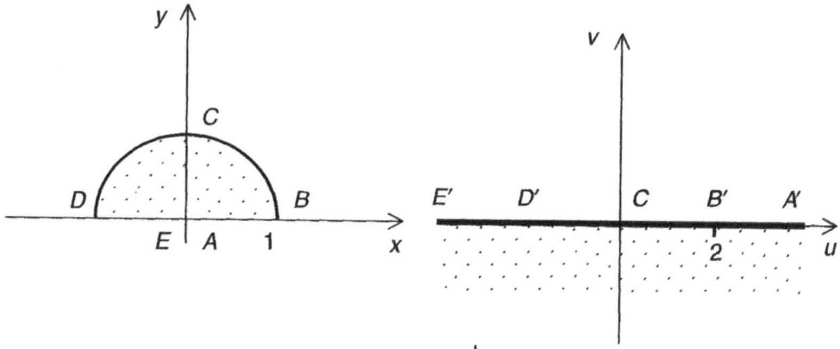

(d) $w = z + \frac{1}{z}$

Figure 5.3 (*Continued*)

The real and imaginary parts of an analytic function of complex variables thus satisfy the two-dimensional *Laplace's equation* and, hence, either one may represent the potential. This fact implies that we can solve for the potential and the corresponding static electric and magnetic fields in a particular microwave structure by choosing an analytical function $w(z)$, determining its real part u and imaginary part v, and taking either u or v as the potential subject to the structure's boundary conditions. For transmission line problems, the obtained potential and electric field can be used to determine the capacitance per unit length of the transmission line and hence the transmission line's characteristic impedance and effective dielectric constant.

We now consider an analytical function $w = f(z)$. Under the transformation f, any infinitesimal linear element dz in the z plane is transformed into a small linear element dw in the w plane, according to the chain rule, as

$$dw = \frac{dw}{dz}\,dz \tag{5.5}$$

provided that dw/dz is nonzero. In general, dw/dz is complex and hence may be expressed in terms of its magnitude a and phase ϕ as

$$\frac{dw}{dz} = ae^{j\phi} \tag{5.6}$$

from which, we can write

$$dw = ae^{j\phi}\,dz \tag{5.7}$$

This result indicates that the transformed element dw is obtained from the original element dz by multiplying its length by a and rotating it through an angle ϕ. Figure 5.4 illustrates this transformation. It hence follows that any infinitesimal area in a neighborhood of the point z would be transformed into a similar

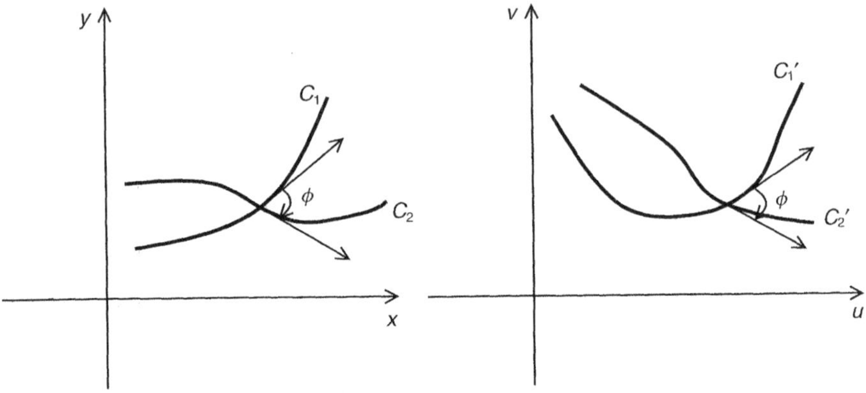

Figure 5.4 A mapping from the curve C in the z plane into the curve C' in the w plane.

infinitesimal area, that is, having approximately the same shape, in a neighborhood of the point w. This kind of mapping preserves the form of the original element and, hence, the name *conformal mapping*. It should be noted that this mapping is only achieved by means of an analytical function defined on a domain whose derivative is never zero. The transformed area is obtained by scaling each linear dimension of the original area by a factor of a and rotating the original element by an angle ϕ. It should be noted that since the scale factor a and the rotational angle ϕ generally change from point to point, a large region in the z plane may be transformed into another region in the w plane that has no resemblance to the original one. It is also recognized that if two curves in the z plane intersect at a particular angle, then their transformed curves in the w plane will also intersect at the same angle, due to the fact that both the transformed curves will be turned through the same angle ϕ. Any conformal mapping would therefore transform orthogonal or parallel curves in the z plane into orthogonal or parallel curves in the w plane, respectively. Figure 5.5 illustrates the transformation of orthogonal curves between the two planes. In mapping region to region, however, parallel curves of the original region may not be transformed into parallel curves, or parallel curves in the transformed region may not correspond to parallel curves in the original region due to the physical constraints of the original region. However, if the curves in the transformed region are orthogonal, then their corresponding curves in the original region must also be orthogonal and vice versa.

As an example of conformal mapping, we consider again the logarithmic function given in Eq. (5.1). By expressing z in the polar form as $z = re^{j\theta}$, we can obtain

$$u = \ln \sqrt{x^2 + y^2} = \ln r \tag{5.8a}$$

$$v = \tan^{-1}\left(\frac{y}{x}\right) = \theta \tag{5.8b}$$

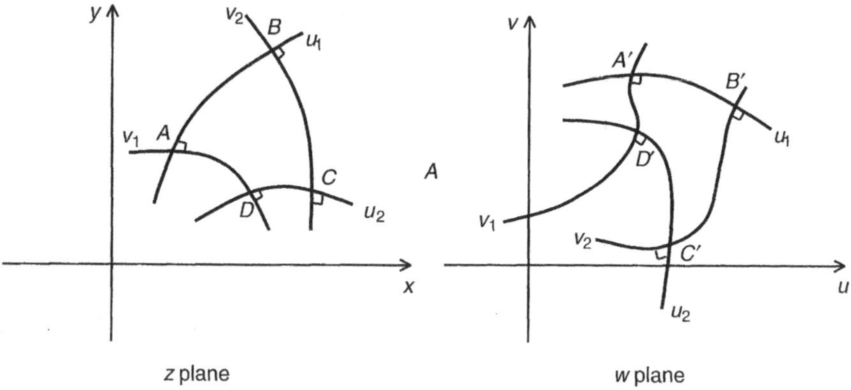

Figure 5.5 A mapping of perpendicular curves between the z and w planes.

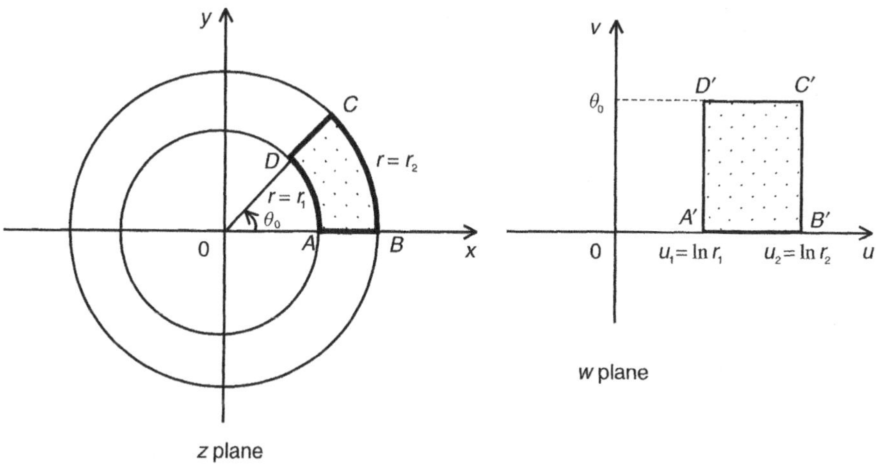

Figure 5.6 The mapping between a circular and a rectangular region.

Equation (5.8) indicates that a circular region in the z plane would transform onto a rectangular region in the w plane as shown in Fig. 5.6. This mapping performs several transformations. It maps the circular region $ABCD$ onto the rectangular region $A'B'C'D'$, the arcs AD and BC into the u-constant segments $A'D'$ and $B'C'$, and the straight-line segments DC and AB into the v-constant segments $D'C'$ and $A'B'$, respectively. The mapping transforms orthogonal curves, such as AB and BC, into orthogonal curves; such as $A'B'$ and $B'C'$, respectively. It also maps the parallel segments AD and BC into the parallel segments $A'D'$ and $B'C'$, respectively. It should be noted that, although the segments DC and AB intersect at an angle $\theta_0 \neq \pi$, the transformed curves $D'C'$ and $A'B'$ are in parallel. This is due to the constraints imposed by the AB and CD or the angle θ_0.

Had the segments DC and AB been transformed as two independent curves, the transformed curves $D'C'$ and $A'B'$ would have intersected at the same angle θ_0. We can also see that the straight lines $u =$ constant and $v =$ constant in the w plane correspond to the circle of radius e^u centered at the origin (0,0) and the straight line passing through the origin and intersecting the circle at a right angle.

In electromagnetic problems, in general, and transmission line problems, in particular, the boundary conditions are normally given by the function itself or by the normal derivative of the function along the boundary. These boundary-value problems are known as *Dirichlet* and *Neumann problems*, respectively. Under conformal mapping these conditions remain unchanged in the transformed region.

Let's consider an analytic function $w = f(z) = u(x, y) + jv(x, y)$ that conformally maps a curve C in the z plane into a curve C' in the w plane. Let a function $h(u, v)$ be defined that is differentiable on C'. If $h(u, v)$ satisfies either of the boundary conditions

$$h(u, v) = d \tag{5.9a}$$

where d is a constant, or

$$\frac{dh(x, y)}{dn} = 0 \tag{5.9b}$$

along C', then the function $H(x, y) = h[u(x, y), v(x, y)]$ also satisfies the corresponding boundary condition along C; namely,

$$H(x, y) = d \tag{5.10a}$$

or

$$\frac{dH(x, y)}{dn} = 0 \tag{5.10b}$$

This property of conformal mapping implies that, instead of solving the boundary-value problem directly in the z plane, we can transform this problem into a simpler one in the w plane, determine its solution, and transfer back to the z plane to obtain the solution of the original problem.

As a demonstration of the above principle, we consider a general two-conductor transmission line, whose transverse plane is in the z plane, embedded in a linear, homogeneous, and isotropic medium of permittivity ε. Under conformal mapping, this two-conductor transmission line is transformed into another two-conductor structure in the w plane. The energy per unit length stored in the electrostatic field of the transmission line in the z plane is given as

$$W_e = \tfrac{1}{2} \iint_S \varepsilon |\mathbf{E}|^2 \, dS \tag{5.11}$$

where $\mathbf{E}$ is the electrostatic field. Replacing $\mathbf{E}$ by $-\nabla_t V$, with V being the potential distribution between the two conductors, expanding the ∇_t operator, and utilizing the Cauchy–Riemann Eq. (5.2a), we can write

$$\begin{aligned} W_e &= \frac{1}{2}\varepsilon \int_x \int_y \left[\left(\frac{\partial V}{\partial x}\right)^2 + \left(\frac{\partial V}{\partial y}\right)^2 \right] dx\, dy \\ &= \frac{1}{2}\varepsilon \int_x \int_y \left[\left(\frac{\partial V}{\partial u}\right)^2 \left(\frac{\partial u}{\partial x}\right)^2 + \left(\frac{\partial V}{\partial v}\right)^2 \left(\frac{\partial v}{\partial y}\right)^2 \right] dx\, dy \\ &= \frac{1}{2}\varepsilon \int_x \int_y \left[\left(\frac{\partial V}{\partial u}\right)^2 + \left(\frac{\partial V}{\partial v}\right)^2 \right] \frac{\partial u}{\partial x}\frac{\partial v}{\partial y}\, dx\, dy \\ &= \frac{1}{2}\varepsilon \int_u \int_v \left[\left(\frac{\partial V}{\partial u}\right)^2 + \left(\frac{\partial V}{\partial v}\right)^2 \right] du\, dv \end{aligned} \tag{5.12}$$

which represents the electrostatic energy per unit length for the transformed structure in the w plane. This energy is also given as

$$W_e = \tfrac{1}{2} C V_o^2 \tag{5.13}$$

where C is the transmission line's capacitance per unit length and V_o stands for the potential difference between the two conductors. The per-unit-length capacitances of the corresponding transmission lines in the z and w planes are thus equal. This result shows that we can determine the capacitance per unit length of the transformed structure in the w plane and use it for that of the original transmission line in the z plane. Note that a transformation back to the z plane is not needed in this case.

The solution for the transformed structure in the w plane would be easier to solve if the original structure is mapped in such a way that either the curves $u =$ constant or the curves $v =$ constant will coincide with the boundary of the transformed structure. This desired mapping $w(z) = u + jv$ may be determined by expressing x and y of the original problem in *parametric forms*. Let $f(x, y) = 0$ be the equation describing a curve C in the z plane. Assume that f can also be expressed in parametric forms as $x = f_1(t)$ and $y = f_2(t)$, where f_1 and f_2 are real functions, and t is the real independent variable whose domain corresponds to the entire curve C. Now consider the following transformation:

$$z = f_1(aw) + jf_2(aw) \tag{5.14}$$

where a is a real constant. Letting $v = 0$ in Eq. (5.14), we obtain

$$x = f_1(au) \tag{5.15a}$$

$$y = f_2(au) \tag{5.15b}$$

Note that both $f_1(au)$ and $f_2(au)$ are real numbers. These equations indicate that the mapping (5.14) would transform the curve $f(x, y) = 0$ in the z plane into the curve $v = 0$ in the w plane. This procedure may enable one to determine an appropriate conformal mapping that can produce a transformed structure that has its boundary coinciding with the curves $u =$ constant or the curves $v =$ constant.

As an example, we consider a circle in the z plane described by

$$f(x, y) = x^2 + y^2 = r^2 \tag{5.16}$$

or

$$x = f_1(t) = r \cos t \tag{5.17a}$$

$$y = f_2(t) = r \sin t \tag{5.17b}$$

in parametric form. The transformation is given from Eq. (5.14) as

$$z = r \cos(aw) + jr \sin(aw) \tag{5.18}$$

which can be rewritten in polar form as

$$z = re^{jaw} \tag{5.19}$$

and hence

$$w = -\frac{j}{a} \ln\left(\frac{z}{r}\right) \tag{5.20}$$

This mapping function would transform a circle of radius r in the z plane into a segment corresponding to $0 \leq u \leq 2\pi/a$ and $v = 0$, which coincides with the curve $v = 0$.

In practice, conformal-mapping problems usually require several transformations in sequence. That is, the original structure is mapped successively to yield a final structure whose solution is known or can easily be found. To illustrate this multiple transformation, we consider the upper plane $y > 0$ and wish to transform it onto a region in the w plane. We perform this transformation in two successive steps. First, we use the mapping defined by

$$w_1 = u_1 + jv_1 = \frac{z - 1}{z + 1} \tag{5.21}$$

to obtain

$$u_1 = \frac{x^2 + y^2 - 1}{(x + 1)^2 + y^2} \tag{5.22a}$$

$$v_1 = \frac{2y}{(x + 1)^2 + y^2} \tag{5.22b}$$

This mapping transforms the upper plane $y > 0$ onto the upper plane $v_1 > 0$ in the w_1 plane. Next, we employ the mapping

$$w = u + jv = \ln w_1 \tag{5.23}$$

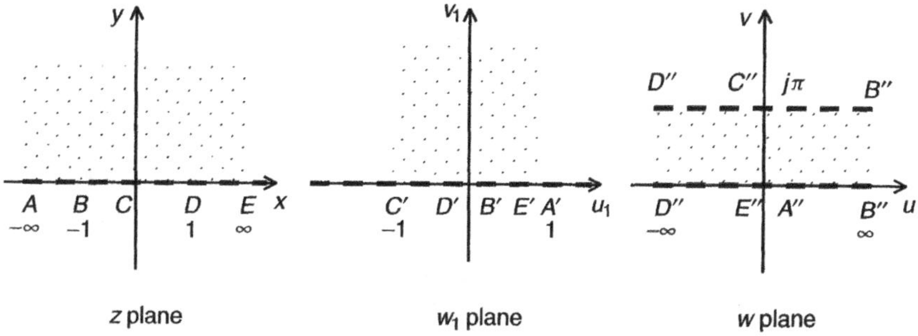

Figure 5.7 Successive mappings from the half-plane $y > 0$ onto the strip $0 < v < \pi$.

which yields, upon expressing w_1 in the polar form of $w_1 = re^{j\theta}$,

$$u = \ln r \tag{5.24a}$$

$$v = j\theta \tag{5.24b}$$

Equation (5.24) shows that the upper plane $v_1 > 0$ is mapped onto the strip $0 < v < \pi$ in the w plane. This composite mapping is shown in Fig. 5.7.

Available transformations such as those listed in Fig. 5.3 are useful for conformal-mapping problems.

5.3 THE SCHWARZ–CHRISTOFFEL TRANSFORMATION

Perhaps the most commonly used conformal mapping in transmission line analysis is the *Schwarz–Christoffel transformation.* This transformation maps the x axis and the upper half of the z plane onto a closed polygon and its interior in the w plane, respectively. We consider a conformal mapping $w = f(z)$ and assume that its derivative with respect to z is given as

$$\begin{aligned} f'(z) = \frac{dw}{dz} &= A(z - x_1)^{-k_1}(z - x_2)^{-k_2} \cdots (z - x_N)^{-k_N} \\ &= A\prod_{n=1}^{N}(z - x_n)^{-k_n} \end{aligned} \tag{5.25}$$

where A is a complex constant, $k_n (n = 1, 2, \ldots, N)$ are real constants, and x_n are real numbers representing points on the x axis and satisfying

$$x_1 < x_2 < \cdots < x_{N-1} < x_N \tag{5.26}$$

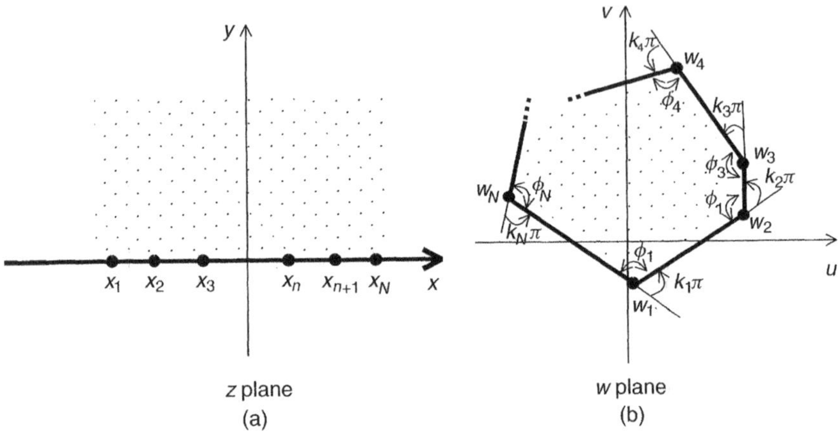

Figure 5.8 The Schwarz–Christoffel transformation.

as shown in Fig. 5.8(a). It should be noted that one or two points of x_n might be at infinity. The argument or angle of $f'(z)$ can be written as

$$\arg f'(z) = \arg A - k_1 \arg(z - x_1) - k_2 \arg(z - x_2) - \cdots - k_N \arg(z - x_N) \tag{5.27}$$

Now we consider a point z lying on the x axis. $z - x_n$ is thus positive or negative when $z > x_n$ or $z < x_n$, respectively, leading to

$$\arg(z - x_n) = \begin{cases} 0, & z > x_n \\ \pi, & z < x_n \end{cases} \tag{5.28}$$

Assuming the point z traverses to the right through different points x_n and letting $\theta_n = \arg f'(z)$ corresponding to $x_n < z < x_{n+1}$, we then have from Eqs. (5.27) and (5.28)

$$\theta_n = \arg A - \pi(k_{n+1} + k_{n+2} + \cdots + k_n) \tag{5.29a}$$

$$\theta_{n+1} = \arg A - \pi(k_{n+2} + k_{n+3} + \cdots + k_n) \tag{5.29b}$$

and hence,

$$\theta_{n+1} - \theta_n = \pi k_{n+1} \tag{5.30}$$

The argument of $f'(z)$ may also be written as

$$\begin{aligned} \arg \frac{dw}{dz} &= \arg \frac{du + j\,dv}{dx} \\ &= \tan^{-1}\left(\frac{dv}{du}\right) \end{aligned} \tag{5.31}$$

which is essentially the angle between the element dw and the real u axis in the w plane. It is now apparent that as the point z traverses the x axis of the z plane through the points x_n, the corresponding point w, transformed by the mapping (5.25), traverses a polygon in the w plane through the vertices $w_n = f(x_n)$ as seen in Fig. 5.8(b). The angle $(\theta_{n+1} - \theta_n)$ of Eq. (5.30) measures the polygon's exterior angle at the vertex w_n. Note that the exterior angles are limited within $\pm\pi$, and so $-1 < k_n < 1$. It is also seen from Eq. (5.30) that as the point z passes through the point x_{n+1}, the direction of the path through which the transformed point w traverses changes by an angle of πk_{n+1} as shown in Fig. 5.8(b). We assume that the sides of the polygon do not cross one another and, hence, the polygon is a closed contour. The positive sense of the polygon is assumed to be counterclockwise. For the closed polygon, its interior angle at the vertex w_{n+1} is defined as the angle ϕ_{n+1} between the two adjacent sides and can thus be obtained from the corresponding exterior angle as

$$\begin{aligned}\phi_{n+1} &= \pi - (\theta_{n+1} - \theta_n) \\ &= \pi(1 - k_{n+1})\end{aligned} \tag{5.32}$$

by making use of Eq. (5.30). Integrating Eq. (5.25) with respect to z and using Eq. (5.32) gives

$$w = f(z) = A \int^z \prod_{n=1}^{N} (z - x_n)^{\phi_n/\pi - 1}\, dz + B \tag{5.33}$$

where B is an arbitrary constant that determines the position of the polygon. The magnitude and angle of the constant A control the size and orientation of the polygon in the w plane. This mapping is known as the Schwarz–Christoffel transformation. This transformation maps the interior points of the upper half of the z plane into the points lying to the left sides of the polygon in the counterclockwise direction. This implies that the upper half of the z plane is mapped onto the interior region of the polygon in the w plane.

In practice, a polygon is normally defined in the w plane for a given problem, and the Schwarz–Christoffel transformation is determined such that the x axis of the z plane is mapped onto the polygon. This process is accomplished by determining the polygon's interior angles ϕ_n and the points x_n and evaluating the integral in Eq. (5.33). All these interior angles are readily obtained from the polygon. For the points x_n, however, only a maximum of three points or three conditions of x_n can be chosen arbitrarily. Hence, for polygons with more than three sides ($N > 3$), some of the points x_n must be determined so that the x axis is transformed into the polygon. This is often difficult. The integration encountered in the transformation may also be quite complicated, further imposing another difficulty in obtaining the solution. For transmission line problems, the x axis of the z plane corresponding to the polygon that represents the transmission line in the w plane always has parts of different potentials. This is due to the fact that a transmission line with two or more conductors has two or more different potentials, respectively; one of these potentials is zero. In these cases, successive mappings would be needed.

5.4 APPLICATIONS OF THE SCHWARZ–CHRISTOFFEL TRANSFORMATION IN TRANSMISISON LINE ANALYSIS

To illustrate the concept of the Schwarz–Christoffel transformation for transmission line analysis, we now use it to analyze a microstrip transmission line with an upper conducting cover as shown in Fig. 5.9. We assume that the strip and ground planes are perfect conductors, the strip is infinitesimally thin, and the ground planes are infinitely large. We also suppose that the interface between the dielectric and air, except where the strip resides, represents a perfect magnetic surface. It should be noted that this assumption is exact only when the upper and lower regions are identical and thus maintaining a perfect symmetry. For the considered microstrip line, this symmetry does not exist and thus error in calculations is expected. Nevertheless, as will be seen, good results for practical microstrip lines are obtained by this conformal mapping.

We can now decompose the total capacitance per unit length of the microstrip line into two independent per-unit-length capacitances: C_o corresponding to the upper (air) and C_r corresponding to the lower (dielectric) planes. Each of these capacitances is determined as though the other capacitance was not present. To determine C_r, we thus only need to consider the lower part of the microstrip line. Because the structure is symmetrical with respect to the central vertical plane and, hence, imposes a perfect magnetic wall there, we can finally consider only a half of the lower part as shown in Fig. 5.10(a). We visualize this lower area as a polygon in the w plane whose vertices are w_1, w_2, w_3, and w_4 and wish to transform it onto the upper half of the z plane as shown in Fig. 5.10(b). It is observed that while the different potentials of the strip and ground plane are on

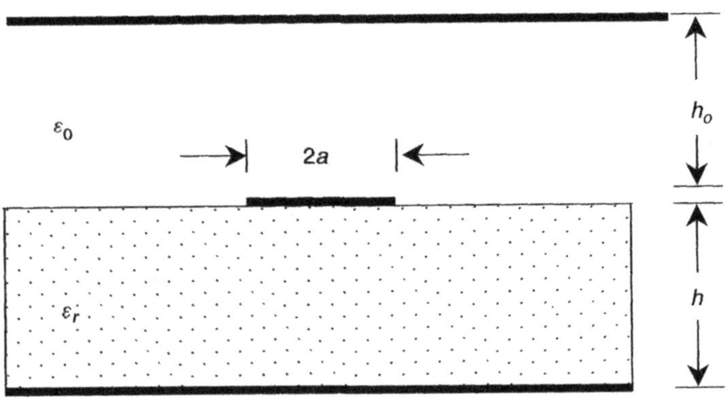

Figure 5.9 Cross section of microstrip line with an upper conducting cover.

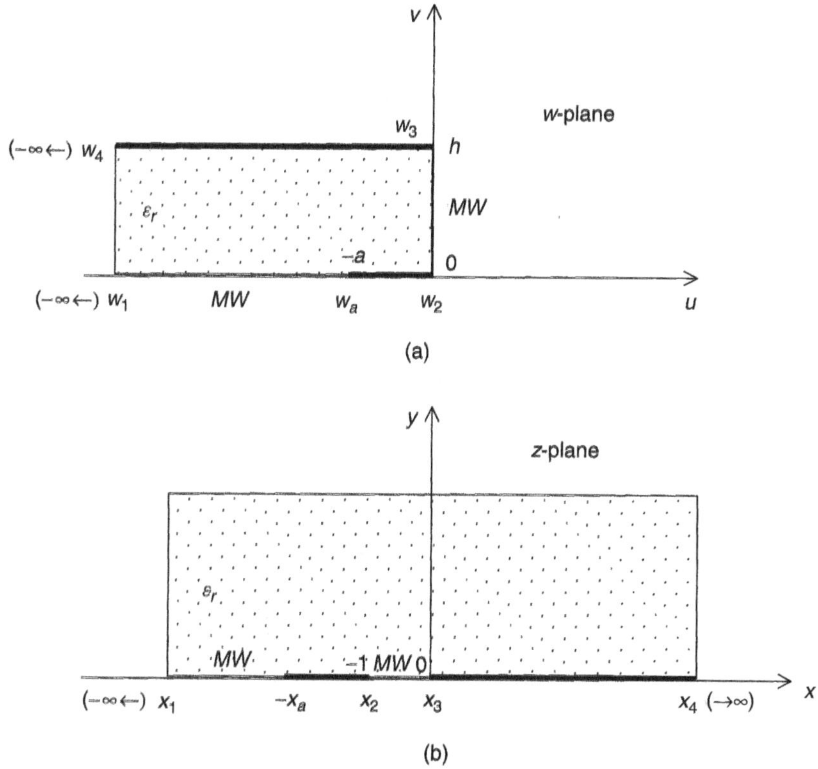

Figure 5.10 The Schwarz–Christoffel mapping of a half of the microstrip's lower part in the w plane (a) into the z plane (b). MW stands for magnetic wall.

different surfaces in the w plane, they appear on the same surface in the z plane. Using the Schwarz–Christoffel transformation (5.33), we can write the mapping from the z plane into the w plane as

$$w = A\int^{z}(z-x_1)^{\phi_1/\pi-1}(z-x_2)^{\phi_2/\pi-1}(z-x_3)^{\phi_3/\pi-1}(z-x_4)^{\phi_4/\pi-1}\,dz + B \tag{5.34}$$

Since the considered polygon is rectangular, the interior angles are $\pi/2$. To cover the entire x axis, we let x_1 and x_4 approach $-\infty$ and $+\infty$. We also choose $x_2 = -1$ and $x_3 = 0$, and rewrite the mapping as

$$\begin{aligned} w &= \frac{A}{\sqrt{x_1 x_4}}\int^{z}\left[(z-x_2)(z-x_3)\left(1-\frac{z}{x_i}\right)\left(1-\frac{z}{x_4}\right)\right]^{-1/2} dz + B \\ &= A_1\int^{z}\frac{dz}{\sqrt{z(z+1)}} + B \end{aligned} \tag{5.35}$$

where $A_1 = A/\sqrt{x_1 x_4}$. Multiplying the numerator and denominator of the integrand by $[z^{-1/2} + (z+1)^{-1/2}][z^{1/2} + (z+1)^{1/2}]$, we obtain

$$\begin{aligned} w &= A_1 \int^z \frac{\left[z^{-1/2} + (z+1)^{-1/2}\right]\left[z^{1/2} + (z+1)^{1/2}\right]}{\left[z^{1/2} + (z+1)^{1/2}\right]^2}\, dz + B \\ &= A_1 \int^z \frac{d\left[z^{1/2} + (z+1)^{1/2}\right]^2}{\left[z^{1/2} + (z+1)^{1/2}\right]^2} + B \\ &= 2A_1 \ln\left[z^{1/2} + (z+1)^{1/2}\right] + B \end{aligned} \tag{5.36}$$

Note that under this mapping we have transformed the points x_n in the z plane into the points w_n in the w plane as follows:

$$x_1 \to -\infty \leftrightarrow w_1 \to -\infty \tag{5.37a}$$

$$x_2 = -1 \leftrightarrow w_2 = 0 \tag{5.37b}$$

$$x_3 = 0 \leftrightarrow w_3 = jh \tag{5.37c}$$

$$x_4 \to \infty \leftrightarrow w_4 \to -\infty + jh \tag{5.37d}$$

Note also that $x_1 < x_2 < x_3 < x_4$ as required by Eq. (5.26). Using the boundary conditions (5.37b) and (5.37c), we obtain

$$2A_1 \ln\sqrt{-1} = -B \tag{5.38a}$$

$$B = jh \tag{5.38b}$$

Replacing -1 in Eq. (5.38a) by $e^{j\pi}$ and solving for A_1 gives

$$A_1 = -\frac{h}{\pi} \tag{5.39}$$

The mapping (5.36) is now given as

$$w = -\frac{2h}{\pi} \ln\left[z^{1/2} + (z+1)^{1/2}\right] + jh \tag{5.40}$$

Using this mapping, we can then write the relation between the two corresponding points $z = -x_a$, where $x_a > 0$, and $w = w_a = -a$ as

$$jx_a^{1/2} + j(x_a - 1)^{1/2} = e^{\pi a/2h} e^{j\pi/2} \tag{5.41}$$

from which we can write

$$(x_a - 1)^{1/2} = e^{\pi a/2h} - x_a^{1/2} \tag{5.42}$$

Taking the square of both sides of Eq. (5.41), we can derive

$$x_a^{1/2} = \frac{e^{\pi a/h} + 1}{2e^{\pi a/2h}} = \frac{e^{\pi a/2h} + e^{-\pi a/2h}}{2} \tag{5.43}$$

and hence

$$x_a = \cosh^2\left(\frac{\pi a}{2h}\right) \tag{5.44}$$

To obtain a structure whose capacitance can be determined easily, we further map the upper half-plane in the z plane of Fig. 5.10(b) onto a polygon in the $w' = u' + jv'$ plane using the Schwarz–Christoffel transformation again. For an easy calculation of the capacitance C_r, we choose a rectangular polygon. This mapping is illustrated in Fig. 5.11. It should be noted that we have transferred

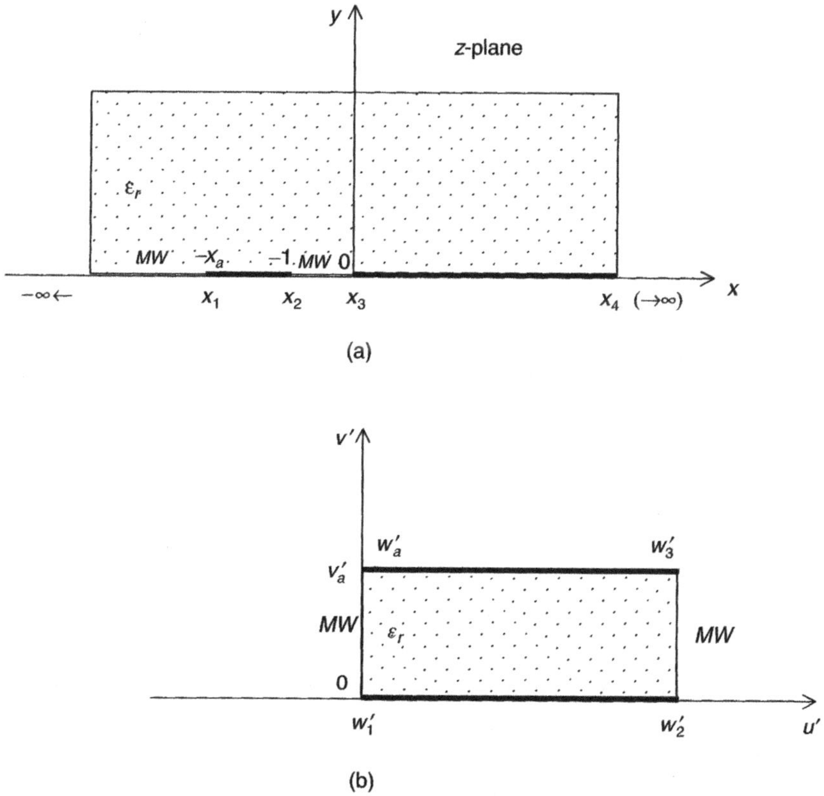

Figure 5.11 Conformal mapping from the upper half in the z plane (a) into the w plane (b) MW denotes magnetic wall.

the different potentials along the x axis to different surfaces in the w' plane. Since the polygon is rectangular we again have each of the interior angles equal to $\pi/2$. For this mapping, the points along the x axis of the z plane are already known from the previous mapping. Note, however, that point x_1 is now chosen to be $-x_a$, while points x_2, x_3, and x_4 are still at -1, 0, and ∞, respectively, as in Fig. 5.10. We choose the four points in the w' plane corresponding to the points x_1, x_2, x_3, and x_4 in the z plane as follows:

$$x_1 = -x_a = -\cosh^2\left(\frac{\pi a}{2h}\right) \leftrightarrow w'_1 = 0 \tag{5.45a}$$

$$x_2 = -1 \leftrightarrow w'_2 = 1 \tag{5.45b}$$

$$x_3 = 0 \leftrightarrow w'_3 = 1 + jv'_a \tag{5.45c}$$

$$x_4 \to \infty \leftrightarrow w'_4 = j'v'_a \tag{5.45d}$$

We can write the Schwarz–Christoffel transformation for this mapping as

$$w' = A' \int^z \frac{dz}{\sqrt{z(z+1)(z+x_a)}} + B' \tag{5.46}$$

where A' and B' are constants. Since B' is as yet arbitrary, we may choose a lower limit for the integration. Setting this limit to zero and letting $z = -t^2$, $A'_1 = 2jA'x_a^{-1/2}$, we have

$$w' = A'_1 \int_0^{\sqrt{-z}} \frac{dt}{\sqrt{(1-t^2)(1-t^2/x_a)}} + B' \tag{5.47}$$

Making use of the inverse elliptic function

$$sn^{-1}(s,k) = \int_0^s \frac{dt}{\sqrt{(1-t^2)(1-k^2t^2)}}, \qquad 0 \le k \le 1 \tag{5.48}$$

we can rewrite Eq. (5.47) as

$$w' = A'_1\ sn^{-1}\left[(-z)^{-1/2}, x_a^{-1/2}\right] + B' \tag{5.49}$$

Substituting the boundary conditions (5.45a)–(5.45c) into Eq. (5.49), we get

$$A'_1\ sn^{-1}(x_a^{1/2}, x_a^{-1/2}) + B' = 0 \tag{5.50a}$$

$$A'_1\ sn^{-1}(1, x_a^{-1/2}) + B' = 1 \tag{5.50b}$$

$$B' = 1 + jv'_a \tag{5.50c}$$

which can then be solved to obtain, upon replacing $k = x_a^{-1/2}$,

$$v_a' = \frac{-j\, sn^{-1}(1,k)}{sn^{-1}(k^{-1},k) - sn^{-1}(1,k)} \tag{5.51}$$

$sn^{-1}(1,k)$ is equal to the *complete elliptic integral of the first kind*, $K(k)$, as

$$sn^{-1}(1,k) = K(k) = \int_0^1 \frac{dt}{\sqrt{(1-t^2)(1-k^2t^2)}} \tag{5.52}$$

Using Wallis formula

$$\int_0^{\pi/2} \sin^n t\, dt = \frac{\pi}{2}\frac{1\times 3\times 5\times\cdots\times(n-1)}{2\times 4\times 6\times\cdots\times n}, \qquad n = \text{even integer} \tag{5.53}$$

we can derive

$$K(k) = \frac{\pi}{2}\left[1 + \left(\frac{1}{2}\right)^2 k^2 + \left(\frac{1\times 3}{2\times 4}\right)^2 k^4 + \left(\frac{1\times 3\times 5}{2\times 4\times 6}\right)^2 k^6 + \cdots\right] \tag{5.54}$$

which can be used to compute $K(k)$ for various values of k.

To simplify Eq. (5.51), we let $k' = \sqrt{1-k^2}$ and obtain the complete integral of the first kind for the modulus k' as

$$\begin{aligned} K(k') &= \int_0^1 \frac{dt}{\sqrt{(1-t^2)(1-t^2+k^2t^2)}} \\ &= -j\int_1^{k^{-1}} \frac{dt}{\sqrt{(1-t^2)(1-k^2t^2)}} \end{aligned} \tag{5.55}$$

The inverse elliptic function $sn^{-1}(k^{-1},k)$ is given as

$$\begin{aligned} sn^{-1}(k^{-1},k) &= \int_0^{k^{-1}} \frac{dt}{\sqrt{(1-t^2)(1-k^2t^2)}} \\ &= \int_0^1 \frac{dt}{\sqrt{(1-t^2)(1-k^2t^2)}} + \int_1^{k^{-1}} \frac{dt}{\sqrt{(1-t^2)(1-k^2t^2)}} \\ &= K(k) - jK(k') \end{aligned} \tag{5.56}$$

upon using Eqs. (5.52) and (5.55). Substituting Eqs. (5.52) and (5.56) into Eq. (5.51) then gives

$$v_a' = \frac{K(k)}{K(k')} \tag{5.57}$$

The upper and lower conducting plates of the polygon in Fig. 5.11(b) have different potentials and constitute a pure parallel-plate capacitor due to the existence of the magnetic walls along the $u' = 0$ and 1 surfaces. As indicated earlier from the result of Eq. (5.13), the capacitance of the transformed structure is equal to that of the original structure. Therefore, the capacitance per unit length of the parallel-plate capacitor is equal to a half of the per-unit-length capcitance C_r corresponding to the microstrip's lower half. Hence, we have

$$C_r = \frac{2\varepsilon_0\varepsilon_r}{v'_a} = 2\varepsilon_0\varepsilon_r\frac{K(k')}{K(k)} \tag{5.58}$$

where

$$k = \operatorname{sech}\left(\frac{\pi a}{2h}\right) \tag{5.59a}$$

$$k' = \sqrt{1-k^2} = \tanh\left(\frac{\pi a}{2h}\right) \tag{5.59b}$$

The per-unit-length capacitance C_o corresponding to the upper half of the microstrip line with air as the dielectric can be derived from Eq. (5.58) as

$$C_o = 2\varepsilon_0\frac{K(k'_o)}{K(k_o)} \tag{5.60}$$

where

$$k_o = \operatorname{sech}\left(\frac{\pi a}{2h_o}\right) \tag{5.61a}$$

$$k'_o = \sqrt{1-k_o^2} = \tanh\left(\frac{\pi a}{2h_o}\right) \tag{5.61b}$$

The total capacitance per unit length of the microstrip line can now be obtained as a summation of the capacitances C_r and C_o as

$$C = 2\varepsilon_0\varepsilon_r\frac{K(k')}{K(k)} + 2\varepsilon_0\frac{K(k'_o)}{K(k_o)} \tag{5.62}$$

The effective dielectric constant is given as

$$\varepsilon_{\text{eff}} = \frac{C}{C_a} \tag{5.63}$$

where C_a is the capacitance per unit length with the dielectric replaced by air. Using Eq. (5.62) we obtain

$$\varepsilon_{\text{eff}} = \frac{\varepsilon_r\dfrac{K(k')}{K(k)} + \dfrac{K(k'_o)}{K(k_o)}}{\dfrac{K(k')}{K(k)} + \dfrac{K(k'_o)}{K(k_o)}} \tag{5.64}$$

When $h = h_o$, the effective dielectric constant becomes

$$\varepsilon_{\text{eff}} = \frac{\varepsilon_r + 1}{2} \tag{5.65}$$

which is independent of the strip width. The characteristic impedance is given as

$$Z_o = \frac{1}{c\sqrt{CC_a}} \tag{5.66}$$

Multiplying the numerator and denominator by C_a and making use of $c = (\varepsilon_0\mu_0)^{-1/2}$, Eq. (5.62), and Eq. (5.63), we get

$$Z_o = \frac{60\pi}{\sqrt{\varepsilon_{\text{eff}}}\left[\dfrac{K(k')}{K(k)} + \dfrac{K(k_o')}{K(k_o)}\right]} \tag{5.67}$$

The ratio $K(k)/K(k')$ can be determined using Eq. (5.54). It can also be conveniently evaluated using the following accurate approximation [3]:

$$\frac{K(k)}{K(k')} = \begin{cases} \dfrac{1}{\pi}\ln\left(2\dfrac{1+\sqrt{k}}{1-\sqrt{k}}\right), & 0.5 \le k^2 \le 1 \\[2ex] \dfrac{\pi}{\ln\left(2\dfrac{1+\sqrt{k'}}{1-\sqrt{k'}}\right)}, & 0 \le k^2 \le 0.5 \end{cases} \tag{5.68}$$

For the conventional open microstrip line with no conducting cover, h_o tends to infinity and hence k_o becomes 1. This results in ε_{eff} equal to ε_r, which is incorrect for the microstrip line. Equations (5.64) and (5.67) cannot therefore be used for the open microstrip line by simply letting h_o approach infinity. This limitation is due to the fact that the assumption of a perfect magnetic surface at the upper dielectric surface is no longer valid as h_o becomes very large. To determine the valid range for these conformal-mapping equations, we compute the effective dielectric constant and characteristic impedance of the microstrip line versus $2a/h$ for different h_o/h and commonly used relative dielectric constants of 2.2, 6.15, and 10.5. We also compare these results with those obtained more accurately by the spectral-domain method described in Chapter 7. We found that Eqs. (5.64) and (5.67) are valid within 6% for h_o/h up to 5. Note that the results when $h_o/h = 5$ can also be used for an open microstrip line. We then have the following conditions for k_o and k_o':

$$k_o = \begin{cases} \operatorname{sech}\left(\dfrac{\pi a}{2h_o}\right), & 0 \le \dfrac{h_o}{h} < 5 \\[2ex] \operatorname{sech}\left(\dfrac{\pi a}{10h}\right), & 5 \le \dfrac{h_o}{h} \end{cases} \tag{5.69a}$$

$$k_o' = \sqrt{1 - k_o^2} = \begin{cases} \tanh\left(\dfrac{\pi a}{2h_o}\right), & 0 \le \dfrac{h_o}{h} < 5 \\ \tanh\left(\dfrac{\pi a}{10h}\right), & 5 \le \dfrac{h_o}{h} \end{cases} \tag{5.69b}$$

When $\varepsilon_r = 1$ and $h_o = h$, the computed results are exact as expected.

5.5 CONFORMAL-MAPPING EQUATIONS FOR COMMON TRANSMISSION LINES

Equations for various commonly used transmission lines, obtained using the conformal-mapping technique are given as follows.

Strip Line (Fig. 5.12) [4]

$$Z_o = \frac{30\pi}{\sqrt{\varepsilon_r}} \frac{K(k')}{K(k)} \tag{5.70a}$$

$$k = \tanh\left(\frac{\pi W}{4a}\right) \tag{5.70b}$$

Coplanar Waveguide (CPW) (Fig. 5.13) [5]

$$\varepsilon_{\text{eff}} = 1 + \frac{\varepsilon_r - 1}{2} \frac{K(k')}{K(k)} \frac{K(k_1)}{K(k_1')} \tag{5.71a}$$

$$Z_o = \frac{30\pi}{\sqrt{\varepsilon_{\text{eff}}}} \frac{K(k')}{K(k)} \tag{5.71b}$$

$$k = \frac{a}{b} \tag{5.71c}$$

$$k_1 = \frac{\sinh(\pi a/2h)}{\sinh(\pi b/2h)} \tag{5.71d}$$

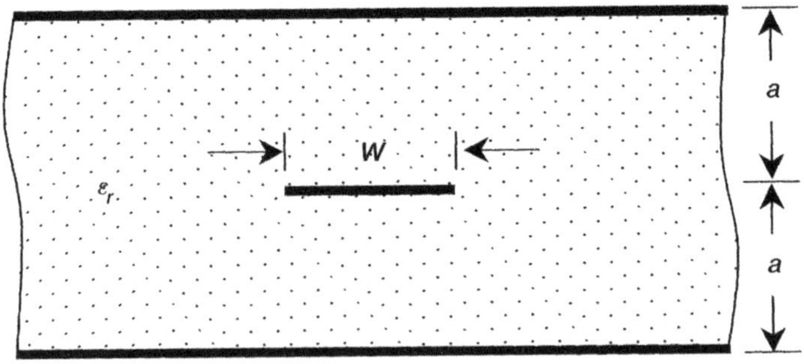

Figure 5.12 Cross section of a strip line.

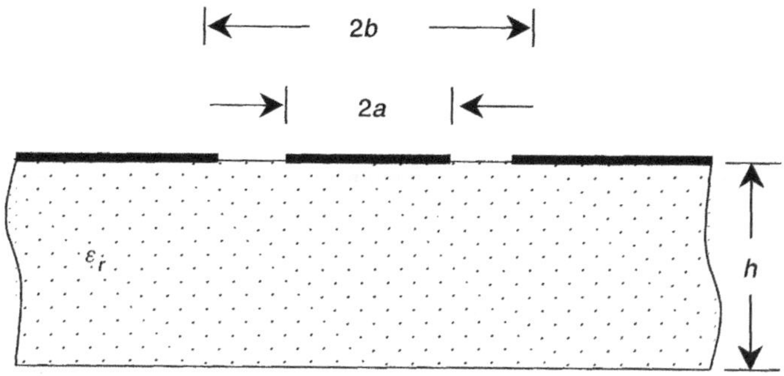

Figure 5.13 Cross section of a CPW.

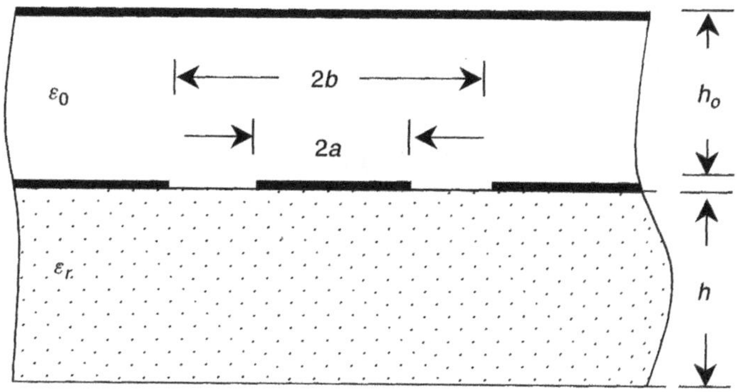

Figure 5.14 Cross section of a CPW with an upper conducting cover.

CPW with Upper Conducting Cover (Fig. 5.14) [6]

$$\varepsilon_{\text{eff}} = 1 + (\varepsilon_r - 1)\frac{\dfrac{K(k_1)}{K(k_1')}}{\dfrac{K(k)}{K(k')} + \dfrac{K(k_2)}{K(k_2')}} \tag{5.72a}$$

$$Z_o = \frac{60\pi}{\sqrt{\varepsilon_{\text{eff}}}} \frac{1}{\dfrac{K(k)}{K(k')} + \dfrac{K(k_2)}{K(k_2')}} \tag{5.72b}$$

$$k = a/b \tag{5.72c}$$

$$k_1 = \frac{\sinh(\pi a/2h)}{\sinh(\pi b/2h)} \tag{5.72d}$$

$$k_2 = \frac{\tanh(\pi a/2h_o)}{\tanh(\pi b/2h_o)} \tag{5.72e}$$

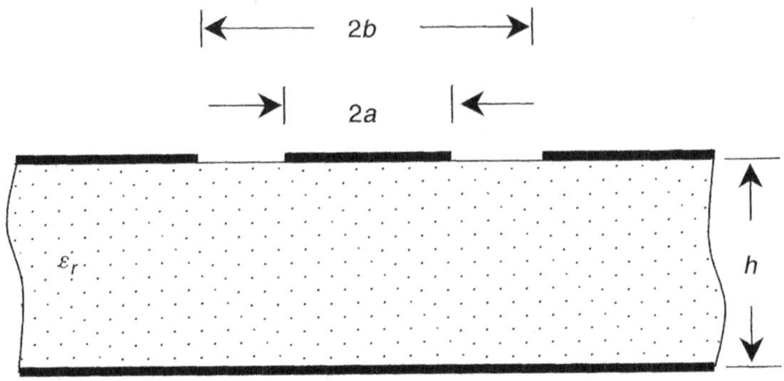

Figure 5.15 Cross section of conductor-backed CPW.

Conductor-Backed CPW (Fig. 5.15) [7]

$$\varepsilon_{\text{eff}} = \frac{1 + \varepsilon_r \dfrac{K(k')}{K(k)} \dfrac{K(k_1)}{K(k_1')}}{1 + \dfrac{K(k')}{K(k)} \dfrac{K(k_1)}{K(k_1')}} \tag{5.73a}$$

$$Z_o = \frac{60\pi}{\sqrt{\varepsilon_{\text{eff}}}} \frac{1}{\dfrac{K(k)}{K(k')} + \dfrac{K(k_1)}{K(k_1')}} \tag{5.73b}$$

$$k = \frac{a}{b} \tag{5.73c}$$

$$k_1 = \frac{\tanh(\pi a/2h)}{\tanh(\pi b/2h)} \tag{5.73d}$$

Conductor-Backed CPW with Upper Conducting Cover (Fig. 5.16) [6]

$$\varepsilon_{\text{eff}} = 1 + (\varepsilon_r - 1) \frac{\dfrac{K(k)}{K(k')}}{\dfrac{K(k)}{K(k')} + \dfrac{K(k_1)}{K(k_1')}} \tag{5.74a}$$

$$Z_o = \frac{60\pi}{\sqrt{\varepsilon_{\text{eff}}}} \frac{1}{\dfrac{K(k)}{K(k')} + \dfrac{K(k_1)}{K(k_1')}} \tag{5.74b}$$

$$k = \frac{\tanh(\pi a/2h)}{\tanh(\pi b/2h)} \tag{5.74c}$$

$$k_1 = \frac{\tanh(\pi a/2h_o)}{\tanh(\pi b/2h_o)} \tag{5.74d}$$

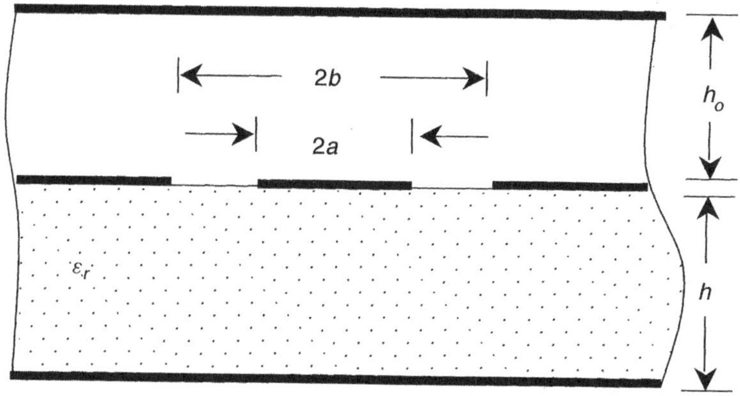

Figure 5.16 Cross section of conductor-backed CPW with an upper conducting cover.

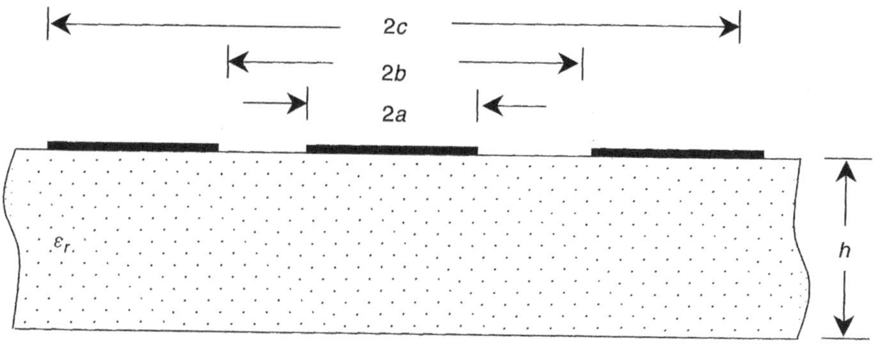

Figure 5.17 Cross section of a CPW with finite ground planes.

CPW with Finite Ground Planes (Fig. 5.17) [6]

$$\varepsilon_{\text{eff}} = 1 + \frac{\varepsilon_r - 1}{2} \frac{K(k)}{K(k')} \frac{K(k_1')}{K(k_1)} \tag{5.75a}$$

$$Z_o = \frac{30\pi}{\sqrt{\varepsilon_{\text{eff}}}} \frac{K(k_1')}{K(k_1)} \tag{5.75b}$$

$$k = \frac{\sinh(\pi a/2h)}{\sinh(\pi b/2h)} \sqrt{\frac{1 - \sinh^2(\pi b/2h)/\sinh^2(\pi c/2h)}{1 - \sinh^2(\pi a/2h)/\sinh^2(\pi c/2h)}} \tag{5.75c}$$

$$k_1 = \frac{a}{b} \sqrt{\frac{1 - (b/c)^2}{1 - (a/c)^2}} \tag{5.75d}$$

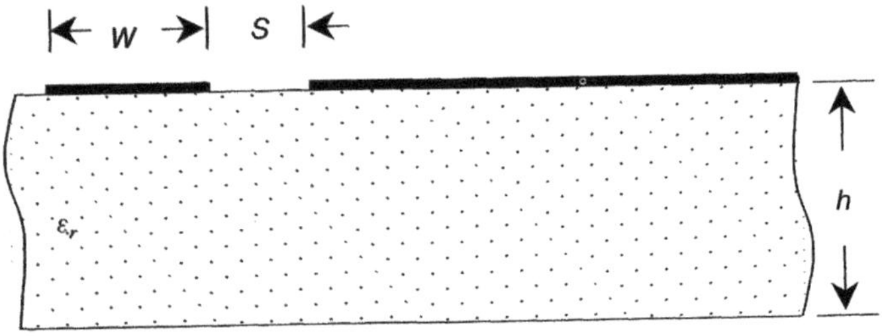

Figure 5.18 Cross section of an asymmetric CPW.

Asymmetrical CPW (Fig. 5.18) [8]

$$\varepsilon_{\text{eff}} = 1 + \frac{\varepsilon_r - 1}{2} \frac{K(k)}{K(k')} \frac{K(k_1)}{K(k_1')} \tag{5.76a}$$

$$Z_o = \frac{60\pi}{\sqrt{\varepsilon_{\text{eff}}}} \frac{K(k)}{K(k')} \tag{5.76b}$$

$$k = \sqrt{\frac{S}{W+S}} \tag{5.76c}$$

$$k_1 = \sqrt{\frac{(t_1 - t_2)(1 - t_2)}{(t_1 + t_2)(1 + t_2)}} \tag{5.76d}$$

$$t_i(i = 1, 2) = \frac{e^{d_i} - 1}{e^{d_i} + 1} \tag{5.76e}$$

$$d_1 = \frac{\pi}{2}\left(\frac{2W}{h} + \frac{S}{h}\right) \tag{5.76f}$$

$$d_2 = \frac{\pi S}{2h} \tag{5.76g}$$

Coplanar Strip (CPS) (Fig. 5.19) [5]

$$\varepsilon_{\text{eff}} = 1 + \frac{\varepsilon_r - 1}{2} \frac{K(k')}{K(k)} \frac{K(k_1)}{K(k_1')} \tag{5.77a}$$

$$Z_o = \frac{120\pi}{\sqrt{\varepsilon_{\text{eff}}}} \frac{K(k)}{K(k')} \tag{5.77b}$$

$$k = \frac{a}{b} \tag{5.77c}$$

$$k_1 = \frac{\sinh(\pi a/2h)}{\sinh(\pi b/2h} \tag{5.77d}$$

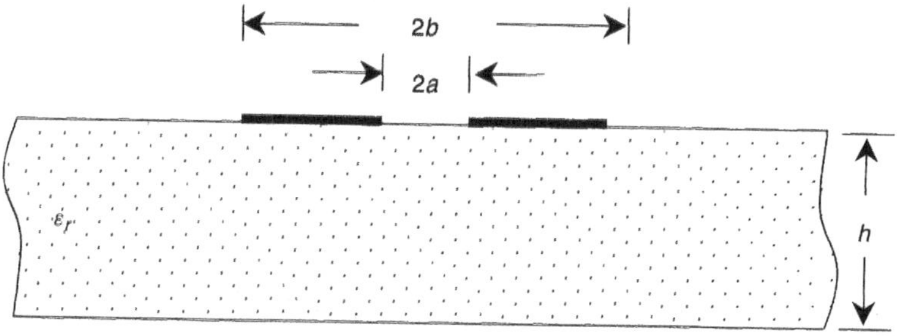

Figure 5.19 Cross section of a CPS.

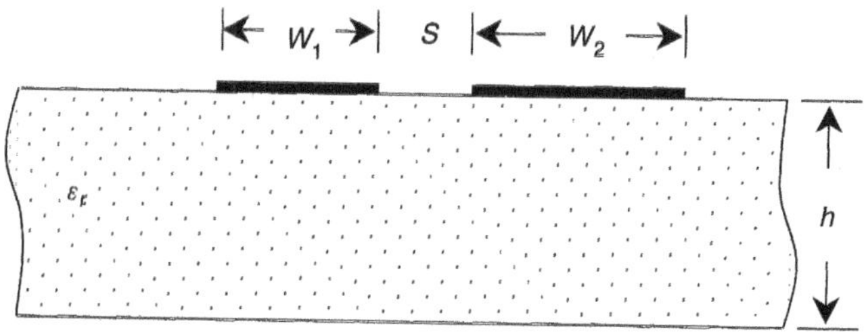

Figure 5.20 Cross section of an asymmetric CPS.

Asymmetrical CPS (Fig. 5.20) [8]

$$\varepsilon_{\text{eff}} = 1 + \frac{\varepsilon_r - 1}{2} \frac{K(k)}{K(k')} \frac{K(k_1)}{K(k_1')} \tag{5.78a}$$

$$Z_o = \frac{60\pi}{\sqrt{\varepsilon_{\text{eff}}}} \frac{K(k)}{K(k')} \tag{5.78b}$$

$$k = \sqrt{1 - \frac{1}{(1 + S/W_1)(1 + S/W_2)}} \tag{5.78c}$$

$$k_1 = \sqrt{\frac{(t_1 - t_2)(t_3 - t_2)}{(t_1 + t_2)(t_3 + t_2)}} \tag{5.78d}$$

$$t_i(i = 1, 2, 3) = \frac{e^{d_i} - 1}{e^{d_i} + 1} \tag{5.78e}$$

$$d_1 = \frac{\pi}{2}\left(\frac{2W_2}{h} + \frac{S}{h}\right) \tag{5.78f}$$

$$d_2 = \frac{\pi S}{2h} \tag{5.78g}$$

$$d_3 = \frac{\pi}{2}\left(\frac{2W_1}{h} + \frac{S}{h}\right) \tag{5.78h}$$

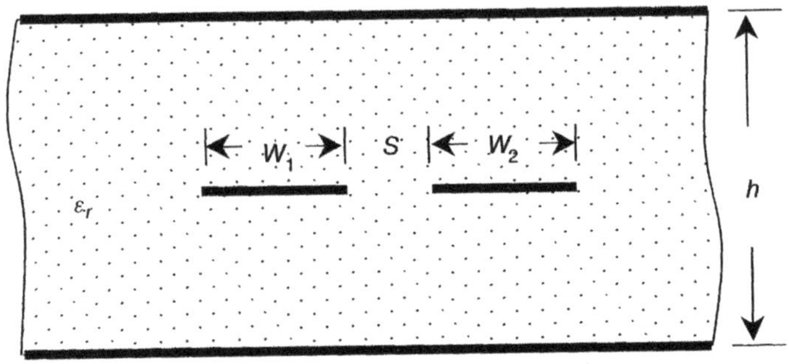

Figure 5.21 Cross section of parallel-coupled strip lines.

Parallel-Coupled Strip Lines (Fig. 5.21) [9]
Even-mode characteristic impedance:

$$Z_{oe} = \frac{30\pi}{\sqrt{\varepsilon_r}} \frac{K(k_e')}{K(k_e)} \tag{5.79a}$$

Odd-mode characteristic impedance:

$$Z_{oo} = \frac{30\pi}{\sqrt{\varepsilon_r}} \frac{K(k_o')}{K(k_o)} \tag{5.79b}$$

$$k_e = \tanh\left(\frac{\pi}{2}\frac{W}{b}\right) \tanh\left(\frac{\pi}{2}\frac{W+S}{b}\right) \tag{5.79c}$$

$$k_o = \tanh\left(\frac{\pi}{2}\frac{W}{b}\right) \coth\left(\frac{\pi}{2}\frac{W+S}{b}\right) \tag{5.79d}$$

In all these equations, the complementary modulus k' is obtained from the modulus k by

$$k' = \sqrt{1-k^2} \tag{5.80}$$

REFERENCES

1. R. E. Collin, *Field Theory of Guided Waves*, 2nd edition, IEEE Press, New York, 1991, pp. 259–273.
2. R. V. Churchill, J. W. Brown, and R. F. Verhey, *Complex Variables and Applications*, 3rd edition, McGraw-Hill, New York, 1976.
3. W. Hilberg, "From Approximations to Exact Relations for Characteristic Impedance," *IEEE Trans. Microwave Theory Tech.*, Vol. MTT-17, pp. 259–265, May 1969.
4. S. B. Cohn, "Characteristic Impedance of Shielded Strip Transmission Line," *IRE Trans. Microwave Theory Tech.*, Vol. MTT-2, pp. 52–55, July 1954.

5. G. Ghione and C. Naldi, "Analytical Formulas for Coplanar Lines in Hybrid and Monolithic MICs," *Electron. Lett.*, Vol. 20, No. 4, pp. 179–181, Feb. 1984.
6. G. Ghione and C. Naldi, "Coplanar Waveguides for MMIC Applications: Effect of Upper Shielding, Conductor Backing, Finite-Extent Ground Planes, and Line-to-Line Coupling," *IEEE Trans. Microwave Theory Tech.*, Vol. MTT-35, No. 3, pp. 260–267, Mar. 1987.
7. G. Ghione and C. Naldi, "Parameters of Coplanar Waveguides with Lower Ground Plane," *Electron. Lett.*, Vol. 19, No. 18, pp. 734–735, Sept. 1983.
8. R. K. Hoffmann, *Handbook of Microwave Integrated Circuits*, Artech House, Dedham, MA, 1987.
9. S. B. Cohn, "Shielded Coupled-Strip Transmission Line," *IRE Trans. Microwave Theory Tech.*, Vol. MTT-3, pp. 29–38, Oct. 1955.

PROBLEMS

5.1 Derive the Cauchy–Riemann equations, (5.2a) and (5.2b), where u and v are the respective real and imaginary parts of an analytical function $w = f(z)$ with z given as $z = x + jy$.

5.2 Consider a coaxial transmission line with the respective inner and outer conductors' radii of a and b. The relative dielectric constant of the medium between the two conductors is ε_r. The conductors are assumed to be perfect, and the medium is lossless. The potentials on the inner and outer conductors are V_o and zero, respectively. Using the conformal mapping defined by the logarithmic function

$$w = A \ln z + B$$

and choosing its real part u as the potential function, derive the potential distribution, electric and magnetic fields, capacitance per unit length, and characteristic impedance of the transmission line. Compare the derived characteristic impedance to the well-known equation:

$$Z_o = \frac{60}{\sqrt{\varepsilon_r}} \ln \frac{b}{a}$$

5.3 Repeat Problem 5.2 with the function's imaginary part v chosen as the potential function.

5.4 Use the mapping function $w = \ln z$ to transform the region in the z plane shown in Fig. P5.1 onto another region in the w plane.

5.5 Prove that, under a conformal mapping, the two families of curves in the z plane corresponding to the curves $u =$ constant and the curves $v =$ constant in the w plane form an orthogonal system.

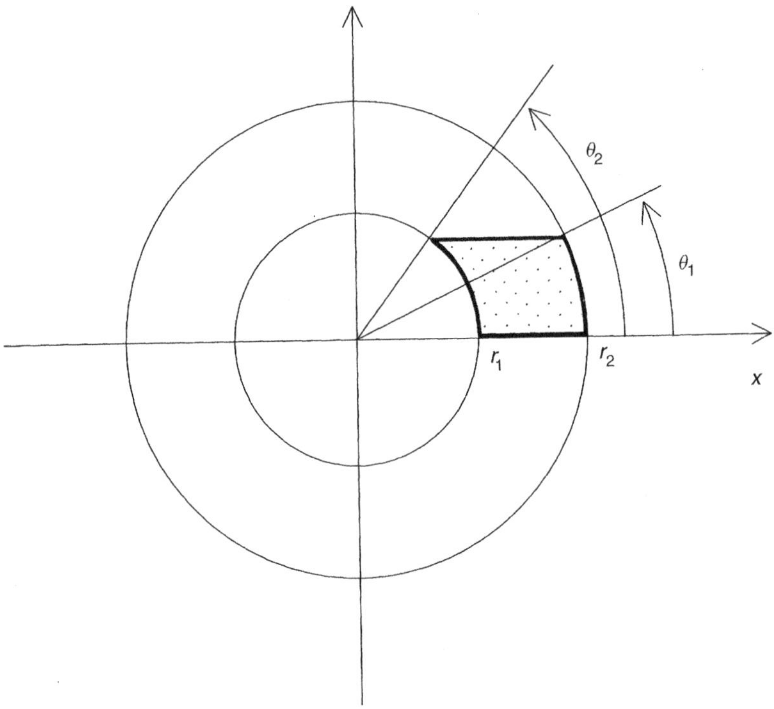

Figure P5.1

5.6 Find a transformation that maps a curve $f(x, y) = 0$ in the z plane into the curve $u = 0$ in the w plane. Use this mapping to transform a circle of radius r in the z plane into the w plane.

5.7 Consider an ellipse described by

$$\frac{x^2}{a^2} + \frac{y^2}{b^2} = r^2$$

where a, b, and r are positive real constants. Determine the transform that maps this ellipse into the curve $v = 0$ in the w plane.

5.8 Verify the mappings that transform the upper plane $y > 0$ onto the strip $0 < v < \pi$, including the correspondence between the marked points and boundaries in Fig. 5.7.

5.9 Consider the following Schwarz–Christoffel transformation:

$$w = j \int_0^z \frac{dz}{\sqrt{z(z+1)(z-1)}}$$

Show that this transformation maps the x axis onto the square with vertices $w_1 = ja$, $w_2 = 0$, $w_3 = a$, and $w_4 = a + ja$, where a is a positive constant given as

$$a = \tfrac{1}{2}B\left(\tfrac{1}{4}, \tfrac{1}{2}\right)$$

with $B(x, y)$ being the *beta function* of two real variables x and y given by

$$B(x, y) = \int_0^1 t^{x-1}(1-t)^{y-1}\,dt, \qquad x > 0,\ y > 0$$

5.10 Verify the transformation between the figures in the z and w planes in Fig. 5.3(a) using the given mapping function.

5.11 Verify the transformation between the figures in the z and w planes in Fig. 5.3(b) using the given mapping function.

5.12 Use the Schwarz–Christoffel transformation to derive the mapping function given in Fig. 5.3(c).

5.13 Derive the capacitance per unit length, effective dielectric constant, and characteristic impedance of the microstrip line, shown in Fig. 5.9, through successive Schwarz–Christoffel transformations. The first mapping between the z and w planes is described in Fig. P5.2.

5.14 Consider an asymmetric strip line shown in Fig. P5.3. We assume that the strip is negligibly thin and a perfect conductor and the ground planes are infinitely large and perfect conductors. Use the Schwarz–Christoffel transformation to derive a closed-form equation for the characteristic impedance of this strip line.

5.15 Calculate the characteristic impedance of the strip line in Problem 5.14 for W/h_2 from 0.1 to 10, ε_r of 2.2 and 10.5, and h_1/h_2 of 1 and 2.

5.16 Consider a CPW with an upper conducting cover as shown in Fig. 5.14. Assume that the central and ground strips have negligible thickness and are perfect conductors, the ground strips are infinitely large, and the upper cover is infinitely large and a perfect conductor. We also suppose that the slots behave as perfect magnetic walls. Use the Schwarz–Christoffel transformation to derive closed-form formulas for the capacitance per unit length, effective dielectric constant, and characteristic impedance of this CPW.

5.17 Calculate the effective dielectric constant and characteristic impedance of the CPW in Problem 5.16 for a/b from 0.1 to 0.8, ε_r of 2.2 and 10.5, h of 0.635 and 1.27 mm, and h_1/h of 1, 2, and 4.

5.18 Consider a conductor-backed coplanar waveguide (CPW) with an upper conducting cover as shown in Fig. 5.16. Assume that the central and ground strips have negligible thickness and are perfect conductors, the

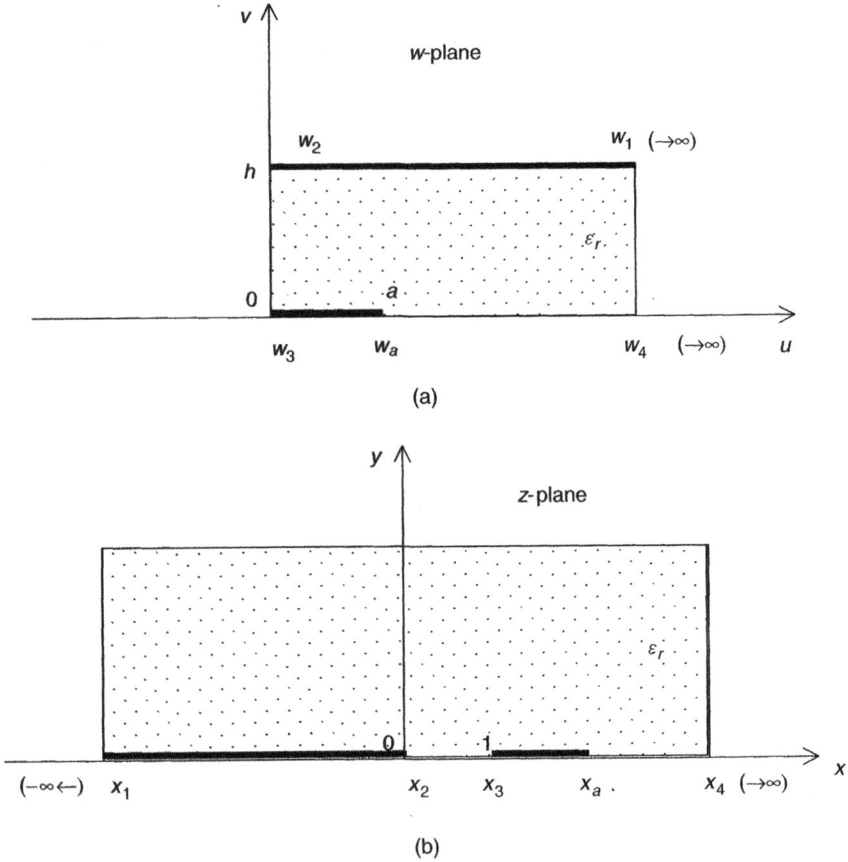

Figure P5.2

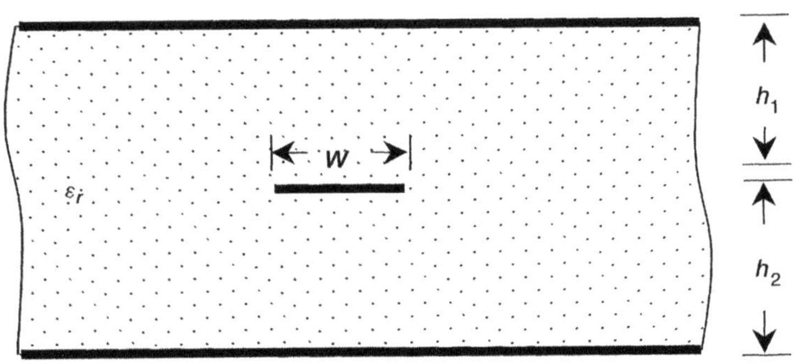

Figure P5.3 Cross section of an asymmetric strip line.

ground strips are infinitely large, and the upper and lower conducting plates are infinitely large and perfect conductors. We also suppose that the slots behave as perfect magnetic walls. Use the Schwarz–Christoffel transformation to derive closed-form formulas for the capacitance per unit length, effective dielectric constant, and characteristic impedance of this CPW.

5.19 Calculate the effective dielectric constant and characteristic impedance of the CPW in Problem 5.18 for a/b from 0.1 to 0.8, ε_r of 2.2 and 10.5, h of 0.635 and 1.27 mm, and h_1/h of 1, 2, and 4.

5.20 Consider a CPW with finite ground planes as shown in Fig. 5.17. Assume that the central and ground strips have negligible thickness and are perfect conductors, and the ground strips are infinitely large. We also suppose that the slots behave as perfect magnetic walls. Use the Schwarz–Christoffel transformation to derive closed-form formulas for the capacitance per unit length, effective dielectric constant, and characteristic impedance of this CPW.

5.21 Calculate the effective dielectric constant and characteristic impedance of the CPW in Problem 5.20 for a/b from 0.1 to 0.8, ε_r of 2.2 and 10.5, h of 0.635 and 1.27 mm, and b/c of 0.2, 0.4, and 0.8.

5.22 Consider an asymmetric CPW as shown in Fig. 5.18. Assume that the central and ground strips have negligible thickness and are perfect conductors, and the ground strip is infinitely large. We also suppose that the upper air–dielectric interface, except where the central and ground strips reside, behaves as a perfect magnetic wall. Use the Schwarz–Christoffel transformation to derive closed-form formulas for the capacitance per unit length, effective dielectric constant, and characteristic impedance of this CPW.

5.23 Calculate the effective dielectric constant and characteristic impedance of the CPW in Problem 5.22 for W/S from 0.1 to 1.5, ε_r of 2.2 and 10.5, and h of 0.635 and 1.27 mm.

5.24 Consider a coplanar strip (CPS) as shown in Fig. 5.19. The strips are assumed to be infinitely thin and perfect conductors. We also suppose that the upper air–dielectric interface, except where the strips reside, appears as a perfect magnetic surface. Use the Schwarz–Christoffel transformation to derive closed-form equations for the effective dielectric constant and characteristic impedance of this CPS.

5.25 Calculate the effective dielectric constant and characteristic impedance of the CPS in Problem 5.24 for a/b from 0.1 to 0.8, ε_r of 2.2 and 10.5, and h of 0.635 and 1.27 mm.

5.26 Consider an asymmetrical CPS as shown in Fig. 5.20. The strips are assumed to be infinitely thin and perfect conductors. We also suppose that the upper air–dielectric interface, except where the strips reside, appears as a perfect magnetic surface. Use the Schwarz–Christoffel transformation to derive closed-form equations for the effective dielectric constant and characteristic impedance of this CPS.

5.27 Calculate the effective dielectric constant and characteristic impedance of the CPS in Problem 5.26 for $S/(S + W_1 + W_2)$ from 0.1 to 0.8, ε_r of 2.2 and 10.5, and h of 0.635 and 1.27 mm.

5.28 Consider two parallel-coupled strip lines as shown in Fig. 5.21. We assume that the strips are negligibly thin and perfect conductors and the ground planes are infinitely large and perfect conductors. Use the Schwarz–Christoffel transformation to derive closed-form equations for the even- and odd-mode capacitances per unit length and characteristic impedances of these parallel-coupled strip lines.

5.29 Calculate the even- and odd-mode characteristic impedances of the parallel-coupled strip lines in Problem 5.28 for W/b from 0.1 to 2, ε_r of 2.2 and 10.5, and $(W + S)/b$ of 0.4, 0.8, and 1.6.

5.30 Consider the microstrip line as shown in Fig. 5.9. Its effective dielectric constant, ε_{eff}, and characteristic impedance, Z_o, can be determined using the conformal-mapping Eqs. (5.64) and (5.69), respectively. Verify Eq. (5.69) by calculating ε_{eff} and Z_o versus $2a/h$ for different h_o/h and $\varepsilon_r = 2.2$, 6.15, and 10.5, and compare them to more accurate results using techniques such as the spectral-domain method presented in Chapter 7.

5.31 Consider a coaxial transmission line, whose cross section is shown in Fig. P5.4. The transmission line is filled with two dielectric materials

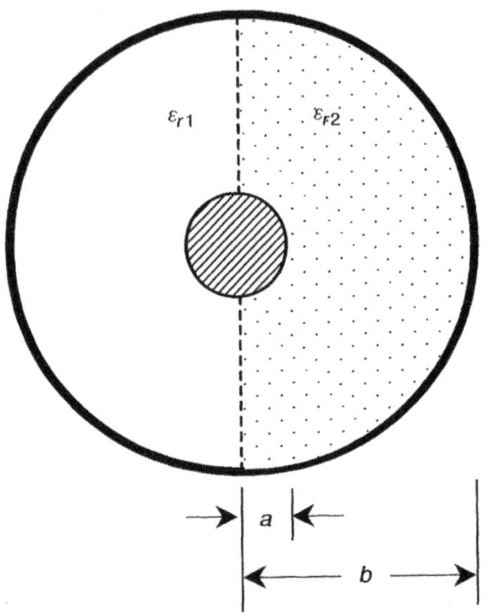

Figure P5.4 Cross section of a coaxial transmission line having two different dielectrics.

having relative dielectric constants of ε_{r1} and ε_{r2}. The inner and outer radii are a and b, respectively. The conductors are assumed to be perfect, and the dielectrics are lossless. Use conformal mapping to derive an expression for the characteristic impedance Z_o. Verify the result with those obtained in Problem 5.2 and by the well-known equation

$$Z_o = \frac{60}{\sqrt{\varepsilon_r}} \ln \frac{b}{a}$$

when $\varepsilon_{r1} = \varepsilon_{r2}$.

CHAPTER SIX

Variational Methods

In electromagnetic problems, in general, and microwave problems, in particular, solutions are normally obtained directly by solving appropriate differential or integral equations. For example, the resonant frequency of a resonator can be determined by solving the wave equation of the electric or magnetic field. On the other hand, *variational methods* operate by indirectly looking for a solution. Generally, a variational method seeks a *functional* that gives a maximum or minimum of a desired quantity. For example, it searches for the charge distribution (*functional*) on a transmission line that produces a maximum of the capacitance per unit length of the transmission line. So a variational method is essentially a maximization or minimization technique.

The main advantage of the variational method is that it produces stationary formulas, which yield results insensitive to the first-order errors in the unknown function. There are, in general, three kinds of variational methods, depending on the technique used to obtain approximate solutions of problems expressed in a variational form: the direct method based on the classical Rayleigh–Ritz or simply Ritz procedure, the indirect method such as Galerkin and least squares, and the semidirect method based on separation of variables. Variational methods can be formulated in both the space and Fourier-transform or spectral domain. The use of the spectral domain simplifies the derivation of the Green's function needed in the analysis, as will be seen later in this chapter and also in Chapter 7 on the spectral-domain method. Applications of variational methods include analysis of transmission lines to obtain characteristic impedances, effective dielectric constants, and losses, analysis of discontinuities, determination of resonant frequencies of resonators, and determination of impedances of antennas and obstacles in waveguides.

In this chapter, we will describe variational methods for analyzing transmission lines in both the space and spectral domain. Coaxial and microstrip transmission lines will be used to illustrate the formulation process. The Ritz procedure will be implemented to obtain numerical results for the characteristic impedances and effective dielectric constants.

6.1 FUNDAMENTALS OF VARIATIONAL METHODS

As indicated earlier, a variational method generally seeks a functional that results in a maximum or minimum of a desired quantity. Let's consider a functional y, which depends on a function $u(x)$ as $y = f[u(x)]$, and assume that the variational method will look for $u(x)$ that gives a maximum or minimum of y. To this end, we assume a function for $u(x)$ and call it a *trial function*. Let $x = x_0$ to be the exact or *true* solution that produces the absolute maximum or minimum of y. Now applying Taylor's series around this *true* $x = x_0$ gives

$$f(x) = f(x_0) + \left.\frac{df}{dx}\right|_{x=x_0}(x - x_0) + \frac{1}{2!}\left.\frac{d^2 f}{dx^2}\right|_{x=x_0} \times (x - x_0)^2 + \cdots + \frac{1}{n!}\left.\frac{d^n f}{dx^n}\right|_{x=x_0}(x - x_0)^n + R_{n+1} \tag{6.1}$$

from which we obtain

$$\Delta f \Delta f(x) - f(x_0) = \left.\frac{df}{dx}\right|_{x=x_0}(x - x_0) + \frac{1}{2!}\left.\frac{d^2 f}{dx^2}\right|_{x=x_0}(x - x_0)^2 + \cdots \tag{6.2}$$

or

$$\Delta f = \delta f(x) + \delta^2 f(x) + \cdots \tag{6.3}$$

where δf is called the first variation of f or the first-order error, and $\delta^2 f$ represents the second variation of f or the second-order error. The functional $y = f[u(x)]$ is called the *stationary* or *variational* expression if $\delta f = 0$ for $x \neq x_0$, and *nonstationary* otherwise. $\delta f = 0$ leads to the derivative of $f(x)$ at $x = x_0$ to be equal to zero, implying that $y(x)$ is a minimum or maximum at $x = x_0$. The value of x corresponding to the minimum or maximum of $y(x)$ is referred to as the upper- or lower-bound solution, respectively. Figure 6.1 illustrates stationary and non stationary solutions for $y = f(x)$. Note that the stationary formula, corresponding to $\delta f = 0$, gives a smaller error in y than does the non stationary formula, which has $\delta f \neq 0$, for the same approximate solution x_1 with a small deviation from the *true* solution x_0.

An important criterion in applying a variational method is that $y = f[u(x)]$ must be *stationary*. In order to prove that, we can let $x = x_0 + \delta x$ and $y = y_0 + \delta y$, where x_0 denotes the *true* solution for x and y_0 is the corresponding value for y, and then we prove that $\delta y = 0$. In this process, we neglect the higher-order terms, which are second-order error.

For problems involving multiple dimensions, we let y be a multidimensional functional as $y = f[u_1(x_1, x_2, \ldots, x_N), u_2(x_1, x_2, \ldots, x_N), \ldots, u_N(x_1, x_2, \ldots, x_N)]$. The variational method then assumes trial functions for $u_i (i = 1, 2, \ldots, N)$. We can obtain *stationary* and *nonstationary* formulas as well as the upper- and lower-bound solutions in the same fashion as for the one-dimensional variational formula discussed earlier.

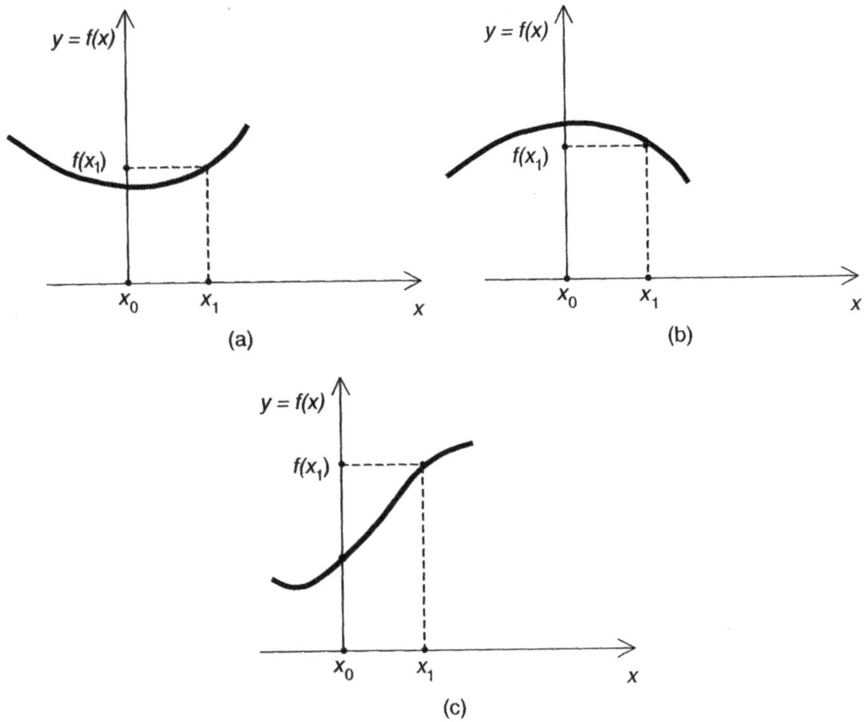

Figure 6.1 Illustration of (a, b) stationary and (c) nonstationary solutions. (a) Upper-bound solution; (b) lower-bound solution. For the same value of x_1, a stationary formula gives a smaller error in y than does the nonstationary formula.

Various techniques can be employed to determine approximate solutions of problems expressed in variational forms. In order to implement these techniques, we must derive the trial function $u(x)$ such that the variational expression y has a minimum or a maximum. $u(x)$ may be selected from a physical consideration; that is, choose $u(x)$ to meet or closely approximate the actual behavior of the parameter it represents. The closers $u(x)$ is to the true value, the better the result for y and the less computation time. The so-called Rayleigh–Ritz or simply Ritz method is perhaps the most commonly used procedure. It is applied by expanding the charge distribution as a sum of known functions whose coefficients are the variational parameters and by solving for the coefficient values by maximizing or minimizing y with respect to them. For the one-dimensional variational problem, the Ritz procedure essentially involves two steps. It first approximates the trial function $u(x)$ by a linear combination of a finite number of functions as

$$u(x) = \sum_{i=1}^{N} a_i u_i(x) \tag{6.4}$$

where $a_i, i = 1, 2, \ldots, N$, are the (independent) unknown constants, referred to as the variational parameters. The a_i need to be determined so that the functional y has a minimum or maximum. The $u_i(x)$ are referred to as the basis functions and should be so selected that Eq. (6.4) satisfies the boundary conditions of the considered physical structure for any choice of a_i. In practical transmission lines, the general behavior of the potential or charge distribution of a transmission line is often known. This can be exploited to choose suitable basis functions $u_i(x)$ so that a linear combination of them, representing the trial function $u(x)$, may approximate closely the potential or charge's actual behavior. In general, the $u_i(x)$ should be selected to form a complete set of orthogonal functions or polynomials. Once the $u_i(x)$ are chosen, we can write y as a function of a_i as $y = f(a_i)$. Finally, the Ritz method continues by finding the variational parameters a_i so that y is maximized or minimized. This is satisfied by taking the derivatives of y with respect to a_i for $i = 1, 2, \ldots, N$ and letting them be zero as

$$\frac{\partial y}{\partial a_i} = 0, \quad i = 1, 2, \ldots, N \tag{6.5}$$

For multidimensional problems, a similar approach is also employed for the Ritz method.

We can see that a variational method produces only approximate solutions for a problem. The main characteristic of this method is that the formula for the desired quantity is *stationary* about the *true* or *exact* solution, and hence the name *stationary* or *variational* formula. This means that a variational expression is relatively insensitive to small changes in the approximation around the *true* solution, which implies that a first-order error in the approximation (i.e., a trial function) produces only a second-order error in the desired solution. A variational formula may produce an upper-bound (maximum) or a lower-bound (minimum) solution of the desired quantity.

6.2 VARIATIONAL EXPRESSIONS FOR THE CAPACITANCE PER UNIT LENGTH OF TRANSMISSION LINES

We learn in Chapter 4 that the static or quasi-static parameters of a transmission line, including the characteristic impedance Z_o and effective dielectric constant ε_{eff}, can be obtained from the transmission line's capacitance per unit length C. Therefore, to determine these parameters using a variational method, variational formulas for C are needed. As for any variational expression, a variational expression for C may produce an upper- or a lower-bound solution for it. In this section, we will derive two kinds of expression for C: the upper bound based on the potential and the lower bound based on the charge density. It will be seen later that the upper and lower bounds to C correspond to the lower and upper bounds for the corresponding characteristic impedance, respectively.

6.2.1 Upper-Bound Variational Expression for C

We begin by noting that the electrostatic energy stored per unit length of a general two-conductor transmission line, shown in Fig. 6.2, is given as

$$\begin{aligned} W_e &= \tfrac{1}{2}\varepsilon_0\varepsilon_r \iint \mathbf{E}\cdot\mathbf{E}^*\,dx\,dy \\ &= \tfrac{1}{2}\varepsilon_0\varepsilon_r \iint \nabla_t\phi\cdot\nabla_t\phi\,dx\,dy \\ &= \tfrac{1}{2}\varepsilon_0\varepsilon_r \iint \left[\left(\frac{\partial\phi}{\partial x}\right)^2 + \left(\frac{\partial\phi}{\partial y}\right)^2\right] dx\,dy \end{aligned} \tag{6.6}$$

where $\mathbf{E}$ is the electric field between the two conductors S_1 and S_2. In practice, one conductor, for example, S_1, is normally held at zero potential representing the ground and the other is kept at a potential V_o. ε_0 denotes the permittivity of free space; ε_r represents the relative dielectric constant of the medium surrounding the two conductors; and $\phi(x, y)$ is the potential at any location in the plane transverse to the direction of propagation. The double integral is carried over the transmission line's cross section. This energy may also be obtained as

$$W_e = \tfrac{1}{2}CV^2 \tag{6.7}$$

where

$$V = \phi_2 - \phi_1 = \int_{S_1}^{S_2} \nabla_t\phi\cdot d\boldsymbol{\ell} \tag{6.8}$$

is the difference in potentials between the two conductors. $d\boldsymbol{\ell}$ represents a differential-length vector along any path from conductor S_1 to conductor S_2.

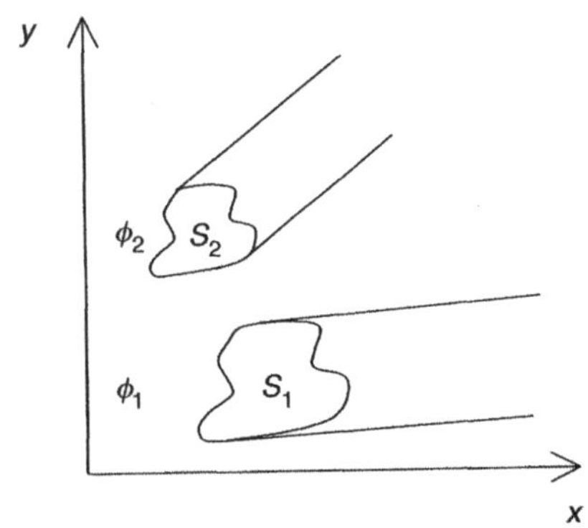

Figure 6.2 General two-conductor transmission line. ϕ_1 and ϕ_2 are the potentials on conductors S_1 and S_2, respectively.

Equating Eq. (6.6) to (6.7) and making use of Eq. (6.8) gives

$$C = \varepsilon_0 \varepsilon_r \frac{\iint |\nabla_t \phi|^2 \, dx \, dy}{\left[\int_{S_1}^{S_2} \nabla_t \phi \cdot \mathbf{d}\ell \right]} \tag{6.9}$$

It can be proved that Eq. (6.9) is stationary. That is, a first-order error in the potential ϕ causes a second-order error in the *true* or *exact* value C_o of C. The *true* solution C_o would result from Eq. (6.9) if we use the *true* value ϕ_o for ϕ. Note that ϕ_o satisfies Laplace's equation exactly. An approximate value for C obtained from Eq. (6.9) corresponding to an approximate ϕ is always greater than C_o. This equation therefore represents an upper-bound expression for C. C_o is the stationary value, which is an absolute minimum, for C. Figure 6.3 shows a possible curve for the capacitance C as a function of ϕ, calculated using the upper-bound variational expression (6.9). The fact that Eq. (6.9) gives upper-bound values to C implies that we should use a trial function for ϕ that results in calculated values of C that are as small as possible. This is an important criterion in selecting a proper trial function for ϕ.

6.2.2 Lower-Bound Variational Expression for C

We consider again the two-conductor transmission line shown in Fig. 6.2, in which conductors S_1 and S_2 are held at zero and V_o potentials, respectively. Now let $\phi(x, y)$ be the potential at any point in the transverse plane due to the (per-unit-length) charge distribution $\rho(x', y')$ located at an arbitrary location (x', y') on conductor S_2. For a unit charge at (x', y'), the resulting potential is basically the Green's function $G(x, y; x', y')$ associated with the transmission line, as discussed

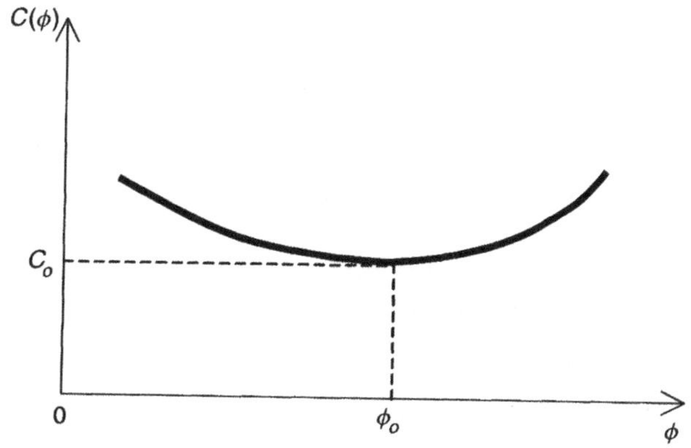

Figure 6.3 Possible values of C calculated using the upper-bound variational expression. The true value C_o represents an absolute minimum.

in Chapter 3. Note that this Green's function is the (potential) response due to a unit (charge) source located at (x', y') on conductor S_2 and can be determined from the physical parameters of a transmission line. The potential ϕ and the Green's function G must satisfy the corresponding Poisson equations subject to appropriate boundary conditions. Using the principle of superposition, it can be shown that the potential at any location (x, y) due to a charge distribution $\rho(x', y')$ at a location (x', y') on conductor S_2 is given as

$$\phi(x, y) = \oint_{S_2} G(x, y; x', y')\rho_s(x', y')\, d\ell' \tag{6.10}$$

where the integral is performed over the surface of conductor S_2 in the transverse plane where the charge is distributed. This potential is equal to V_o for locations on S_2. That is,

$$V_o = \oint_{S_2} G(x, y; x', y')\rho_s(x', y')\, d\ell' \tag{6.11}$$

where (x, y) indicates a location on S_2 only. The capacitance per unit length of the transmission line is obtained as

$$C = \frac{Q}{V_o} = \frac{Q^2}{QV_o} \tag{6.12}$$

where

$$Q = \oint_{S_2} \rho_s(x, y)\, d\ell \tag{6.13}$$

denotes the total charge on S_2. Substituting Eqs. (6.10) and (6.13) into Eq. (6.12), we have

$$\frac{1}{C} = \frac{\oint_{S_2}\oint_{S_2} G(x, y; x', y')\rho_s(x, y)\rho_s(x', y')\, d\ell\, d\ell'}{\left[\oint_{S_2} \rho_s(x, y)\, d\ell\right]^2} \tag{6.14}$$

which is stationary. Note that the *true* value C_o for C can be obtained from Eq. (6.14) if we use the *true* value ρ_o for ρ. It can be proved easily that an approximate value for C calculated using Eq. (6.14) corresponding to an approximate ρ is always smaller than the *true* C_o. This suggests that this equation is a lower-bound expression for C. C_o is the stationary value, which is an absolute maximum, for C in this case. Figure 6.4 illustrates possible values of the capacitance C as a function of the charge distribution ρ, calculated using the lower-bound variational expression (6.14). The lower-bound characteristic of Eq. (6.14) presents a useful criterion in selecting an appropriate trial function for ρ, in which we should choose that trial function leading to calculated results of C as large as possible.

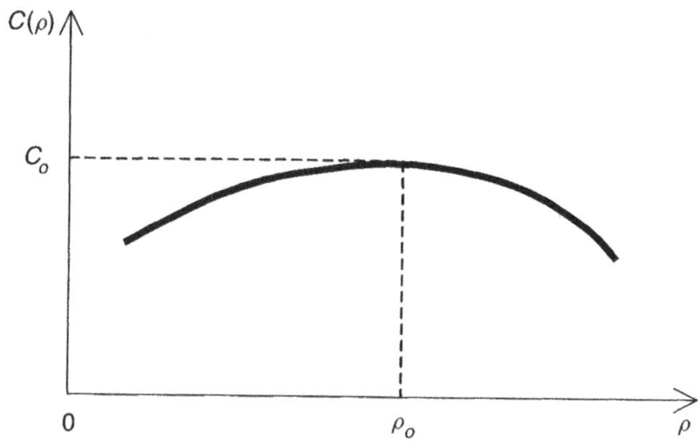

Figure 6.4 Possible values of C calculated using the upper-bound variational expression. The true value C_o represents an absolute maximum.

6.2.3 Determination of C, Z_o, and ε_{eff}

The capacitance per unit length C can be determined using either the upper- or lower-bound expression. The choice of a particular formula depends on individual problems. In general, if the potential ϕ is easily approximated, then one should use the upper-bound formula. If, however, ρ can be determined easily, then using the lower-bound equation is more appropriate. For complex problems such as multilayer, multiconductor transmission lines, both upper- and lower-bound expressions may be used together for the best accuracy.

As discussed in Section 6.1, in order to implement a variational method using the upper- or lower-bound expression, we must derive a trial function for ϕ or ρ, respectively. Let $u(x)$ be the (unknown) trial function representing either ϕ or ρ, and let $y = f[u(x)]$ represent either the variational expression (6.9) for C or (6.14) for $1/C$, respectively. We must determine $u(x)$ so that y has a minimum or a maximum. $u(x)$ should be selected to meet or closely approximate the actual behavior of ϕ or ρ. Often, we determine $u(x)$ using Eq. (6.4) based on the Ritz procedure as

$$u(x) = \sum_{i=1}^{N} a_i u_i(x) \tag{6.15}$$

where $u_i(x)$, $i = 1, 2, \ldots, N$, are basis functions. The a_i are the unknown variational parameters. Once the $u_i(x)$ are chosen, we can write y as a function of a_i as $y = f(a_i)$. We take the derivatives of y with respect to a_i for $i = 1, 2, \ldots, N$ and let them be zero to maximize or minimize y as

$$\frac{\partial y}{\partial a_i} = 0, \quad i = 1, 2, \ldots, N \tag{6.16}$$

This produces a system of N homogeneous linear equations, from which we can solve for C as discussed in the Appendix at the end of this chapter. It should be noted that a more accurate result for C is obtained by using more variational parameters a_i but at the expense of more computation time.

Once C is determined, we can then calculate the characteristic impedance and effective dielectric constant as

$$Z_o = \frac{1}{c\sqrt{CC_a}} \tag{6.17}$$

and

$$\varepsilon_{\text{eff}} = \frac{C}{C_a} \tag{6.18}$$

respectively, where c is the free-space velocity and C_a is the capacitance per unit length with the dielectrics removed. From Eq. (6.17) we can see that the upper and lower bounds to C correspond to the lower and upper bounds to Z_o.

It should be noted again that an exact value for C and, hence, Z_o and ε_{eff}, can be obtained only if the potential ϕ or charge distribution ρ is given exactly. In practice, only approximate values for C, Z_o, and ε_{eff} may be determined because values for ϕ and ρ are obtained approximately through trial functions.

6.3 FORMULATION OF VARIATIONAL METHODS IN THE SPACE DOMAIN

The formulation process is best illustrated via practical transmission lines. We will first implement the upper-bound variational expression to determine the per-unit-length capacitance and characteristic impedance of a coaxial transmission line. We will then analyze a three-layer microstrip line based on the lower-bound variational expression.

6.3.1 Variational Formulation Using Upper-Bound Expression

Figure 6.5 shows the considered coaxial transmission line. The conductors are assumed to be perfect conductors, and the substrate is lossless. Without lost of generality we assume that the potentials on the outer conductor (S_1) and inner conductor (S_2) are zero and V_o, respectively.

Before beginning the analysis, we should note that, for this problem, the *true* or *exact* values for the potential at any location corresponding to a radius r in the transverse plane, ϕ_o, and the per-unit-length capacitance C_o can be derived as

$$\phi_o = \frac{V_o}{\ln(a/b)} \ln\left(\frac{r}{b}\right) \tag{6.19}$$

where a and b are the radii of the inner and outer conductors, respectively, and

$$C_o = \frac{2\pi\varepsilon_0\varepsilon_r}{\ln(b/a)} \tag{6.20}$$

Figure 6.5 Coaxial transmission line. ε_r is the substrate's relative dielectric constant; a and b are radii of the inner and outer conductors, respectively.

where ε_r represents the substrate's relative dielectric constant. These equations may be used to verify the accuracy of results computed here.

Now using the upper-bound variational expression (6.9) and following Collin [1], we express C as

$$C = \varepsilon_0 \varepsilon_r \frac{\iint |\nabla_t \phi|^2 \, dx \, dy}{\left[\int_{S_1}^{S_2} \nabla_t \phi \cdot \mathbf{d}\boldsymbol{\ell} \right]^2} \tag{6.21}$$

Using the Ritz method described in Section 6.1, we first assume a trial function for ϕ as

$$\phi(r) = a_0 + a_1 r + a_2 r^2 \tag{6.22}$$

where a_1 and a_2 are the (unknown) variational parameters. Next, taking the partial derivatives of Eq. (6.22) in the cylindrical-coordinate system and using them in the transverse operator ∇_t, we can write

$$\nabla_t \phi = (a_1 + 2a_2 r)\, \mathbf{a}_r \tag{6.23}$$

Substituting Eq. (6.23) into (6.21) gives

$$\begin{aligned} C &= \varepsilon_0 \varepsilon_r \frac{\int_0^{2\pi} \int_a^b (a_1 + 2a_2 r)^2 r \, dr \, d\theta}{\left[\int_a^b (a_1 + 2a_2 r) \, dr \right]^2} \\ &= \frac{2\pi \varepsilon_0 \varepsilon_r \left[\dfrac{b^2 - a^2}{2} a_1^2 + 4 \dfrac{b^3 - a^3}{3} a_1 a_2 + (b^4 - a^4)\, a_2^2 \right]}{(b-a)^2 a_1^2 + 2(b^2 - a^2)(b-a)\, a_1 a_2 + (b^2 - a^2)^2\, a_2^2} \end{aligned} \tag{6.24}$$

The Ritz procedure continues by letting the partial derivatives of C with respect to a_1 and a_2 be zero. This will produce the minimum value for C and lead to a system of homogeneous equations,

$$\left[\frac{2C}{\pi\varepsilon}(b-a)^2 - 2(b^2-a^2)\right]a_1 + \left[\frac{2C}{\pi\varepsilon}(b^2-a^2)(b-a) - \frac{8}{3}(b^3-a^3)\right]a_2 = 0$$

$$\left[\frac{2C}{\pi\varepsilon}(b^2-a^2)(b-a) - \frac{8}{3}(b^3-a^3)\right] + \left[\frac{2C}{\pi\varepsilon}(b^2-a^2) - 4(b^4-a^4)\right]a_2 = 0 \tag{6.25}$$

as outlined in Section 6.2.3. Note that these equations contain (unknown) C in the coefficient matrix, so the (unknown) variational parameters a_1 and a_2 cannot be solved prior to the determination of C. To solve for C, we require that the determinant of the coefficient matrix equal zero, which is necessary in order to have nontrivial solutions for these equations. Enforcing this condition yields a characteristic equation of C, from which we can solve for C directly. Assuming b is equal to $2a$, the calculated value for C is $2.889\pi\varepsilon_0\varepsilon_r$, which is only 0.1% different from the exact value of $2.886\pi\varepsilon_0\varepsilon_r$ for C calculated from Eq. (6.20). This small error is achieved even though the assumed trial function for ϕ does not follow closely the true ϕ_o given in Eq. (6.19). This is the principal advantage of variational methods. We can, of course, obtain more accurate results for C by employing more terms in the trial function of ϕ but with an increase in the computation time.

Once C is obtained, we can then calculate the characteristic impedance, using Eq. (6.17):

$$Z_o = \frac{1}{c\sqrt{CC_a}} \tag{6.26}$$

We can also solve Eq. (6.25) for a_1 and a_2 using the calculated value of C. Applying the boundary condition for the inner conductor we obtain

$$V_o = a_0 + a_1 b + a_2 b^2 \tag{6.27}$$

which can be solved for a_0 by making use of the results obtained for a_1 and a_2. The potential at any location corresponding to a radius r in the transverse plane can now be determined completely from Eq. (6.22).

6.3.2 Variational Formulation Using Lower-Bound Expression

The formulation of the variational method based on the lower-bound expression is illustrated using a three-layer shielded microstrip line shown in Fig. 6.6. The enclosure and conducting strip are assumed to be perfect conductors. The strip thickness is assumed negligible, and the dielectric substrates are assumed lossless. This structure includes the open and shielded single-layer microstrip line, suspended strip line, strip line, and two-layer microstrip line. For example, the (open single-layer) microstrip line is obtained by letting h_1 and a be very large,

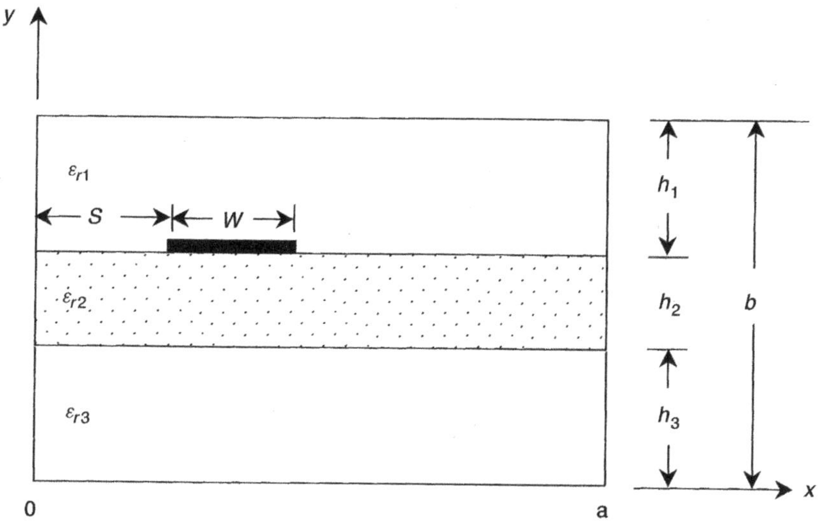

Figure 6.6 Cross section of a shielded three-layer microstrip line.

$h_3 = 0$, and $\varepsilon_{r1} = 1$. This structure, which is embedded in inhomogeneous media, supports a quasi-TEM mode.

For this problem, the upper-bound variational expression would require a two-dimensional trial function for the potential ϕ. On the other hand, only a one-dimensional trial function for the charge distribution ρ would be needed, due to the fact that the strip thickness is assumed infinitely thin. A two-dimensional function is more difficult to derive and consumes more computation time than that of one dimension. The upper-bound expression based on the charge distribution is therefore preferred over the lower-bound formula for the considered structure. Using the lower-bound variational expression (6.14), we now express the inverse of the per-unit-length capacitance of the transmission line $1/C$ as

$$\frac{1}{C} = \frac{\oint_S \oint G(x;x')\rho_s(x)\rho_s(x')\,dx\,dx'}{\left[\oint_s \rho_s(x)\,dx\right]^2} \tag{6.28}$$

where S represents the surface of the conducting strip along the x direction, and the integral is taken over the strip's surface in the transverse plane. $\rho_s(x)$ is the (per-unit-length) surface charge density at the upper interface where the strip resides; it exists only on the strip and solely depends on x for this problem. $G(x;x')$ is the Green's function, which is symmetric in x and x' due to reciprocity. Equation (6.28) is valid for any transmission line whose strips are infinitesimally thin.

To calculate C and, hence, Z_o and ε_{eff}, we need to determine the Green's function G and a suitable trial function for the charge density ρ. The Green's function is given as [2]

$$G(x;x') = \sum_{n=1}^{\infty} g_n g(x) g(x') \tag{6.29}$$

where

$$g(n) = \frac{2}{n\,\pi\,Y} \tag{6.30}$$

$$g(x) = \sin(\alpha_n x) \tag{6.31}$$

$$g(x') = \sin(\alpha_n x') \tag{6.32}$$

$$\alpha_n = \frac{n\,\pi}{a}, \quad n = 1, 2, \ldots \tag{6.33}$$

$$Y = \varepsilon_0 \varepsilon_{r2} \frac{\varepsilon_{r1} \coth(\alpha_n h_1) + \varepsilon_{r2} \tanh(\alpha_n h_2)}{\varepsilon_{r2} + \varepsilon_{r1} \coth(\alpha_n (h_1) \tanh(\alpha_n h_2)} + \varepsilon_0 \varepsilon_{r3} \coth(\alpha_n h_3) \tag{6.34}$$

ε_{ri} and h_i, $i = 1, 2, 3$, are the relative dielectric constant and thickness of the ih dielectric layer. a denotes the transmission line's enclosure width. This Green's function can also be derived using the procedure described in Chapter 3 (see Problem 3.19).

Now using the Ritz method, we let the charge density on the strip be

$$\rho_s(x) = \sum_{i=1}^{N} a_i \rho_{si}(x) \tag{6.35}$$

where a_i, $i = 1, 2, \ldots, N$, are the unknown variational parameters, and $\rho_{si}(x)$ are the basis functions, chosen to approximate the charge density as

$$\rho_{si}(x) = \begin{cases} \dfrac{\cos\left((i-1)\pi \dfrac{x-S}{W}\right)}{\sqrt{1 - \left(\dfrac{2(x-S)-W}{W}\right)^2}}, & S \le x \le S + W \\ 0, & \text{otherwise} \end{cases} \tag{6.36}$$

Sketches of these charge basis functions are shown in Fig. 6.7. Note that the basis function ρ_{s1}, corresponding to $i = 1$, is minimum at the center of the strip and increases rapidly near the strip edges and, thus, approximates closely the actual charge distribution on the strip. We can therefore expect that good results for C may be obtained even if only one basis function ($N = 1$) is used. Proceeding with the Ritz procedure, we take the partial derivatives of $1/C$ in

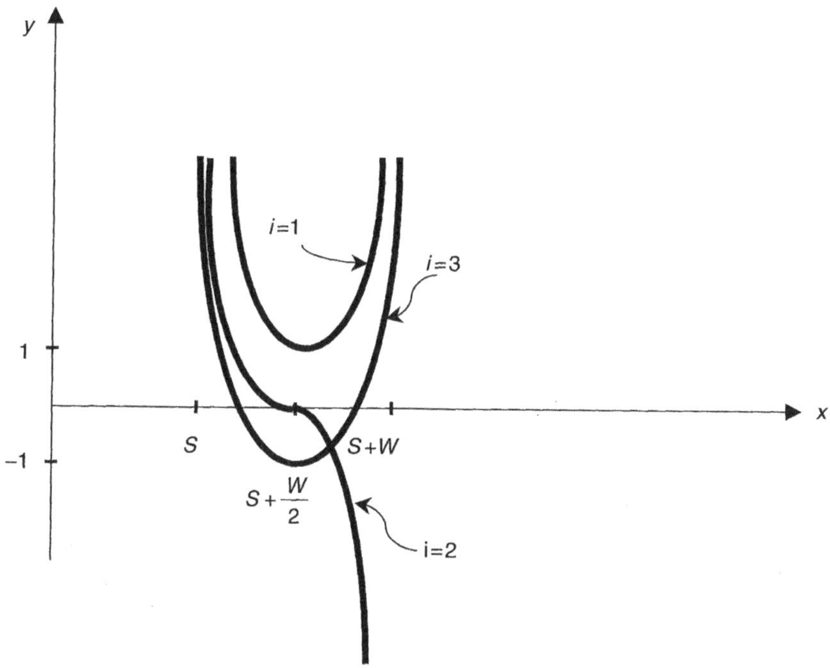

Figure 6.7 Sketches of the charge basis functions.

Eq. (6.24) with respect to a_1 and a_2 and let them be zero. This will produce the maximum value for $1/C$ and lead to a system of homogeneous equations as outlined in Section 6.2.3. In the Appendix at the end of this chapter, we derive two different systems of homogeneous equations with and without $1/C$ in the coefficient matrices. The equations that contain $1/C$ are given in Eq. (A6.14) as

$$\sum_{i=1}^{N}\left(\sum_{n=1}^{\infty} g_n\, SQ_{ni}\, SQ_{nj} - \frac{1}{C}Q_iQ_j\right) a_i = 0; \quad j = 1, 2, \ldots, N \tag{6.37}$$

where g_n is given in Eqs. (6.30) and (6.34), and the total charge on the strip, Q_i, is given from Eq. (A6.6) as

$$Q_i = \int_S^{S+W} \rho_{si}(x)\, dx \tag{6.38}$$

Upon substitution of the charge basis function $\rho_{si}(x)$ from Eq. (6.32), Q_i becomes

$$Q_i = \frac{\pi W}{2} J_0\left(\frac{(i-1)\pi}{2}\right) \cos\left(\frac{(i-1)\pi}{2}\right) \tag{6.39}$$

where J_0 stands for the zero-order Bessel function of the first kind. The other parameter, SQ_i, is obtained from Eq. (A6.7) as

$$SQ_{ni} = \int_S^{S+W} g(x)\rho_{si}(x)\,dx \tag{6.40}$$

Substituting Eqs. (6.31) and (6.36) into Eq. (6.40), we obtain, after several manipulations,

$$SQ_{ni} = \frac{\pi W}{4}\left[J_0\left(\frac{\alpha_n W + (i-1)\pi}{2}\right)\sin\left(\alpha_n S + \frac{\alpha_n W + (i-1)\pi}{2}\right)\right.$$
$$\left. + J_0\left(\frac{\alpha_n W - (i-1)\pi}{2}\right)\sin\left(\alpha_n S + \frac{\alpha_n W - (i-1)\pi}{2}\right)\right] \tag{6.41}$$

Now using Eqs. (6.39) and (6.41) in Eq. (6.37), we can obtain the final N linear equations in N unknowns $a_i, i = 1, 2, \ldots, N$. These equations contain the (unknown) C in the coefficient matrix, so the (unknown) variational parameters a_i cannot be solved prior to the determination of C. However, in order for the nontrivial solutions of a_i to exist, the determinant of the coefficient matrix must be zero. Doing so results in a characteristic equation of C, from which we can solve for C directly. Once C is determined, we can then calculate the characteristic impedance and effective dielectric constant using Eqs. (6.17) and (6.18), respectively, as

$$Z_0 = \frac{1}{c\sqrt{CC_a}} \tag{6.42}$$

$$\varepsilon_{\text{eff}} = \frac{C}{C_a} \tag{6.43}$$

Using the calculated value of C, we can also solve the system of equations for a_i and then determine the charge density $\rho(x)$.

The system of equations that does not contain $1/C$ is given in Eq. (A6.16) as

$$\sum_{i=1}^{N}\sum_{n=1}^{\infty}[g_n\, SQ_{nj}\cdot Q_i - g_n\, SQ_{ni}\cdot Q_j]\,a_i = 0; \quad j = 1, 2, \ldots, N \tag{6.44}$$

Substituting Eqs. (6.39) and (6.41) into Eq. (6.44), we obtain the final system of N homogeneous equations in N unknowns $a_i, i = 1, 2, \ldots, N$, from which we can solve for a_i. Using the results for a_i we can calculate C obtained in

Eq. (A6.5) as

$$\frac{1}{C} = \frac{\sum_{n=1}^{\infty} g_n \left(\sum_{i=1}^{N} a_i S Q_{ni} \right)^2}{\left(\sum_{i=1}^{N} a_i Q_i \right)^2} \tag{6.45}$$

Once C is determined, we can of course compute the characteristic impedance and effective dielectric constant.

If a simple expression such as

$$\rho_s(x) = \begin{cases} x^2, & S \le x \le S + W \\ 0, & \text{otherwise} \end{cases} \tag{6.46}$$

is used for $\rho(x)$, we can derive a closed-form expression for C.

It should be noted again that the closer $\rho(x)$ is to the true value ρ_o the better the result for C. Also, a more accurate result can be obtained by using more variational parameters a_i, which is, of course, achieved at the expense of increased computation time.

Perhaps the most difficult step in implementing a variational method based on the lower-bound expression is the derivation of the Green's function $G(x, y; x', y')$. This Green's function, however, can be found easily if we perform the variational analysis in the Fourier-transform or spectral domain as described in the following section.

6.4 VARIATIONAL METHODS IN THE SPECTRAL DOMAIN

As for the case of variational methods in the space domain, in order to determine the static or quasi-static parameters of a transmission line using variational methods in the spectral domain, we need to have variational formulas for the per-unit-length capacitance C in the spectral domain. Here, we will derive the lower-bound expression for C in the spectral domain and then formulate the variational process.

6.4.1 Lower-Bound Variational Expression for C in the Spectral Domain

We begin by considering a general two-conductor transmission line as shown in Fig. 6.2. The potentials on conductors S_1 and S_2 are assumed to be zero and V_o, respectively. The lower-bound variational expression for C in the space domain given in Eq. (6.14) can be rewritten as

$$\frac{1}{C} = \frac{\oint_{S_2} \rho_s(x, y)\phi(x, y)\, d\ell}{Q^2} \tag{6.47}$$

where $\phi(x, y)$ is the potential at any point in the transverse plane given in Eq. (6.10) and $Q(x, y)$ is the total charge on conductor S_2 given in Eq. (6.13). We will now perform the analysis in the spectral domain using the following Fourier transform definition

$$\tilde{f}(\alpha, y) = \int_{-\infty}^{\infty} f(x, y)e^{j\alpha x}\, dx \tag{6.48}$$

where α is the Fourier transform variable and the tilde ($\sim$) indicates the Fourier-transformed quantity. This Fourier transform along with Parseval's theorem are discussed in Appendix A at the end of Chapter 7. Applying Parseval's theorem

$$\int_{-\infty}^{\infty} f(x, y)g^*(x, y)\, dx = \frac{1}{2\pi} \int_{-\infty}^{\infty} \tilde{f}(\alpha, y)\tilde{g}^*(\alpha, y)\, d\alpha \tag{6.49}$$

to the right-hand side of Eq. (6.47) leads to

$$\frac{1}{C} = \frac{1}{2\pi Q^2} \int_{-\infty}^{\infty} \tilde{\rho}_s(\alpha, y)\tilde{\phi}(\alpha, y)\, d\alpha \tag{6.50}$$

Note that Eqs. (6.49) and (6.50) are general and specifically apply to (open) transmission lines. For a transmission line enclosed in a conducting channel, the Fourier transform is finite and corresponds to the discrete Fourier transform variable α_n, which must be determined based on the transmission line's boundary conditions. For instance, for the three-layer shielded microstrip line shown in Fig. 6.6, the Fourier transform can be defined as

$$\tilde{f}(\alpha_n, y) = \frac{2}{a} \int_0^a f(x, y) \sin \alpha_n x\, dx \tag{6.51}$$

where $\alpha_n = n\pi/a$, $n = 1, 2, \ldots$. This Fourier transform, given in Appendix A of Chapter 7, is also used in the spectral-domain formulation presented in Chapter 7. The corresponding Parseval's relation is

$$\int_0^a f(x, y)g^*(x, y)\, dx = \frac{a}{2} \sum_{n=1}^{\infty} \tilde{f}(\alpha_n, y)\, \tilde{g}^*(\alpha_n, y) \tag{6.52}$$

and Eq. (6.50) becomes

$$\frac{1}{C} = \frac{a}{2Q^2} \sum_{n=1}^{\infty} \tilde{\rho}_s(\alpha_n, y)\tilde{\phi}(\alpha_n, y) \tag{6.53}$$

Similar to the case of the (space-domain) Green's function described in Section 6.2.2, the Green's function in the spectral domain, $\tilde{G}(\alpha)$, is defined as the resultant potential due to a unit charge located on conductor S_2 and so may be written as

$$\tilde{G}(\alpha)\frac{\tilde{\rho}_s(\alpha, y)}{\varepsilon_0} = \tilde{\phi}(\alpha, y) \tag{6.54}$$

Equation (6.50) can now be rewritten, upon making use of Eq. (6.54), as

$$\frac{1}{C} = \frac{1}{2\pi\varepsilon_0 Q^2}\int_{-\infty}^{\infty} \tilde{\rho}_s^2(\alpha, y)\tilde{G}(\alpha, y)\, d\alpha \tag{6.55}$$

For the case of the three-layer shielded microstrip line shown in Fig. 6.6, using Eq. (6.54) in (6.53) gives

$$\frac{1}{C} = \frac{a}{2\varepsilon_0 Q^2}\sum_{n=1}^{\infty} \tilde{\rho}_s^2(\alpha_n, y)\tilde{G}(\alpha_n, y) \tag{6.56}$$

Both Eqs. (6.55) and (6.56) are stationary. Because these equations are derived directly from the lower-bound variational formula, they also represent lower-bound expressions for C in the spectral domain. This lower-bound characteristic is useful in selecting an appropriate trial function for the charge distribution $\rho(x, y)$, in which we should choose a trial function that results in as large a calculated value of C as possible.

6.4.2 Determination of C, Z_o, and ε_{eff}

To obtain numerical results for C using a spectral-domain variational expression such as Eq. (6.55) and, hence, Z_o and ε_{eff}, we must determine a trial function for the charge density $\rho(x, y)$. This trial function must be so chosen that the resultant capacitance per unit length C is maximized. Often, we use the Ritz method and express the charge density as a truncated summation of known basis functions in the space domain as

$$\rho_s(x) \cong \sum_{i=1}^{N} a_i \rho_{si}(x) \tag{6.57}$$

where $\rho_i(x)$, $i = 1, 2, \ldots, N$, are the basis functions chosen to describe the charge distribution on the strip and are nonzero only on the strip; a_i are the unknown coefficients known as variational parameters; and N denotes the number of basis functions used to approximate the strip's charge density. In the spectral domain, the charge density given in Eq. (6.57) becomes

$$\tilde{\rho}_s(\alpha) \cong \sum_{i=1}^{N} a_i \tilde{\rho}_{si}(\alpha) \tag{6.58}$$

Criteria for choosing the basis functions were already described earlier. In addition to these standards, the basis functions used in the spectral-domain variational method should have closed-form Fourier transforms to reduce the computation time. Using Eq. (6.58) in (6.55) then gives us an expression for $1/C$ whose only unknowns are a_i. Continuing with the Ritz method, we set the derivatives of $1/C$ with respect to a_i for $i = 1, 2, \ldots, N$ to zero. This results in a system of N homogeneous linear equations, from which we can solve for C. The derivation of these equations is given in the Appendix at the end of this chapter.

Once C is determined, we can compute the characteristic impedance and effective dielectric constant of the transmission line using Eqs. (6.17) and (6.18), respectively. Equation (6.17) produces a value that is an upper bound to the true characteristic impedance. The obtained characteristic impedance and effective dielectric constant are also approximate due to the fact that the charge density on the strip is chosen approximately.

6.4.3 Formulation

The formulation process of the variational method based on the lower-bound expression in the spectral domain is illustrated by using the same three-layer shielded microstrip line shown in Fig. 6.6. We will present details in determining the per-unit-length capacitance, characteristic impedance, and effective dielectric constant of the transmission line. The process is very similar to the one we present for the quasi-static spectral-domain method in Chapter 7.

We begin with the two-dimensional Laplace equation for the electric potential in a plane transverse to the direction of wave propagation,

$$\frac{\partial^2 \phi_i(x, y)}{\partial x^2} + \frac{\partial^2 \phi_i(x, y)}{\partial y^2} = 0, \quad i = 1, 2, 3 \tag{6.59}$$

where $\phi_i(x, y)$ is the unknown potential in the ith region. The boundary conditions for the considered structure are

$$\phi_i(0, y) = \phi_i(a, y) = 0 \tag{6.60}$$

$$\phi_1(x, b) = \phi_3(x, 0) = 0 \tag{6.61}$$

$$\phi_1(x, h_2 + h_3) = \phi_2(x, h_2 + h_3) = V^U(x) \tag{6.62}$$

$$\phi_2(x, h_3) = \phi_3(x, h_3) \tag{6.63}$$

$$\varepsilon_{r2} \left.\frac{\partial \phi_2}{\partial y}\right|_{y=h_2+h_3} - \varepsilon_{r1} \left.\frac{\partial \phi_1}{\partial y}\right|_{y=h_2+h_3} = \frac{\rho(x)}{\varepsilon_0} \tag{6.64}$$

$$\varepsilon_{r3} \left.\frac{\partial \phi_3}{\partial y}\right|_{y=h_3} - \varepsilon_{r2} \left.\frac{\partial \phi_2}{\partial y}\right|_{y=h_3} = 0 \tag{6.65}$$

where ε_0 is the free-space permittivity and ε_{ri}, $i = 1, 2, 3$, is the relative dielectric constant of the ith layer; $V^U(x)$ denotes the potential at the upper interface, where the strip resides, and can be expressed as

$$V^U(x) = V_o(x) + V(x) \tag{6.66}$$

$V_o(x) = V_o$ on the strip and zero elsewhere; we choose the value of V_o. $V(x)$ is the unknown potential at the upper interface and is zero on the strip. $\rho(x)$ represents the unknown charge density on the strip and is nonzero only on the strip.

To perform the variational method in the spectral domain for the considered problem, we define the Fourier transform $\tilde{f}(\alpha_n, y)$ of $f(x, y)$ as

$$\tilde{f}(\alpha_n, y) = \frac{2}{a}\int_0^a f(x, y)\sin\alpha_n x\,dx \tag{6.67}$$

where $\alpha_n = n\pi/a$, $n = 1, 2, 3, \ldots,$ denoting the spectral order or term. This Fourier transform is the sine transform used for odd functions; it is chosen here because the functions $\phi_i(x, y)$ are required to be zero at $x = 0$ and a. As can be verified, the choice for the Fourier transform variable α_n will cause the boundary conditions of Eq. (6.60) on $\phi_i(x, y)$ to be met automatically.

Laplace's equation in the Fourier-transform or spectral domain can now be obtained, upon applying Eq. (6.67) to (6.59), as

$$\frac{\partial^2\tilde{\phi}_i}{\partial y^2} - \alpha_n^2\,\tilde{\phi}_i = 0 \tag{6.68}$$

This equation is derived in Section 7.2. The boundary conditions of the considered structure in the spectral domain are obtained by Fourier transforming Eqs. (6.61)–(6.65) as

$$\tilde{\phi}_1(\alpha_n, b) = \tilde{\phi}_3(\alpha_n, 0) = 0 \tag{6.69}$$

$$\tilde{\phi}_1(\alpha_n, h_2 + h_3) = \tilde{\phi}_2(\alpha_n, h_2 + h_3) = \tilde{V}^U(\alpha_n) \tag{6.70}$$

$$\tilde{\phi}_2(\alpha_n, h_3) = \tilde{\phi}_3(\alpha_n, h_3) \tag{6.71}$$

$$\varepsilon_{r2}\left.\frac{\partial\tilde{\phi}_2}{\partial y}\right|_{y=h_2+h_3} - \varepsilon_{r1}\left.\frac{\partial\tilde{\phi}_1}{\partial y}\right|_{y=h_2+h_3} = \frac{\tilde{\rho}_s(\alpha_n)}{\varepsilon_0} \tag{6.72}$$

$$\varepsilon_{r3}\left.\frac{\partial\tilde{\phi}_3}{\partial y}\right|_{y=h_3} - \varepsilon_{r2}\left.\frac{\partial\tilde{\phi}_2}{\partial y}\right|_{y=h_3} = 0 \tag{6.73}$$

Note that the boundary condition (6.60) is not transformed since it is already satisfied by the choice of α_n. The potential along the upper interface is given in the spectral domain as

$$\tilde{V}^U(\alpha_n) = \tilde{V}_o(\alpha_n) + \tilde{V}(\alpha_n) \tag{6.74}$$

The solution of Eq. (6.68), which satisfies the boundary conditions (6.69)–(6.73), is derived in Chapter 7 as

$$\tilde{G}(\alpha_n)\tilde{\rho}_s(\alpha_n) = \tilde{V}^U(\alpha_n) \tag{6.75}$$

where

$$\tilde{G}(\alpha_n) = \frac{\tanh\alpha_n h_1\left(\dfrac{\varepsilon_{r3}}{\varepsilon_{r2}}\tanh\alpha_n h_2 + \tanh\alpha_n h_3\right)}{\varepsilon_0\alpha_n\tanh\alpha_n h_2\left[\varepsilon_{r3}\left(\dfrac{\varepsilon_{r1}}{\varepsilon_{r2}} + \tanh\alpha_n h_1\coth\alpha_n h_2\right) + \tanh\alpha_n h_3(\varepsilon_{r1}\coth\alpha_n h_2 + \varepsilon_{r2}\tanh\alpha_n h_1)\right]} \tag{6.76}$$

This function is equal to the potential $\tilde{V}^U(\alpha_n)$ when the charge $\tilde{\rho}(\alpha_n)$ is set to one and, thus, represents the Green's function in the spectral domain. $\tilde{V}^U(\alpha_n)$ is given in Eq. (6.74) as

$$\tilde{V}^U(\alpha_n) = \tilde{V}_0(\alpha_n) + \tilde{V}(\alpha_n) \tag{6.77}$$

$\tilde{V}_0(\alpha_n)$ and $\tilde{V}(\alpha_n)$ stand for the Fourier transforms of $V_o(x)$ and $V(x)$, respectively. $V_o(x)$ is chosen and is thus known. $V(x)$ is given as

$$\tilde{V}(\alpha_n) = \frac{2}{a}\int_0^a V(x)\sin\alpha_n x\,dx \tag{6.78}$$

with $\alpha_n = n\pi/a$, $n = 1, 2, 3\ldots.$ It should be noted here that the process of deriving Eq. (6.75), and hence Eq. (6.76), would be simpler if we assume h_1 and h_2 are always nonzero. Furthermore, the spectral-domain Green's function is independent of the dimensions along the x axis. It is basically the total potential, corresponding to a unit charge, along the interface where the conductor resides. It is recognized that the Green's function in the spectral domain is easier to derive than that in the space domain.

Up to now, the formulation is exact and the same as that for the spectral-domain method described in Section 7.2. The reciprocal of the per-unit-length capacitance, $1/C$, is derived in Eq. (6.56) as

$$\frac{1}{C} = \frac{a}{2\varepsilon_0 Q^2}\sum_{n=1}^{\infty}\tilde{\rho}_s^2(\alpha_n, y)\,\tilde{G}(\alpha_n, y) \tag{6.79}$$

where $\tilde{G}(\alpha_n)$ is given in Eq. (6.36) and

$$Q = \int_S \rho_s(x)\,dx \tag{6.80}$$

To solve for C, we therefore need to determine the strip's charge density $\rho(x)$. Using the Ritz method, we express the charge density as a truncated summation of basis functions in the space domain as

$$\rho(x) \cong \sum_{i=1}^{N} a_i \rho_i(x) \tag{6.81}$$

where $\rho_i(x)$, $i = 1, 2, \ldots,$ are the basis functions describing the charge distribution on the strip and are nonzero only on the strip. These basis functions strongly influence the numerical efficiency of the solution process and the accuracy of the solutions and thus should be chosen carefully following the criteria discussed previously. The a_i are the unknown coefficients referred to as the variational parameters; and N denotes the number of basis functions used for the strip's charge density. In the spectral domain,

$$\tilde{\rho}(\alpha_n) \cong \sum_{i=1}^{N} a_i \tilde{\rho}_i(\alpha_n) \tag{6.82}$$

The basis functions used here are

$$\rho_i(x) = \begin{cases} \dfrac{\cos\left((i-1)\pi \dfrac{x-S}{W}\right)}{\sqrt{1-\left(\dfrac{2(x-S)-W}{W}\right)^2}}, & S \le x \le S+W \\ 0, & \text{otherwise} \end{cases} \tag{6.83}$$

which are the same as those used for the space-domain variational method of Section 6.3.2. The Fourier transforms of these basis functions are obtained using Eq. (6.67) as

$$\tilde{\rho}_{si}(\alpha_n) = \frac{\pi}{2}\frac{W}{a}[\sin(k_1+k_2)J_0(k_1+k_3) + \sin(k_2-k_1)J_0(k_1-k_3)] \tag{6.84}$$

$$k_1 = (j-1)\frac{\pi}{2} \tag{6.85}$$

$$k_2 = \alpha_n \left(S + \frac{W}{2}\right) \tag{6.86}$$

$$k_3 = \alpha_n \frac{W}{2} \tag{6.87}$$

where J_0 stands for the zero-order Bessel function of the first kind.

Substitute Eq. (6.82) into (6.56), take the derivatives with respect to a_i, and let them be zero according to the Ritz method. This can result in two systems of linear homogeneous equations—one containing $1/C$ and one not. The system of equations that does not contain $1/C$ is given by Eq. (A6.21), in the Appendix at the end of this chapter, as

$$\sum_{i=1}^{N} a_i \left(\sum_{n=1}^{\infty} \tilde{\rho}_{si}(\alpha_n)\tilde{G}(\alpha_n)\tilde{\rho}_{sj}(\alpha_n) \right) = 0 \tag{6.88}$$

Now using Eqs. (6.84) and (6.36) in Eq. (6.88), we can obtain the final N linear equations in N unknowns $a_i, i = 1, 2, \ldots, N$, from which we can solve for a_i. Using the results for a_i we can calculate C obtained in Eq. (A6.19) of the Appendix at the end of this chapter as

$$\frac{1}{C} = \frac{a}{2\varepsilon_0 Q^2} \sum_{i=1}^{N} \sum_{j=1}^{\infty} a_i a_j \sum_{n=1}^{\infty} \tilde{\rho}_{si}(\alpha_n)\tilde{G}(\alpha_n)\tilde{\rho}_{sj}(\alpha_n) \tag{6.89}$$

Once C is determined, it can then be used to calculate the characteristic impedance and effective dielectric constant as

$$Z_0 = \frac{1}{c\sqrt{CC_a}} \tag{6.90}$$

and

$$\varepsilon_{\text{eff}} = \frac{C}{C_a} \tag{6.91}$$

respectively, where c is the free-space velocity and C_a is the capacitace per unit length with the dielectrics removed. Because of the use of an approximate charge density for the strip, C evaluated using Eq. (6.79) only gives an approximate result. However, as a characteristic of the variational method, this expression is stationary; that is, a first-order error in the choice of the charge density basis functions ρ_i will only produce a second-order error. Therefore, Eq. (6.79) should give a value for C that is very close to the exact result, provided that good choices for the basis functions are used. It can be proved that C from Eq. (6.79) is always smaller than the exact result and, hence, this equation represents the lower-bound expression for C.

It should be noted that both the number of basis functions and the number of spectral terms, used in Eq. (6.79) for determining C, affect the accuracy of the numerical results. The larger these numbers the more accurate the results, but at the expense of increased computation time. For most engineering purposes, three basis functions and 200 spectral terms are sufficient.

REFERENCES

1. R. E. Collin, *Field Theory of Guided Waves*, 2nd edition, IEEE Press, New York, 1991, Chap. 4.

2. B. Bhat and S. K. Koul, "Unified Approach to Solve a Class of Strip and Microstrip-like Transmission Lines," *IEEE Trans. Microwave Theory Tech.*, Vol. MTT-30, pp. 679–686, May 1982.

3. E. Yamashita and R. Mittra, "Variational Method for the Analysis of Microstrip Lines," *IEEE Trans. Microwave Theory Tech.*, Vol. MTT-16, pp. 251–256, Apr. 1968.

PROBLEMS

6.1 Prove that Eq. (6.9) is stationary.

6.2 Prove that C, given by Eq. (6.9), produces an upper bound to the true value C_o.

6.3 Prove that Eq. (6.14) is stationary.

6.4 Prove that C, given by Eq. (6.14), produces a lower bound to the true value C_o.

6.5 Derive Eq. (6.41).

6.6 Derive a system of linear equations using Eqs. (6.37), (6.39), and (6.41), whose unknowns are a_i, $i = 1, 2, \ldots, N$, and the corresponding characteristic equation for C.

6.7 Consider a three-layer shielded microstrip as shown in Fig. 6.6. The enclosure and conducting strip are assumed to be perfect conductors. The strip thickness is assumed negligible, and the dielectric substrates are assumed lossless. The charge basis functions are given in Eq. (6.36).

(a) Formulate the lower-bound variational analysis for determining the characteristic impedance, Z_o, and effective dielectric constant, ε_{eff}.

(b) Write a computer program to calculate Z_o and ε_{eff}. Calculate and plot Z_o and ε_{eff} versus W/h_2 from 0.1 to 5, $\varepsilon_{r2} = 2.2, h_1 = h_3 = 0.66$ mm, $h_2 = 0.254$ mm, $a = 2.54$ mm, and $S = 0.25(a - W)$ and $0.5(a - W)$. Compare results using one to five basis functions.

6.8 Use the program developed in Problem 6.7 to calculate and plot Z_o and ε_{eff} versus W/h from 0.05 to 5 for the microstrip line, as shown in Fig. 4.2, having $\varepsilon_r = 2.2$ and 10.5 and $h = 0.635$ and 1.27 mm. The strip thickness is negligible. Compare results with those calculated using the program developed in Problem 6.15. Provide an assessment of the basis functions used.

6.9 Consider the strip line as shown in Fig. P3.2 of Problem 3.18. It is assumed that the strip is a perfect conductor and has zero thickness. Also, the ground planes are assumed to be infinitely wide and perfect conductors. The dielectric is assumed to be perfect.

(a) Derive the Green's function in closed form (see Problem 3.18).

(b) Derive the upper-bound expression for the characteristic impedance Z_o, assuming the charge distribution is

$$\rho_s(x) = \begin{cases} x^2, & x \in \text{strip} \\ 0, & \text{otherwise} \end{cases}$$

(c) Calculate and plot Z_o versus the normalized strip width, W/a, from 0.1 to 5 for $\varepsilon_r = 2.2$ and 10.5 and $a = 0.635$, and 1.27 mm. Compare results to those obtained in Problem 4.18.

6.10 Consider a shielded microstrip line as shown in Fig. 3.2. The enclosure and conducting strip are assumed to be perfect conductors. The strip thickness is assumed negligible, and the dielectric substrates are assumed lossless.

(a) Derive expressions for the capacitance per unit length, C, characteristic impedance, Z_o, and effective dielectric constant, ε_{eff}, using a variational method, assuming the following charge distribution:

$$\rho_s(x) = \begin{cases} A, & \dfrac{a-W}{2} \le x \le \dfrac{a+W}{2} \\ 0, & \text{otherwise} \end{cases}$$

where A is a constant.

(b) Calculate and plot Z_o and ε_{eff} versus W/h from 0.05 to 5 for $\varepsilon_r = 2.2$ and 10.5 and $h = 0.635$ and 1.27 mm. Compare results to those obtained using the program developed in Problem 6.7 and provide an evaluation of the basis functions used.

(c) Determine the potential at $x = 5a/6$ and $y = h$.

6.11 Repeat Problem 6.10, parts (a) and (b), using

$$\rho_s(x) = \begin{cases} \dfrac{1}{\sqrt{1 - \left(\dfrac{|x - a/2|}{W/2}\right)^2}}, & \dfrac{a-W}{2} \le x \le \dfrac{a+W}{2} \\ 0, & \text{otherwise} \end{cases}$$

Compare the calculated results with those obtained in parts (a) and (b) of Problem 6.10 and provide an assessment of the basis functions used.

6.12 Determine Eq. (A6.14) for the microstrip line shown in Fig. 6.6. The charge basis functions are given in Eq. (6.36). The resultant equations contain the zero-order Bessel function of the first kind,

$$J_0(\alpha) = \frac{1}{\pi} \int_{-1}^{1} \frac{e^{-j\alpha x}}{\sqrt{1 - x^2}} dx$$

6.13 Prove that both Eqs. (6.55) and (6.56) are stationary.

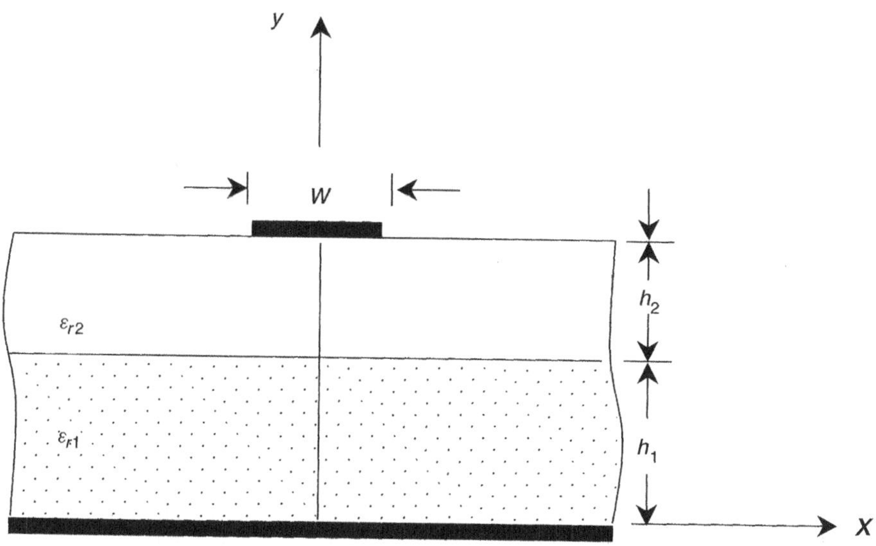

Figure P6.1 Cross section of a two-layer microstrip line.

6.14 Consider a three-layer shielded microstrip as shown in Fig. 6.6. The enclosure is assumed to be a perfect conductor. Determine the Fourier transform variable α_n in Eq. (6.67).

6.15 Consider a microstrip line as shown in Fig. P6.1. It is assumed that the strip is a perfect conductor and has zero thickness. Also, the ground plane is assumed to be infinitely wide and a perfect conductor. The dielectrics are assumed to be perfect.

(a) Derive the Green's function in the spectral domain.

(b) Formulate the lower-bound variational analysis in the spectral domain for determining the capacitance per unit length, characteristic impedance, Z_o, and effective dielectric constant, ε_{eff}, using the following charge distribution:

$$\rho_s(x) = \begin{cases} \dfrac{1}{\sqrt{1 - \left(\dfrac{|x|}{W/2}\right)^2}}, & x \leq \dfrac{W}{2} \\ 0, & \text{otherwise} \end{cases}$$

(c) Write a computer program to calculate Z_o and ε_{eff}. Compute and plot Z_o and ε_{eff} versus W/h_2 from 0.1 to 5 for $\varepsilon_{r1} = 10.5$, $h_1 = 1.27$ mm and $\varepsilon_{r1} = 2.2$, $h_2 = 0.508$ mm.

6.16 Consider a three-layer shielded microstrip as shown in Fig. 6.6. The enclosure and conducting strip are assumed to be perfect conductors. The strip thickness is assumed negligible, and the dielectric substrates are assumed lossless. Assume the following distribution for the charge:

$$\rho_s(x) = \begin{cases} x, & S \le x \le S + W \\ 0, & \text{otherwise} \end{cases}$$

(a) Derive the lower-bound expression for $1/C$ in the spectral domain, and then Z_o and ε_{eff}.

(b) Calculate and plot Z_o and ε_{eff} versus W/h_2 from 0.1 to 5, for $\varepsilon_{r2} = 2.2$, $h_1 = h_3 = 0.026$ in., $h_2 = 0.254$ mm, $a = 2.54$ mm, and $S = 0.25(a - W)$ and $0.5(a - W)$. Compare results to those obtained in Problem 6.7 and provide an assessment of the basis functions used.

6.17 Consider a shielded CPW as shown in Fig. 7.1. The enclosure and ground and central strips are assumed to be perfect conductor. The metallization thickness of the ground and central strips is negligible, and the dielectric substrates are considered lossless. Assume that the basis functions for the charge distributions on the central strip, ρ_{s1i}, and ground strips, ρ_{s2i} and ρ_{s3i}, are given by Eqs. (7.47)–(7.49).

(a) Formulate the variational analysis in the spectral domain for determining the characteristic impedance, Z_o and effective dielectric constant, ε_{eff}.

(b) Write a computer program to calculate Z_o and ε_{eff}. Compute and plot Z_o and ε_{eff} versus W/h_2 from 0.1 to 5, for $\varepsilon_{r1} = 1$, $\varepsilon_{r2} = 2.2$, $\varepsilon_{r3} = 10.5$, $h_1 = h_3 = 0.66$ mm, $h_2 = 0.254$ mm, $a = 2.54$ mm, $S_1 = S_2 = 0.254$ mm, and $G_1 = G_2$. Compare results using one to five basis functions.

6.18 Use the program developed in Problem 6.17 to calculate and plot Z_o and ε_{eff} for the conventional CPW shown in Fig. 4.3(a) versus the dimension ratio a/b from 0.1 to 0.9 for a relative dielectric constant, ε_r, of 2.2 and 10.5 and a normalized substrate thickness, h/b, of 0.1, 0.5, 4, and 10. Assume the ground and central strips have zero thickness. Compare the results to those obtained in Problem 4.10.

6.19 Repeat Problem 6.18 for the conductor-backed CPW with zero strip thickness shown in Fig. 4.3(b). Compare the results to those obtained in Problem 4.11.

6.20 Consider a CPS as shown in Fig. 4.4. The strips are assumed to be perfect conductor with negligible thickness, and the dielectric substrate is assumed lossless. Assume that the basis functions for the charge distributions on the strips are given by

$$\rho_{s1i}(x) = \begin{cases} \dfrac{\cos\left((i-1)\pi\dfrac{x+b}{b-a}\right)}{\sqrt{1-\left(\dfrac{2(x+b)-(b-a)}{b-a}\right)^2}}, & -b \le x \le -a \\ 0, & \text{otherwise} \end{cases}$$

$$\rho_{s2i}(x) = \begin{cases} \dfrac{\cos\left((i-1)\pi\dfrac{x-a}{b-a}\right)}{\sqrt{1-\left(\dfrac{2(x-a)-(b-a)}{b-a}\right)^2}}, & a \le x \le b \\ 0, & \text{otherwise} \end{cases}$$

(a) Formulate the variational analysis in the spectral domain for determining the characteristic impedance, Z_o, and effective dielectric constant, ε_{eff}.

(b) Write a computer program to calculate Z_o and ε_{eff}. Compute and plot Z_o and ε_{eff} versus a/b from 0.1 to 0.9, for a relative dielectric constant, ε_r, of 2.2 and 10.5, and a normalized substrate thickness, h/b, of 0.1, 0.5, 4, and 10. Compare results using one to five basis functions, and to those obtained in Problem 4.15.

6.21 Consider the strip line of Problem 6.9. Assume that the basis functions for the charge distribution on the strip are given as

$$\rho_{si}(x) = \begin{cases} \dfrac{\cos\left((i-1)\pi\dfrac{x+W}{2W}\right)}{\sqrt{1-\left(\dfrac{(x+W)-W}{W}\right)^2}}, & -W \le x \le W \\ 0, & \text{otherwise} \end{cases}$$

(a) Formulate the variational analysis in the spectral domain for determining the characteristic impedance, Z_o.

(b) Write a computer program to calculate Z_o. Compute and plot Z_o versus the normalized strip width, W/d, from 0.1 to 5 for $\varepsilon_r = 2.2$ and 10.5 and $d = 0.635$ and 1.27 mm. Compare results using one to five basis functions, and to those obtained in Problems 4.18 and 6.9.

6.22 Consider a shielded CPW as shown in Fig. 7.1. The enclosure and ground and central strips are assumed to be perfect conductors. The metallization thickness of the ground and central strips is negligible, and the dielectric substrates are considered lossless. Assume that the charge distributions on the central strip, ρ_{s1}, and ground strips, ρ_{s2} and ρ_{s3}, are defined only on the strips as follows:

$$\rho_s(x) = \frac{1}{\pi\sqrt{\left(\frac{W}{2}\right)^2 - \left(x - G_1 - S_1 - \frac{W}{2}\right)^2}}, \quad G_1 + S_1 < x < G_1 + S_1 + W$$

$$\rho_{s2}(x) = \frac{x}{2G_1\sqrt{G_1^2 - x^2}}, \quad 0 < x < G_1$$

$$\rho_{s3}(x) = \frac{a - x}{2G_2\sqrt{G_2^2 - (a - x)^2}}, \quad a - G_2 < x < a$$

(a) Formulate the variational analysis in the space domain for determining the characteristic impedance, Z_o, and effective dielectric constant, ε_{eff}.

(b) Write a computer program to calculate Z_o and ε_{eff}. Compute and plot Z_o and ε_{eff} versus W/h_2 from 0.1 to 5, for $\varepsilon_{r1} = 1$, $\varepsilon_{r2} = 2.2$, $\varepsilon_{r3} = 10.5$, $h_1 = h_3 = 0.66$ mm, $h_2 = 0.254$ mm, $a = 2.54$ mm, $S_1 = S_2 = 0.254$ mm, and $G_1 = G_2$. Compare results to those in Problem 6.17.

6.23 Repeat Problem 6.22 in the spectral domain.

APPENDIX: SYSTEMS OF HOMOGENEOUS EQUATIONS FROM THE LOWER-BOUND VARIATIONAL FORMULATION

The lower-bound variational expression for the capacitance per unit length of a transmission line from Eq. (6.28) is repeated here as

$$\frac{1}{C} = \frac{\oint_S \oint G(x; x')\rho_s(x)\rho_s(x')\,dx\,dx'}{\left[\oint_S \rho_s(x)\,dx\right]^2} \tag{A6.1}$$

Here we assume the transmission line has an infinitely thin strip. The Green's function $G(x; x')$ may be assumed to be of the form

$$G(x; x') \underline{\Delta} \sum_{n=1}^{\infty} g_n g(x) g(x') \tag{A6.2}$$

where $g(x)$ and $g(x')$ are identical in form, and g_n depends on the transmission line parameters. $G(x;x')$ is given in Eqs. (6.29)–(6.34) for the three-layer microstrip line considered in Section 6.3.1. The charge distribution $\rho(x)$ on the strip can be approximated using the Ritz method as

$$\rho(x) = \sum_{i=1}^{N} a_i \rho_i(x) \tag{A6.3}$$

Now substituting Eq. (A6.2) into (A6.1), we obtain

$$\begin{aligned} \frac{1}{C} &= \frac{\oint_S \oint \left[\sum_{n=1}^{\infty} g_n g(x) g(x') \right] \rho_s(x) \rho_s(x')\, dx\, dx'}{\left[\oint_S \rho_s(x)\, dx \right]^2} \\ &= \frac{\sum_{n=1}^{\infty} g_n \left[\oint_S g(x) \rho_s(x)\, dx \right]^2}{\left[\oint_S \rho_s(x)\, dx \right]^2} \end{aligned} \tag{A6.4}$$

Using Eq. (A6.3) in (A6.4) then yields

$$\frac{1}{C} = \frac{\sum_{n=1}^{\infty} g_n \left(\sum_{i=1}^{N} a_i\, SQ_{ni} \right)^2}{\left(\sum_{i=1}^{N} a_i\, Q_i \right)^2} \tag{A6.5}$$

where

$$Q_i = \oint_S \rho_{si}(x)\, dx, \qquad i = 1, 2, \ldots, N \tag{A6.6}$$

$$SQ_{ni} = \oint_S g(x) \rho_{si}(x)\, dx, \quad i = 1, 2, \ldots, N \text{ and } n = 1, 2, \ldots, \infty \tag{A6.7}$$

Note that g_n, Q_i, and SQ_{ni} can all be computed for a particular transmission line. The capacitance given by Eq. (A6.5) is now a function only of the unknowns $a_i, i = 1, 2, \ldots, N$.

Let $1/C$ in Eq. (A6.5) be

$$\frac{1}{C} = \frac{H(a_1, a_2, \ldots, a_N)}{\mathrm{I}(a_1, a_2, \ldots, a_N)} \tag{A6.8}$$

where

$$H = \sum_{n=1}^{\infty} g_n \left(\sum_{i=1}^{N} a_i SQ_{ni} \right)^2 \tag{A6.9}$$

and

$$I = \left(\sum_{i=1}^{N} a_i Q_i \right)^2 \tag{A6.10}$$

Taking the partial derivative of Eq. (A6.8) with respect to a_j and dividing the denominator and numerator by I then yields

$$\frac{\partial}{\partial a_j}\left(\frac{1}{C}\right) = \frac{1}{I}\left(\frac{\partial H}{\partial a_j} - \frac{1}{C}\frac{\partial I}{\partial a_j}\right) \tag{A6.11}$$

where

$$\frac{\partial H}{\partial a_j} = 2\sum_{i=1}^{N} a_i \sum_{n=1}^{\infty} g_n SQ_{ni} SQ_{nj} \tag{A6.12}$$

and

$$\frac{\partial I}{\partial a_j} = 2\sum_{i=1}^{N} a_i Q_i Q_j \tag{A6.13}$$

To maximize the value of C in accordance with the lower-bound formula, we let the partial derivatives of $1/C$ with respect to a_j in Eq. (A6.11) be zero. This leads to

$$\sum_{i=1}^{N}\left(\sum_{n=1}^{\infty} g_n\, S\,Q_{ni}\, S\,Q_{nj} - \frac{1}{C} Q_i\, Q_j\right) a_i = 0, \quad j = 1, 2, \ldots, N \tag{A6.14}$$

which is a system of N linear homogeneous equations in N unknowns a_i, $i = 1, 2, \ldots, N$. In order to have nontrivial solutions, the coefficient matrix must be singular. To satisfy this requirement, we can set the determinant of the matrix to zero, from which we can solve directly $1/C$.

Another way to solve for $1/C$ is described as follows. We take the derivative of Eq. (A6.8) and let it be zero to obtain

$$I\frac{\partial H}{\partial a_j} - H\frac{\partial I}{\partial a_j} = 0 \tag{A6.15}$$

Substituting Eqs. (A6.9), (A6.10), (A6.12), and (A6.13) into Eq. (A6.15), we obtain

$$\sum_{i=1}^{N}\sum_{n=1}^{\infty}(g_n SQ_{nj} \cdot Q_i - g_n SQ_{ni} \cdot Q_j)\, a_i = 0, \quad j = 1, 2, \ldots, N \tag{A6.16}$$

which represents a system of N linear homogeneous equations in N unknowns $a_i, i = 1, 2, \ldots, N$. Note that these equations do not contain $1/C$ as does Eq. (A6.14). We can now solve Eq. (A6.16) for a_i and substitute the results into Eq. (A6.5) along with values of the parameters Q_i and SQ_{ni} from Eqs. (A6.6) and (A6.7), respectively, to determine a value for C.

A system of equations similar to Eqs. (A6.14) and (A6.16) can also be obtained in the spectral domain. For instance, the system of equations that does not contain $1/C$, corresponding to the shielded three-layer microstrip line considered in Section 6.4.2, is derived as follows. We rewrite the expression for $1/C$ in Eq. (6.56) as

$$\frac{1}{C} = \frac{a}{2\varepsilon_0 Q^2} \sum_{n=1}^{\infty} \tilde{\rho}_s(\alpha_n, y)\tilde{G}(\alpha_n, y)\tilde{\rho}_s(\alpha_n, y) \tag{A6.17}$$

where the spectral-domain Green's function $\tilde{G}(\alpha_n)$ is given in Eq. (6.76), and the charge density $\tilde{\rho}_s(\alpha_n)$ is expanded in Eq. (6.58) as

$$\tilde{\rho}_s(\alpha_n) \cong \sum_{i=1}^{N} a_i \tilde{\rho}_{si}(\alpha_n) \tag{A6.18}$$

Substituting Eq. (A6.18) into (A6.17), we obtain, with reordering of summations,

$$\frac{1}{C} = \frac{a}{2\varepsilon_0 Q^2} \sum_{i=1}^{N} \sum_{j=1}^{\infty} a_i a_j \sum_{n=1}^{\infty} \tilde{\rho}_{si}(\alpha_n)\tilde{G}(\alpha_n)\tilde{\rho}_{sj}(\alpha_n) \tag{A6.19}$$

Now we apply the Ritz method to obtain the unknown coefficients a_i by setting

$$\frac{\partial(1/C)}{\partial a_m} = 0, \qquad m = 1, 2, \ldots, N \tag{A6.20}$$

This yields a set of linear homogeneous equations,

$$\sum_{i=1}^{N} a_i \left(\sum_{n=1}^{\infty} \tilde{\rho}_{si}(\alpha_n)\tilde{G}(\alpha_n)\tilde{\rho}_{sj}(\alpha_n) \right) = 0 \tag{A6.21}$$

It is worthwhile to note that the coefficient matrix in Eq. (A6.21) is independent of the total charge on the strip. Also, the matrix is symmetric as expected by reciprocity. We can now solve Eq. (A6.21) for a_i and substitute the results back into Eq. (A.19) along with the Green's function from Eq. (6.76) to determine the capacitance C.

CHAPTER SEVEN

Spectral-Domain Method

Many numerical methods exist to date for analyzing microwave and millimeter-wave passive structures. Among them, the spectral-domain analysis (SDA) is one of the most popular ones. It was developed in 1974 [1]. A SDA version for quasi-static analysis was also presented [2]. SDA is basically a Fourier-transformed version of the integral equation method. However, as compared to the conventional space-domain integral equation method, the SDA has several advantages. Its formulation results in a system of coupled algebraic equations instead of coupled integral equations. Closed-form expressions can easily be obtained for the Green's functions. In addition, incorporation of physical conditions of analyzed structures via the so-called basis functions is achieved, and the obtained solutions are stationary. These features make the SDA numerically simpler and more efficient than the conventional integral equation method. SDA has been used extensively in analyzing planar transmission lines (e.g., [3]). In this chapter, we will present a detailed formulation of the SDA for planar transmission lines in both quasi-static and dynamic domains. Other applications of SDA, to resonators and antenna and scattering problems, can be found in Itoh [4], Zhang and Itoh [5], and Scott [6], respectively.

7.1 FORMULATION OF THE QUASI-STATIC SPECTRAL-DOMAIN ANALYSIS

Figure 7.1 shows a cross section of the three-layer coplanar waveguide (CPW) to be used in illustrating the quasi-static SDA formulation. The central and ground strips are assumed to be perfect electric conductors of zero thickness, uniform and infinite in the z direction. The dielectric substrates are assumed to be lossless. The enclosure or channel is assumed to be a perfect electric conductor and is used to simplify some of the analysis and computations, but the resulting analysis can be used for some transmission line structures discussed in Chapter 4 by choosing appropriate parameters. For instance, the conductor-backed CPW is obtained by

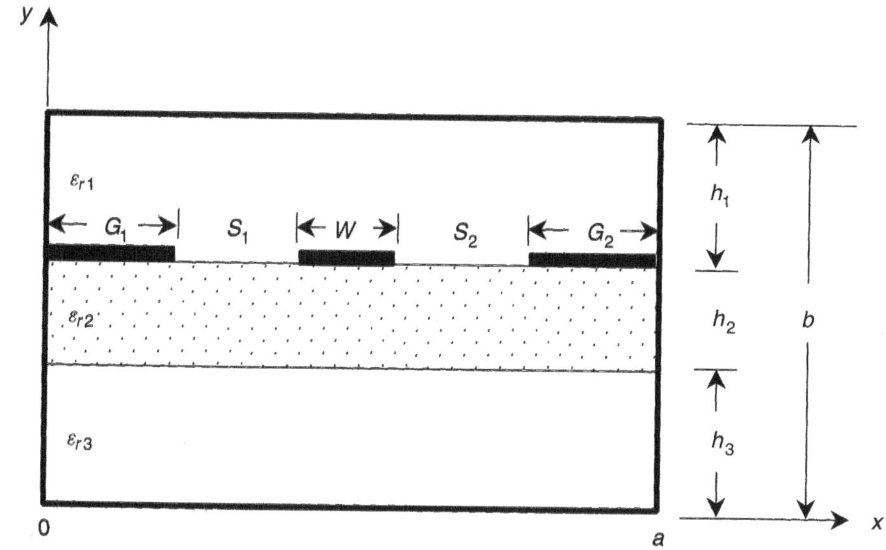

Figure 7.1 Cross section of the asymmetric CPW used for illustrating the quasi-static SDA.

letting $h_3 = 0$, $S_1 = S_2$, h_1 and a be large, and $\varepsilon_{r1} = 1$; the microstrip line is obtained by letting $h_3 = 0$, $G_1 = G_2 = 0$, h_1 and a be large, and $\varepsilon_{r1} = 1$; the strip line is realized when we let $h_3 = 0$, $G_1 = G_2 = 0$, $h_1 = h_2$, a be large, and $\varepsilon_{r1} = \varepsilon_{r2}$. Due to the difference of multiple dielectrics surrounding the metallic strips, the dominant propagating mode is quasi-TEM.

A quasi-static analysis solves the two-dimensional Laplace equation for the electric potential in a plane transverse to the direction of wave propagation subject to appropriate boundary conditions in the space domain,

$$\frac{\partial^2 \phi_i(x, y)}{\partial x^2} + \frac{\partial^2 \phi_i(x, y)}{\partial y^2} = 0, \quad i = 1, 2, 3 \tag{7.1}$$

where $\phi_i(x, y)$ is the unknown potential in the ith region. The quasi-static SDA, on the other hand, solves Laplace's equation by applying a moment method, Galerkin's technique, discussed in Appendix B at the end of this chapter, in the Fourier-transform or spectral domain. The analysis obtains the charge density on the central strip, and from this the per-unit-length (PUL) capacitance is obtained. The PUL capacitance can then be used to determine the effective dielectric constant and characteristic impedance of the transmission line. The boundary conditions are derived from the fact that $\phi(x, y)$ is continuous everywhere and $\mathbf{n} \cdot (\mathbf{D}_2 - \mathbf{D}_1) = \rho_s$ at the upper interface (between the first and second layers.) $\mathbf{D}_i$, $i = 1, 2$, is the electric flux density in region i, $\mathbf{n}$ is the unit vector normal to the interface, and ρ_s stands for the surface charge density. These boundary conditions are

$$\phi_i(0, y) = \phi_i(a, y) = 0 \tag{7.2}$$

$$\phi_1(x, b) = \phi_3(x, 0) = 0 \tag{7.3}$$

$$\phi_1(x, h_2 + h_3) = \phi_2(x, h_2 + h_3) = V^U(x) \tag{7.4}$$

$$\phi_2(x, h_3) = \phi_3(x, h_3) \tag{7.5}$$

$$\varepsilon_{r2} \left.\frac{\partial \phi_2}{\partial y}\right|_{y=h_2+h_3} - \varepsilon_{r1} \left.\frac{\partial \phi_1}{\partial y}\right|_{y=h_2+h_3} = \frac{\rho_s(x)}{\varepsilon_0} \tag{7.6}$$

$$\varepsilon_{r3} \left.\frac{\partial \phi_3}{\partial y}\right|_{y=h_3} - \varepsilon_{r2} \left.\frac{\partial \phi_2}{\partial y}\right|_{y=h_3} = 0 \tag{7.7}$$

where ε_0 is the free-space permittivity and ε_{ri}, $i = 1, 2, 3$, is the relative dielectric constant of the ith layer. $V^U(x)$ denotes the potential at the upper interface and can be expressed as

$$V^U(x) = V_o(x) + V(x) \tag{7.8}$$

$V_o(x) = V_o$ on the central strip and zero elsewhere; we choose the value of the V_o. The ground strips are assumed to be at zero potential. $V(x)$ is the unknown potential on the two slots and is zero on the central and ground strips. The charge density at the upper interface, ρ_s, can be described as

$$\rho_s(x) = \rho_{s1}(x) + \rho_{s2}(x) + \rho_{s3}(x) \tag{7.9}$$

where $\rho_{s1}(x)$, $\rho_{s2}(x)$, and $\rho_{s3}(x)$ are unknown charge densities on the central, left, and right ground strips, respectively, and are nonzero only on the corresponding strips.

To perform the SDA, a Fourier transform is needed. For the considered problem, we define the Fourier transform $\tilde{f}(\alpha_n, y)$ of $f(x, y)$ as follows:

$$\tilde{f}(\alpha_n, y) = \frac{2}{a} \int_0^a f(x, y) \sin \alpha_n x \, dx \tag{7.10}$$

where $\alpha_n = n\pi/a$, $n = 1, 2, 3, \ldots$, denoting the spectral order or term. This choice for the Fourier-transform variable α_n will cause the boundary condition of Eq. (7.2) on $\phi_i(x, y)$ to be met automatically. This Fourier transform along with the corresponding Parseval's theorem is given in Appendix A at the end of this chapter.

Multiply both sides of Eq. (7.1) and taking the integral with respect to x from 0 to a gives

$$\int_0^a \frac{\partial^2 \phi_i(x, y)}{\partial x^2} \sin \alpha_n x \, dx + \int_0^a \frac{\partial^2 \phi_i(x, y)}{\partial y^2} \sin \alpha_n x \, dx = 0 \tag{7.11}$$

Using the following derivative property of the Fourier transform,

$$\int_{-\infty}^{\infty} \frac{d^k f(x)}{dx^k} e^{j\alpha x}\, dx = (-j\alpha)^k \int_{-\infty}^{\infty} f(x) e^{j\alpha x}\, dx \tag{7.12}$$

we obtain

$$\begin{aligned} \int_0^a \frac{d^k f(x)}{dx^k} \sin \alpha x\, dx &= \alpha^k \int_0^a f(x) \cos \alpha x\, dx, \qquad k \text{ odd} \\ &= -\alpha^k \int_0^a f(x) \sin \alpha x\, dx, \quad k \text{ even} \end{aligned} \tag{7.13}$$

assuming $f(x)$ is real.

The first term on the left-hand side of Eq. (7.11) is the Fourier transform of $\partial^2\phi_i(x, y)/\partial x^2$. Using Eq. (7.13) in this term gives

$$\begin{aligned} \int_0^a \frac{\partial^2 \phi_i(x, y)}{\partial x^2} \sin \alpha_n x\, dx &= -\alpha_n^2 \int_0^a \phi_i(x, y) \sin \alpha_n x\, dx \\ &= -\alpha_n^2 \tilde{\phi}_i(\alpha_n, y) \end{aligned} \tag{7.14}$$

The second term is obtained, upon using Eq. (7.10), as

$$\int_0^a \frac{\partial^2 \phi_i(x, y)}{\partial y^2} \sin \alpha_n x\, dx = \frac{\partial^2 \tilde{\phi}_i(\alpha_n, y)}{\partial y^2} \tag{7.15}$$

Substituting Eqs. (7.14) and (7.15) into Eq. (7.11) then yields

$$\frac{\partial^2 \tilde{\phi}_i}{\partial y^2} - \alpha_n^2 \tilde{\phi}_i = 0 \tag{7.16}$$

which represents Laplace's equation in the Fourier-transform or spectral domain. Fourier-transforming Eqs. (7.3)–(7.9), we obtain

$$\tilde{\phi}_1(\alpha_n, b) = \tilde{\phi}_3(\alpha_n, 0) = 0 \tag{7.17}$$

$$\tilde{\phi}_1(\alpha_n, h_2 + h_3) = \tilde{\phi}_2(\alpha_n, h_2 + h_3) = \tilde{V}^U(\alpha_n) \tag{7.18}$$

$$\tilde{\phi}_2(\alpha_n, h_3) = \tilde{\phi}_3(\alpha_n, h_3) \tag{7.19}$$

$$\varepsilon_{r2} \left.\frac{\partial \tilde{\phi}_2}{\partial y}\right|_{y=h_2+h_3} - \varepsilon_{r1} \left.\frac{\partial \tilde{\phi}_1}{\partial y}\right|_{y=h_2+h_3} = \frac{\tilde{\rho}_s(\alpha_n)}{\varepsilon_0} \tag{7.20}$$

$$\varepsilon_{r3} \left.\frac{\partial \tilde{\phi}_3}{\partial y}\right|_{y=h_3} - \varepsilon_{r2} \left.\frac{\partial \tilde{\phi}_2}{\partial y}\right|_{y=h_3} = 0 \tag{7.21}$$

which are the boundary conditions of the considered CPW structure in the spectral domain, and

$$\tilde{V}^U(\alpha_n) = \tilde{V}_o(\alpha_n) + \tilde{V}(\alpha_n) \tag{7.22}$$

$$\tilde{\rho}_s(\alpha_n) = \tilde{\rho}_{s1}(\alpha_n) + \tilde{\rho}_{s2}(\alpha_n) + \tilde{\rho}_{s3}(\alpha_n) \tag{7.23}$$

The boundary condition (7.2) is not transformed since it is already satisfied by the choice of α_n. The solution of Eq. (7.16) is well known. A judicious choice yields the following forms:

$$\tilde{\phi}_1(\alpha_n, y) = A \sinh \alpha_n (b - y) + E \cosh \alpha_n (b - y) \tag{7.24}$$

$$\tilde{\phi}_2(\alpha_n, y) = B \sinh \alpha_n (y - h_3) + C \cosh \alpha_n (y - h_3) \tag{7.25}$$

$$\tilde{\phi}_3(\alpha_n, y) = D \sinh \alpha_n y + F \cosh \alpha_n y \tag{7.26}$$

where A, B, C, D, E, and F are unknown constants. In order to satisfy the first boundary condition (7.17), E and F must be equal to zero. This fact will be used implicitly in subsequent applications of the boundary conditions. Now substituting Eqs. (7.25) and (7.26) into Eq. (7.19) yields

$$D \sinh \alpha_n h_3 = C \tag{7.27}$$

Applying Eq. (7.21) to Eqs. (7.25) and (7.26), we obtain

$$\alpha_n \varepsilon_{r2} B - \alpha_n \varepsilon_{r3} D \cosh \alpha_n h_3 = 0 \tag{7.28}$$

Using Eq. (7.18) in Eqs. (7.24) and (7.25) gives

$$A \sinh \alpha_n h_1 = B \sinh \alpha_n h_2 + C \cosh \alpha_n h_2 \tag{7.29}$$

Substituting Eqs. (7.24) and (7.25) into Eq. (7.20), we have

$$\alpha_n A \cosh \alpha_n h_1 + \alpha_n \varepsilon_{r2} (B \cosh \alpha_n h_2 + C \sinh \alpha_n h_2) = \frac{\tilde{\rho}_s(\alpha_n)}{\varepsilon_0} \tag{7.30}$$

Now manipulating Eqs. (7.27)–(7.30) and making use of Eq. (7.18), we can solve for $\tilde{V}^U(\alpha_n)$ in terms of $\tilde{\rho}_s(\alpha_n)$ as

$$\tilde{G}(\alpha_n)\tilde{\rho}_s(\alpha_n) = \tilde{V}^U(\alpha_n) \tag{7.31}$$

where $\tilde{G}(\alpha_n)$ is called the Green's function in the spectral domain and is given as

$$\tilde{G}(\alpha_n) = \frac{\tanh \alpha_n h_1 \left(\dfrac{\varepsilon_{r3}}{\varepsilon_{r2}} \tanh \alpha_n h_2 + \tanh \alpha_n h_3 \right)}{\begin{array}{c}\varepsilon_0 \alpha_n \tanh \alpha_n h_2 \left[\varepsilon_{r3}(\varepsilon_{r1}/\varepsilon_{r2} + \tanh \alpha_n h_1 \coth \alpha_n h_2) \right. \\ \left. + \tanh \alpha_n h_3 (\varepsilon_{r1} \coth \alpha_n h_2 + \varepsilon_{r2} \tanh \alpha_n h_1) \right]\end{array}} \tag{7.32}$$

$\tilde{V}^U(\alpha_n)$ is given in Eq. (7.22) as

$$\tilde{V}^U(\alpha_n) = \tilde{V}_o(\alpha_n) + \tilde{V}(\alpha_n)$$

$\tilde{V}_o(\alpha_n)$ and $\tilde{V}(\alpha_n)$ stand for the Fourier transforms of $V_o(x)$ and $V(x)$, respectively. $V_o(x)$ is chosen and is thus known. $V(x)$ is given as

$$\tilde{V}(\alpha_n) = \frac{2}{a}\int_0^a V(x)\sin\alpha_n x\,dx \tag{7.33}$$

with $\alpha_n = n\pi/a$, $n = 1, 2, 3, \ldots$. It should be noted here that the process of deriving Eq. (7.31) and hence (7.32) would be simpler if we assume h_1 and h_2 are always nonzero. Furthermore, the spectral-domain Green's function is independent of the dimensions along the x axis. It is basically the total potential, corresponding to a unit charge, along the interface where the conductors reside. As seen in Chapter 6 on variational methods, the Green's function in the spectral domain is easier to derive than that in the space domain.

Up to now, the formulation is exact. There are two unknowns in Eq. (7.31)—$\tilde{\rho}_s(\alpha_n)$ and $\tilde{V}(\alpha_n)$. We now apply Galerkin's technique in the spectral domain to eliminate the unknown voltage $\tilde{V}$ so that Eq. (7.31) can be solved. We begin by expressing each strip's charge density as a truncated summation of basis functions in the space domain as

$$\rho_s(x) \cong \sum_{j=1}^{3}\sum_{i=1}^{N_j} d_{ji}\rho_{sji}(x) \tag{7.34}$$

where $\rho_{sji}(x)$ describes the charge distribution on the jth strip and is nonzero only on that strip; d_{ji} is the unknown coefficient; and N_j denotes the number of basis functions used for the jth strip's charge density. In the spectral domain,

$$\tilde{\rho}_s(\alpha_n) \cong \sum_{j=1}^{3}\sum_{i=1}^{N_j} d_{ji}\tilde{\rho}_{sji}(\alpha_n) \tag{7.35}$$

Substitute Eq. (7.35) into (7.31) and take the *inner product* of the resultant equation with respect to $\tilde{\rho}_{sji}(\alpha_n)$ for $j = 1, 2, 3$ and $i = 1, 2, \ldots, N_j$. That is, multiply the resultant equation by $\tilde{\rho}_{sji}(\alpha_n)$, $j = 1, 2, 3$ and $i = 1, 2, \ldots, N_j$, and sum over α_n, letting the spectral order n go from 1 to infinity. This results in a system of coupled linear algebraic equations, referred to as the Galerkin or Rayleigh–Ritz equations:

$$\sum_{i=1}^{N_1} P_{11}^{ij} d_{1i} + \sum_{i=1}^{N_2} P_{12}^{ij} d_{2i} + \sum_{i=1}^{N_3} P_{13}^{ij} d_{3i} = Q_1^j, \quad j = 1, 2, \ldots, N_1 \tag{7.36}$$

$$\sum_{i=1}^{N_1} P_{21}^{ij} d_{1i} + \sum_{i=1}^{N_2} P_{22}^{ij} d_{2i} + \sum_{i=1}^{N_3} P_{23}^{ij} d_{3i} = Q_2^j, \quad j = 1, 2, \ldots, N_2 \tag{7.37}$$

$$\sum_{i=1}^{N_1} P_{31}^{ij} d_{1i} + \sum_{i=1}^{N_2} P_{32}^{ij} d_{2i} + \sum_{i=1}^{N_3} P_{33}^{ij} d_{3i} = Q_3^j, \quad j = 1, 2, \ldots, N_3 \tag{7.38}$$

where

$$P_{kl}^{ij} = \langle \tilde{\rho}_{skj}, \tilde{G}\tilde{\rho}_{sli} \rangle = \sum_{n=1}^{\infty} \tilde{\rho}_{skj} \tilde{G} \tilde{\rho}_{sli}, \quad k = 1, 2, 3 \text{ and } l = 1, 2, 3$$

$$Q_k^j = \langle \tilde{\rho}_{skj}, \tilde{V}_o + \tilde{V} \rangle = \langle \tilde{\rho}_{skj}, \tilde{V}_o \rangle + \langle \tilde{\rho}_{skj}, \tilde{V} \rangle$$

$$= \sum_{n=1}^{\infty} \tilde{\rho}_{skj} \tilde{V}_o + \sum_{n=1}^{\infty} \tilde{\rho}_{skj} \tilde{V}, \quad k = 1, 2, 3$$

The notation $\langle \cdot \rangle$ indicates an inner product discussed in Appendix B at the end of this chapter. Applying Parseval's theorem

$$\int_0^a f(x, y) g^*(x, y)\, dx = \frac{a}{2} \sum_{n=1}^{\infty} \tilde{f}(\alpha_n, y) \tilde{g}^*(\alpha_n, y) \tag{7.39}$$

to the right-hand sides of Eqs. (7.36)–(7.38) produces

$$\sum_{i=1}^{N_1} P_{11}^{ij} d_{1i} + \sum_{i=1}^{N_2} P_{12}^{ij} d_{2i} + \sum_{i=1}^{N_3} P_{13}^{ij} d_{3i}$$
$$= \frac{2}{a} V_o \int_{G_1+S_1}^{G_1+S_1+W} \rho_{s1j}(x)\, dx, \quad j = 1, 2, \ldots, N_1 \tag{7.40}$$

$$\sum_{i=1}^{N_1} P_{21}^{ij} d_{1i} + \sum_{i=1}^{N_2} P_{22}^{ij} d_{2i} + \sum_{i=1}^{N_3} P_{23}^{ij} d_{3i} = 0, \quad j = 1, 2, \ldots, N_2 \tag{7.41}$$

$$\sum_{i=1}^{N_1} P_{31}^{ij} d_{1i} + \sum_{i=1}^{N_2} P_{32}^{ij} d_{2i} + \sum_{i=1}^{N_3} P_{33}^{ij} d_{3i} = 0, \quad j = 1, 2, \ldots, N_3 \tag{7.42}$$

The infinity upper limit of the spectral order n in the summations represents the number of spectral terms that need to be carried out. In practice, this number is truncated to a finite value N to reduce the computation time. Parseval's theorem eliminates the summations involving the unknown voltage $\tilde{V}$ in Eqs. (7.36)–(7.38), due to the fact that the charge densities and voltages are nonzero in complementary regions along the plane of the strips in the space domain. Therefore, the use of Galerkin's technique allows us to eliminate the unknown $\tilde{V}$. Furthermore, two summations involving known voltage $\tilde{V}_o$ in Eqs. (7.36)–(7.38) are eliminated. Equations (7.40)–(7.42) can now be solved for the unknown coefficients d_{ji} of the charge density basis functions ρ_{sij}, where $i = 1, 2, \ldots, N_i$ and $j = 1, 2, 3$. The transmission line's PUL capacitance is given as

$$C = \frac{Q_o}{V_o} = \frac{\int_0^a \rho_s(x)\,dx}{V_o} = \frac{\sum_{i=1}^{N_1} d_{1i} \int_{G_1+S_1}^{G_1+S_1+W} \rho_{s1i}(x)\,dx}{V_o} \tag{7.43}$$

It should be observed that the integral in the numerator of Eq. (7.43) is the same as the integral on the right-hand side of Eq. (7.40). So if we let the right-hand side of Eq. (7.40), which is already calculated, to be R_j, then we can rewrite Eq. (7.43) as

$$C = \frac{a}{2V_o^2} \sum_{i=1}^{N_1} d_{1i} R_i \tag{7.44}$$

Note that R_j is equal to Q_1^j in Eq. (7.36). C evaluated using Eq. (7.44) only gives an approximate result. However, analogous with variational methods in Chapter 6, Eq. (7.44) is a stationary expression; that is, a first-order error in the choice of the charge density basis functions ρ_{sij} will produce only a second-order error. Therefore, Eq. (7.44) should give a value for C that is very close to the exact result provided that good choices for the basis functions are used. It can also be proved that C from Eq. (7.44) is always smaller than the exact result and, hence, Eq. (7.44) represents the lower-bound expression for C in much the same way as the lower-bound variational expression discussed in Chapter 6. C can then be used to calculate the characteristic impedance and effective dielectric constant as

$$Z_o = \frac{1}{c\sqrt{CC_a}} \tag{7.45}$$

and

$$\varepsilon_{\text{eff}} = \frac{C}{C_a} \tag{7.46}$$

respectively, where c is the free-space velocity and C_a is the capacitace per unit length with the dielectrics removed.

To obtain numerical results for C and, hence, Z_o and ε_{eff}, we need to choose basis functions for the charge densities, $\rho_{sji}(x)$, with $j = 1, 2, 3$ for the central strip and left and right ground strips, respectively. These basis functions influence strongly the numerical efficiency of the solution process and the accuracy of the solutions. Computation time can be reduced significantly if the chosen basis functions closely describe actual behaviors of the charge distributions and have closed-form Fourier transforms. In addition, the basis functions should form complete sets, so that solution accuracy can be enhanced by increasing the number of basis functions. This is an important criterion and should be implemented strictly when choosing the basis functions in the SDA solution. Furthermore, they should be twice continuously differentiable to avoid spurious solutions. Moreover, as C obtained from Eq. (7.44) represents a lower bound to the true capacitance value similar to that from the lower-bound variational expression in Chapter 6, the charge basis functions should be selected so that the calculated value for C will be as large as possible. The same criteria are also used for the choice of basis functions needed for the variational methods in Chapter 6. The basis functions used here for the considered problem have the form

$$\rho_{s1i}(x) = \frac{\cos\left((i-1)\pi\dfrac{(x-S_1-G_1)}{W}\right)}{\sqrt{1-\left(\dfrac{2(x-S_1-G_1)-W}{W}\right)^2}} \tag{7.47}$$

$$\rho_{s2i}(x) = \frac{\cos\left(\left(i-\dfrac{1}{2}\right)\pi\dfrac{(x-G_1)}{G_1}\right)}{\sqrt{1-(x/G_1)^2}} \tag{7.48}$$

$$\rho_{s3i}(x) = \frac{\cos\left(\left(i-\dfrac{1}{2}\right)\pi\dfrac{(x-a+G_2)}{G_2}\right)}{\sqrt{1-\left(\dfrac{a-x}{G_2}\right)^2}} \tag{7.49}$$

These basis functions are defined only on the strips. Sketches of the charge basis functions (7.47) for the central strip are shown in Fig. 7.2. Note that the basis function ρ_{s11} corresponding to $i = 1$ is very large near the central strip's edges and minimum at the strip's center, and ρ_{s21} and ρ_{s31} are also very large near the edges of the left and right ground strips. These functions thus approximate closely the actual charge distributions on the central and ground strips. We then expect that good results for C may be obtained even if only one basis function is used for each of the charges on these strips. The Fourier transforms of these basis functions are obtained using Eq. (7.10) as

$$\begin{aligned}\tilde{\rho}_{s1i}(\alpha_n) = \frac{\pi W}{2a}\Bigg\{ & J_0\left(\left|\frac{(i-1)\pi+\alpha_n W}{2}\right|\right) \\ & \times \sin\left[\frac{(i-1)\pi}{2}+\alpha_n\left(S_1+G_1+\frac{W}{2}\right)\right] - J_0\left(\left|\frac{(i-1)\pi-\alpha_n W}{2}\right|\right) \\ & \times \sin\left[\frac{(i-1)\pi}{2}-\alpha_n\left(S_1+G_1+\frac{W}{2}\right)\right]\Bigg\}\end{aligned} \tag{7.50}$$

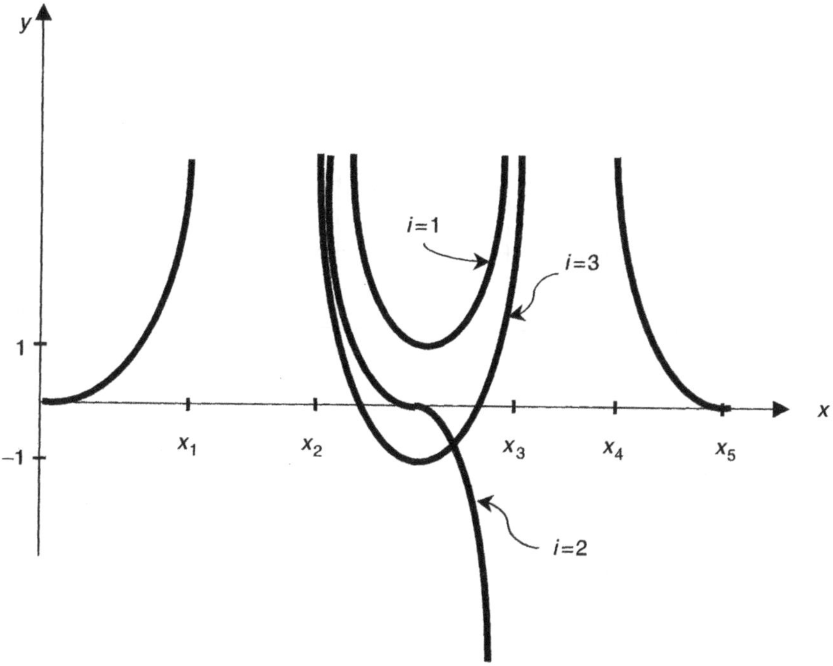

Figure 7.2 Sketches of the charge distributions' basis functions. $x_1 = G_1, x_2 = x_1 + S_1$, $x_3 = x_2 + W, x_4 = x_3 + S_2$, and $x_5 = x_4 + G_2$.

$$\tilde{\rho}_{s2i}(\alpha_n) = (-1)^i \frac{\pi G_1}{2a} \left\{ J_0 \left(\left| \left(i - \frac{1}{2} \right) \pi + \alpha_n G_1 \right| \right) - J_0 \left(\left| \left(i - \frac{1}{2} \right) \pi - \alpha_n G_1 \right| \right) \right\} \tag{7.51}$$

$$\tilde{\rho}_{s3i}(\alpha_n) = (-1)^{i+n+1} \frac{\pi G_2}{2a} \left\{ J_0 \left(\left| \left(i - \frac{1}{2} \right) \pi + \alpha_n G_2 \right| \right) - J_0 \left(\left| \left(i - \frac{1}{2} \right) \pi - \alpha_n G_2 \right| \right) \right\} \tag{7.52}$$

where J_0 stands for the zero-order Bessel function of the first kind. The parameter R_i, representing the right-hand side of Eq. (7.40), that appears in Eq. (7.44) can be obtained upon using Eq. (7.47) as

$$R_i = \frac{\pi W V_o}{a} J_0 \left(\frac{i-1}{2} \pi \right) \cos \left(\frac{i-1}{2} \pi \right) \tag{7.53}$$

A remark needs to be made at this point that both the numbers of basis functions N_j and spectral terms N used in Eq. (7.40)–(7.42) affect the accuracy of the

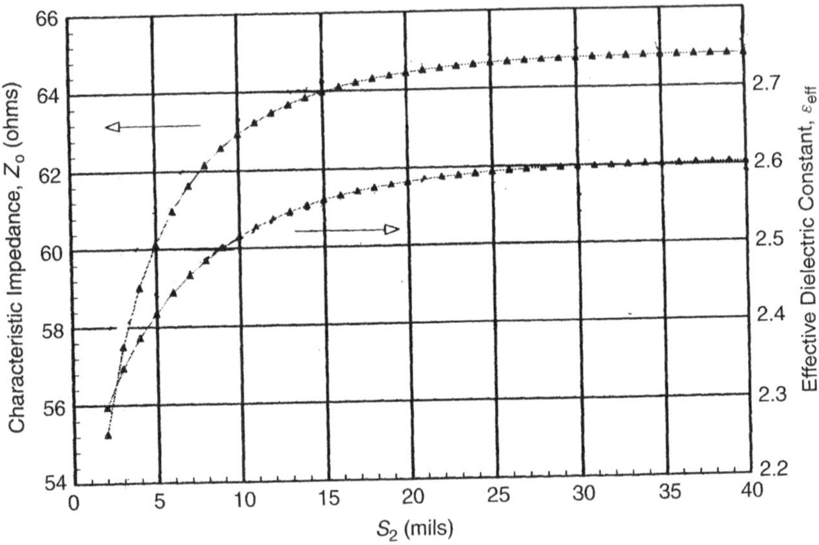

Figure 7.3 Calculated characteristic impedance and effective dielectric constant of the CPW using the quasi-static SDA. $a = 2.54$ mm, $2W = G = 0.508$ mm, $h_1 = h_3 = 4h_2 = 0.508$ mm, $\varepsilon_{r1} = 1$, $\varepsilon_{r2} = 2.2$, and $\varepsilon_{r3} = 10.5$.

numerical results. The larger these numbers the more accurate the results, but at the expense of increased computation time. For most engineering purposes, three basis functions ($N_j = 3$) and 200 spectral terms ($N = 200$) are sufficient.

As a demonstration of the quasi-static SDA, we show in Fig. 7.3 the calculated values of the characteristic impedance and effective dielectric constant for the considered CPW versus the right gap.

7.2 FORMULATION OF THE DYNAMIC SPECTRAL-DOMAIN ANALYSIS

A three-layer CPW is also used to illustrate the dynamic SDA formulation. However, to simplify the analysis without loss of generality, we consider only a symmetrical CPW shown in Fig. 7.4. Analysis of the asymmetrical CPW shown in Fig. 7.1 using the dynamic SDA can be found in Rahman and Nguyen [7] and Nguyen [8]. The dynamic SDA allows us to examine all the eigenmodes existing in the transmission line structure. For the considered CPW, these modes consist of both TE and TM fields. Although SDA can produce results for all of the real and complex eigenmodes for this structure [9], in this section we are restricted to the analysis of the real eigenmodes, including the dominant (CPW) mode, for the purpose of illustrating the SDA. The dominant CPW mode is the quasi-TEM mode discussed in Section 7.1, which is the principal propagating mode in integrated circuits and is the one in which we are most interested in this book.

A dynamic analysis solves the wave equations or Helmholtz equations for the electric and magnetic fields or the electric and magnetic potentials in the space

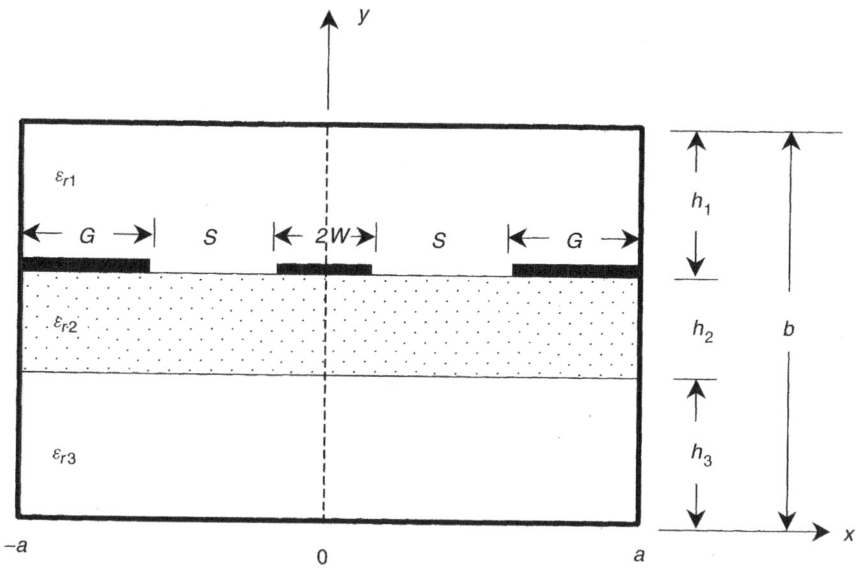

Figure 7.4 Cross section of the CPW used in the dynamic SDA.

domain subject to proper boundary conditions. The dynamic SDA, instead, solves these equations in the spectral domain using Galerkin's technique, discussed in Appendix B at the end of this chapter. The analysis can obtain the propagation constants, effective dielectric constants, and characteristic impedances of the transmission line for all of the eigenmodes. In essence, its formulation process is similar to that for the quasi-static case.

Let $\psi_i^e(x, y)$ and $\psi_i^h(x, y)$ represent the scalar electric and magnetic potentials associated with the TM and TE modes, respectively, in the ith region ($i = 1, 2, 3$) in the space domain. The wave equations of these potentials are obtained from Eqs. (2.46b) and (2.46d) as

$$\nabla_t^2 \psi_i^p(x, y) - (\beta^2 - k_i^2)\psi_i^p(x, y) = 0; \quad p = e, h \tag{7.54}$$

where

$$\nabla_t^2 = \frac{\partial^2}{\partial x^2} + \frac{\partial^2}{\partial y^2} \tag{7.55}$$

β is the propagation constant and $k_i = \omega\sqrt{\varepsilon_i \mu_i}$ denotes the wave number in region i. ω is the angular frequency, and ε_i and μ_i represent the permittivity and permeability, respectively, of medium i.

The boundary conditions for the considered CPW are as follows.

For $-a \le x \le a$, $y = h_3$:

$$E_{x3}(x, h_3) = E_{x2}(x, h_3) \tag{7.56}$$

$$E_{z3}(x, h_3) = E_{z2}(x, h_3) \tag{7.57}$$

$$H_{x3}(x, h_3) = H_{x2}(x, h_3) \tag{7.58}$$

$$H_{z3}(x, h_3) = H_{z2}(x, h_3) \tag{7.59}$$

For $-a \le x \le a$, $y = h_2 + h_3$:

$$E_{x2}(x, h_2 + h_3) = E_{x1}(x, h_2 + h_3) = E_x(x) \tag{7.60}$$

$$E_{z2}(x, h_2 + h_3) = E_{z1}(x, h_2 + h_3) = E_z(x) \tag{7.61}$$

$$H_{x2}(x, h_2 + h_3) - H_{x1}(x, h_2 + h_3) = J_z(x) \tag{7.62}$$

$$H_{z2}(x, h_2 + h_3) - H_{z1}(x, h_2 + h_3) = -J_x(x) \tag{7.63}$$

For $-a \le x \le a$, $y = 0$ and b:

$$E_{x3}(x, y) = E_{x1}(x, y) = E_{z3}(x, y) = E_{z1}(x, y) = 0 \tag{7.64}$$

For $x = \pm a$ and $0 \le y \le b$:

$$E_{z3}(x, y) = E_{z1}(x, y) = 0 \tag{7.65}$$

where E_i and H_i, $i = 1, 2, 3$, stand for the electric and magnetic fields respectively, in region i. $E_x(x)$ and $E_z(x)$ are the x and z components of the unknown electric field on the two slots; they are nonzero on the slots and zero elsewhere. $J_x(x)$ and $J_z(x)$ denote the total respective unknown x- and z-directed current densities on the central and ground strips, and they are nonzero only on these strips. The last boundary condition (7.65) is only used to determine the variable α_n for the Fourier transform defined later.

In order to carry out the SDA, we now need to define an appropriate Fourier transform. For the considered structure, we define the Fourier transform of $f(x, y)$ as

$$\tilde{f}(\alpha_n, y) = \int_{-a}^{a} f(x, y) e^{j\alpha_n x} dx \tag{7.66}$$

or

$$\tilde{f}(\alpha_n, y) = 2 \int_{0}^{a} f(x, y) \sin \alpha_n x \, dx \tag{7.67}$$

and

$$\tilde{f}(\alpha_n, y) = 2 \int_{0}^{a} f(x, y) \cos \alpha_n x \, dx \tag{7.68}$$

for even and odd functions with respect to x, respectively. α_n is chosen to satisfy the boundary conditions at the central plane ($x = 0$) and the two side walls

$(x = \pm a)$ and thus is given by

$$\alpha_n = \begin{cases} \dfrac{n\pi}{a}, & \text{odd function} \\ \left(n - \dfrac{1}{2}\right)\dfrac{\pi}{a}, & \text{even function} \end{cases} \tag{7.69}$$

In actual computations, however, the choice of α_n does not make any difference because a large number of the spectral terms is normally used. So either expression for α_n may be used. Discussion of these Fourier transforms and their associated Parseval's theorem are given in Appendix A at the end of this chapter. There are two kinds of modes existing in CPW: even and odd modes. The even mode is the dominant mode, normally referred to as the CPW mode. This mode corresponds to $E_z(x)$ and $H_z(x)$, which are even and odd functions with respect to x, respectively. The odd mode is normally referred to as the slot line mode and corresponds to odd $E_z(x)$ and even $H_z(x)$. The z-directed magnetic and electric fields at the plane of symmetry $(x = 0)$ are zero for the even and odd modes, respectively.

Now following the same approach in Section 7.1 for the quasi-static SDA formulation, we take the Fourier transform of Eq. (7.54) and use the differentiation property to obtain

$$\frac{\partial^2}{\partial y^2}\tilde{\psi}_i^p(\alpha_n, y) - \gamma_i^2\tilde{\psi}_i^p(\alpha_n, y) = 0; \quad p = e, h \text{ and } i = 1, 2, 3 \tag{7.70}$$

where

$$\gamma_i^2 = \beta^2 + \alpha_n^2 - k_i^2 \tag{7.71}$$

$\tilde{\psi}_i^e(\alpha_n, y)$ and $\tilde{\psi}_i^h(\alpha_n, y)$ are the scalar electric and magnetic potentials, respectively, in the spectral domain. Equation (7.70) is the wave equation or Helmholtz equation of these potentials in the Fourier-transform or spectral domain. Taking the Fourier transform of Eqs. (7.56)–(7.64) gives the boundary conditions of the considered CPW in the spectral domain as

$$\tilde{E}_{x3}(\alpha_n, h_3) = \tilde{E}_{x2}(\alpha_n, h_3) \tag{7.72}$$

$$\tilde{E}_{z3}(\alpha_n, h_3) = \tilde{E}_{z2}(\alpha_n, h_3) \tag{7.73}$$

$$\tilde{H}_{x3}(\alpha_n, h_3) = \tilde{H}_{x2}(\alpha_n, h_3) \tag{7.74}$$

$$\tilde{H}_{z3}(\alpha_n, h_3) = \tilde{H}_{z2}(\alpha_n, h_3) \tag{7.75}$$

$$\tilde{E}_{x2}(\alpha_n, h_2 + h_3) = \tilde{E}_{x1}(\alpha_n, h_2 + h_3) = \tilde{E}_x(\alpha_n) \tag{7.76}$$

$$\tilde{E}_{z2}(\alpha_n, h_2 + h_3) = \tilde{E}_{z1}(\alpha_n, h_2 + h_3) = \tilde{E}_z(\alpha_n) \tag{7.77}$$

$$\tilde{H}_{x2}(\alpha_n, h_2 + h_3) - \tilde{H}_{x1}(\alpha_n, h_2 + h_3) = \tilde{J}_z(\alpha_n) \tag{7.78}$$

$$\tilde{H}_{z2}(\alpha_n, h_2 + h_3) - \tilde{H}_{z1}(\alpha_n, h_2 + h_3) = -\tilde{J}_x(\alpha_n) \tag{7.79}$$

$$\tilde{E}_{x3}(\alpha_n, 0) = \tilde{E}_{x1}(\alpha_n, 0) = \tilde{E}_{z3}(\alpha_n, 0) = \tilde{E}_{z1}(\alpha_n, 0) = 0 \tag{7.80}$$

$$\tilde{E}_{x3}(\alpha_n, b) = \tilde{E}_{x1}(\alpha_n, b) = \tilde{E}_{z3}(\alpha_n, b) = \tilde{E}_{z1}(\alpha_n, b) = 0 \tag{7.81}$$

The fields in each region i may be given in terms of the potential functions as

$$E_{zi}(x, y, z) = j\frac{k_i^2 - \beta^2}{\beta}\psi_i^e(x, y)e^{-j\beta z} \tag{7.82}$$

$$H_{zi}(x, y, z) = j\frac{k_i^2 - \beta^2}{\beta}\psi_i^h(x, y)e^{-j\beta z} \tag{7.83}$$

$$E_{ti}(x, y, z) = \left(\nabla_t\psi_i^e(x, y) - \frac{\omega\mu_i}{\beta}\mathbf{a}_z \times \nabla_t\psi_i^h(x, y)\right)e^{-j\beta z} \tag{7.84}$$

$$H_{ti}(x, y, z) = \left(\nabla_t\psi_i^h(x, y) + \frac{\omega\varepsilon_i}{\beta}\mathbf{a}_z \times \nabla_t\psi_i^e(x, y)\right)e^{-j\beta z} \tag{7.85}$$

where the subscript t indicates the transverse (x or y) component, and

$$\nabla_t = \frac{\partial}{\partial x}\mathbf{a}_x + \frac{\partial}{\partial y}\mathbf{a}_y \tag{7.86}$$

In the spectral domain, they are

$$\tilde{E}_{xi}(\alpha_n, y) = -j\alpha_n\tilde{\psi}_i^e(\alpha_n, y) + \frac{\omega\mu_i}{\beta}\frac{\partial\tilde{\psi}_i^h(\alpha_n, y)}{\partial y} \tag{7.87}$$

$$\tilde{E}_{yi}(\alpha_n, y) = j\frac{\alpha_n\omega\mu_i}{\beta}\tilde{\psi}_i^h(\alpha_n, y) + \frac{\partial\tilde{\psi}_i^e(\alpha_n, y)}{\partial y} \tag{7.88}$$

$$\tilde{E}_{zi}(\alpha_n, y) = j\frac{k_i^2 - \beta^2}{\beta}\tilde{\psi}_i^e(\alpha_n, y) \tag{7.89}$$

$$\tilde{H}_{xi}(\alpha_n, y) = -j\alpha_n\tilde{\psi}_i^h(\alpha_n, y) + \frac{\omega\varepsilon_i}{\beta}\frac{\partial\tilde{\psi}_i^e(\alpha_n, y)}{\partial y} \tag{7.90}$$

$$\tilde{H}_{yi}(\alpha_n, y) = -j\frac{\alpha_n\omega\varepsilon_i}{\beta}\tilde{\psi}_i^e(\alpha_n, y) + \frac{\partial\tilde{\psi}_i^h(\alpha_n, y)}{\partial y} \tag{7.91}$$

$$\tilde{H}_{zi}(\alpha_n, y) = j\frac{k_i^2 - \beta^2}{\beta}\tilde{\psi}_i^h(\alpha_n, y) \tag{7.92}$$

General solutions of Eq. (7.70) are

$$\tilde{\psi}_1^e(\alpha_n, y) = A_e \sinh\gamma_1(b - y) + E_e \cosh\gamma_1(b - y) \tag{7.93}$$

$$\tilde{\psi}_1^h(\alpha_n, y) = A_h \cosh\gamma_1(b - y) + E_h \sinh\gamma_1(b - y) \tag{7.94}$$

$$\tilde{\psi}_2^e(\alpha_n, y) = B_e \sinh \gamma_2(y - h_3) + C_e \cosh \gamma_2(y - h_3) \tag{7.95}$$

$$\tilde{\psi}_2^h(\alpha_n, y) = B_h \sinh \gamma_2(y - h_3) + C_h \cosh \gamma_2(y - h_3) \tag{7.96}$$

$$\tilde{\psi}_3^e(\alpha_n, y) = D_e \sinh \gamma_3 y + F_e \cosh \gamma_3 y \tag{7.97}$$

$$\tilde{\psi}_3^h(\alpha_n, y) = D_h \cosh \gamma_3 y + F_h \sinh \gamma_3 y \tag{7.98}$$

where $A_{e,h}$, $B_{e,h}$, $C_{e,h}$, $D_{e,h}$, $E_{e,h}$, and $F_{e,h}$ are unknown constants. In order to satisfy the last two boundary conditions, Eqs. (7.80) and (7.81), $E_{e,h}$, and $F_{e,h}$ must all be equal to zero. This fact will be used implicitly in later applications of the remaining boundary conditions. The fields in the three regions in the spectral domain can now be derived by substituting Eqs. (7.93)–(7.98) into (7.87)–(7.92) as

$$\tilde{E}_{x1}(\alpha_n, y) = -j\alpha_n A_e \sinh \gamma_1(b - y) - \frac{\omega\mu_1\gamma_1}{\beta} A_h \sinh \gamma_1(b - y) \tag{7.99}$$

$$\tilde{E}_{x2}(\alpha_n, y) = -j\alpha_n [B_e \sinh \gamma_2(y - h_3) + C_e \cosh \gamma_2(y - h_3)] + \frac{\omega\mu_2\gamma_2}{\beta}[B_h \cosh \gamma_2(y - h_3) + C_h \sinh \gamma_2(y - h_3)] \tag{7.100}$$

$$\tilde{E}_{x3}(\alpha_n, y) = -j\alpha_n D_e \sinh \gamma_3 y + \frac{\omega\mu_3\gamma_3}{\beta} D_h \sinh \gamma_3 y \tag{7.101}$$

$$\tilde{E}_{z1}(\alpha_n, y) = jA_e \frac{k_1^2 - \beta^2}{\beta} \sinh \gamma_1(b - y) \tag{7.102}$$

$$\tilde{E}_{z2}(\alpha_n, y) = j\frac{k_2^2 - \beta^2}{\beta}[B_e \sinh \gamma_2(y - h_3) + C_e \cosh \gamma_2(y - h_3)] \tag{7.103}$$

$$\tilde{E}_{z3}(\alpha_n, y) = jD_e \frac{k_3^2 - \beta^2}{\beta} \sinh \gamma_3 y \tag{7.104}$$

$$\tilde{H}_{x1}(\alpha_n, y) = -j\alpha_n A_h \cosh \gamma_1(b - y) + \frac{\omega\varepsilon_1\gamma_1}{\beta} A_e \cosh \gamma_1(b - y) \tag{7.105}$$

$$\tilde{H}_{x2}(\alpha_n, y) = -j\alpha_n [B_h \sinh \gamma_2(y - h_3) + C_h \cosh \gamma_2(y - h_3)] - \frac{\omega\varepsilon_2\gamma_2}{\beta}[B_e \cosh \gamma_2(y - h_3) + C_e \sinh \gamma_2(y - h_3)] \tag{7.106}$$

$$\tilde{H}_{x3}(\alpha_n, y) = -j\alpha_n D_h \cosh \gamma_3 y - \frac{\omega\varepsilon_3\gamma_3}{\beta} D_e \cosh \gamma_3 y \tag{7.107}$$

$$\tilde{H}_{z1}(\alpha_n, y) = jA_h \frac{k_1^2 - \beta^2}{\beta} \cosh \gamma_1(b - y) \tag{7.108}$$

$$\tilde{H}_{z2}(\alpha_n, y) = j\frac{k_2^2 - \beta^2}{\beta}[B_h \sinh \gamma_2(y - h_3) + C_h \cosh \gamma_2(y - h_3)] \tag{7.109}$$

$$\tilde{H}_{z3}(\alpha_n, y) = jD_h \frac{k_3^2 - \beta^2}{\beta} \cosh \gamma_3 y \tag{7.110}$$

Now we apply the boundary conditions (7.72)–(7.79). Applying Eq. (7.72) and using Eqs. (7.100) and (7.101) yields

$$-j\alpha_n D_e \sinh \alpha_n h_3 + \frac{\omega\mu_3\gamma_3}{\beta} D_h \sinh \alpha_n h_3 = -j\alpha_n C_e + \frac{\omega\mu_2\gamma_2}{\beta} B_h \quad (7.111a)$$

Using Eq. (7.73) in Eqs. (7.103) and (7.104), we obtain

$$\left(k_3^2 - \beta^2\right) D_e \sinh \gamma_3 h_3 = \left(k_2^2 - \beta^2\right) C_e \quad (7.111b)$$

Applying Eq. (7.74) to Eqs. (7.106) and (7.107), we get

$$-j\alpha_n D_h \cosh \alpha_n h_3 + \frac{\omega\varepsilon_3\gamma_3}{\beta} D_e \cosh \alpha_n h_3 = -j\alpha_n C_h - \frac{\omega\varepsilon_2\gamma_2}{\beta} B_e \quad (7.112)$$

Substituting Eqs. (7.109) and (7.110) into Eq. (7.75) gives

$$\left(k_3^2 - \beta^2\right) D_h \cosh \gamma_3 h_3 = \left(k_2^2 - \beta^2\right) C_h \quad (7.113)$$

Substituting Eqs. (7.99) and (7.100) into Eq. (7.76), we obtain

$$\begin{aligned} &- j\alpha_n (B_e \sinh \gamma_2 h_2 + C_e \cosh \gamma_2 h_2) + \frac{\omega\mu_2\gamma_2}{\beta}(B_h \cosh \gamma_2 h_2 + C_h \sinh \gamma_2 h_2) \\ &= -j\alpha_n A_e \sinh \gamma_1 h_1 - \frac{\omega\mu_1\gamma_1}{\beta} A_h \sinh \gamma_1 h_1 = \tilde{E}_x(\alpha_n) \end{aligned} \quad (7.114)$$

Substituting Eqs. (7.102) and (7.103) into Eq. (7.77) yields

$$\left(k_2^2 - \beta^2\right)(B_e \sinh \gamma_2 h_2 + C_e \cosh \gamma_2 h_2) = \left(k_1^2 - \beta^2\right) A_e \sinh \gamma_1 h_1 = -j\beta\tilde{E}_z(\alpha_n) \quad (7.115)$$

Substituting Eqs. (7.105) and (7.106) into Eq. (7.78) leads to

$$\begin{aligned} &- j\alpha_n (B_h \sinh \gamma_2 h_2 + C_h \cosh \gamma_2 h_2) - \frac{\omega\varepsilon_2\gamma_2}{\beta}(B_e \cosh \gamma_2 h_2 + C_e \sinh \gamma_2 h_2) \\ &+ j\alpha_n A_h \cosh \gamma_1 h_1 - \frac{\omega\varepsilon_1\gamma_1}{\beta} A_e \cosh \gamma_1 h_1 = \tilde{J}_z(\alpha_n) \end{aligned} \quad (7.116)$$

Finally, applying Eq. (7.79) in Eqs. (7.108) and (7.109), we have

$$\left(k_2^2 - \beta^2\right)(B_h \sinh \gamma_2 h_2 + C_h \cosh \gamma_2 h_2) - \left(k_1^2 - \beta^2\right) A_h \cosh \gamma_1 h_1 = -j\beta\tilde{J}_x(\alpha_n) \quad (7.117)$$

From Eqs. (7.111a)–(7.117) we can solve for $\tilde{J}_x(\alpha_n)$ and $\tilde{J}_z(\alpha_n)$ in terms of $\tilde{E}_x(\alpha_n)$ and $\tilde{E}_z(\alpha_n)$ as

$$\tilde{G}_{11}(\alpha_n, \beta)\tilde{E}_x(\alpha_n) + \tilde{G}_{12}(\alpha_n, \beta)\tilde{E}_z(\alpha_n) = \tilde{J}_x(\alpha_n) \quad (7.118)$$

$$\tilde{G}_{21}(\alpha_n, \beta)\tilde{E}_x(\alpha_n) + \tilde{G}_{22}(\alpha_n, \beta)\tilde{E}_z(\alpha_n) = \tilde{J}_z(\alpha_n) \quad (7.119)$$

where $\tilde{G}_{ij}(\alpha_n, \beta)$, $i, j = 1, 2$, are the Green's functions in the spectral domain and are given as

$$\tilde{G}_{11} = \frac{1}{\alpha_n^2 + \beta^2}\left(\alpha_n^2 \tilde{G}_e - \beta^2 \tilde{G}_h\right) \tag{7.120}$$

$$\tilde{G}_{21} = -\tilde{G}_{12} = -\frac{j\alpha_n \beta}{\alpha_n^2 + \beta^2}\left(\tilde{G}_e + \tilde{G}_h\right) \tag{7.121}$$

$$\tilde{G}_{22} = \frac{1}{\alpha_n^2 + \beta^2}\left(\beta^2 \tilde{G}_e - \alpha_n^2 \tilde{G}_h\right) \tag{7.122}$$

$$\tilde{G}_e = -\frac{k_0}{\alpha_n^{(2)}}\left(\frac{\varepsilon_{r2}\alpha_n^{(3)} \tanh\left(\alpha_n^{(2)} h_2\right) \tanh\left(\alpha_n^{(3)} h_3\right) - \alpha_n^{(2)}\varepsilon_{r3}}{\alpha_n^{(3)} \tanh\left(\alpha_n^{(3)} h_3\right) + \dfrac{\alpha_n^{(2)}\varepsilon_{r3} \tanh\left(\alpha_n^{(2)} h_2\right)}{\varepsilon_{r2}}}\right) + \frac{k_0 \varepsilon_{r1}}{\alpha_n^{(1)} \tanh\left(\alpha_n^{(1)} h_1\right)} \tag{7.123}$$

$$\tilde{G}_h = -\frac{1}{k_0}\left(\frac{1 - \alpha_n^{(2)} \tanh\left(\alpha_n^{(2)} h_2\right) \dfrac{\tanh\left(\alpha_n^{(3)} h_3\right)}{\alpha_n^{(3)}}}{\dfrac{\tanh\left(\alpha_n^{(2)} h_2\right)}{\alpha_n^{(2)}} + \dfrac{\tanh\left(\alpha_n^{(3)} h_3\right)}{\alpha_n^{(3)}}}\right) - \frac{\alpha_n^{(1)}}{k_0 \tanh\left(\alpha_n^{(1)} h_1\right)} \tag{7.124}$$

$$\alpha_n^{(1)} = \sqrt{k_1^2 - \beta^2 - \alpha_n^2} \tag{7.125}$$

$$\alpha_n^{(2)} = \sqrt{k_2^2 - \beta^2 - \alpha_n^2} \tag{7.126}$$

$$\alpha_n^{(3)} = \sqrt{k_3^2 - \beta^2 - \alpha_n^2} \tag{7.127}$$

with k_0 being the free-space wave number. Equations (7.118) and (7.119) are the Fourier transforms of the conventional coupled integral equations in the space domain. As for the case of quasi-static SDA, the spectral-domain Green's functions obtained here are also independent of the dimensions along the x axis. These functions are easier to obtain than those involved in the coupled integral equations encountered in the space domain.

There are four unknowns in Eqs. (7.118) and (7.119); they are $\tilde{E}_x$, $\tilde{E}_z$, $\tilde{J}_x$, and $\tilde{J}_z$ along the plane where the conductors are located. The analysis, up to this stage, is exact. We now apply Galerkin's technique in the spectral domain to cancel the two unknowns $\tilde{J}_x$ and $\tilde{J}_z$ and solve Eqs. (7.118) and (7.119) approximately. To this end, we express the slots' electric fields as truncated summations of basis functions in the space domain as

$$E_x(x) \cong \sum_{m=0}^{M} c_m E_{xm}(x) \tag{7.128}$$

$$E_z(x) \cong \sum_{k=1}^{K} d_k E_{zk}(x) \tag{7.129}$$

where c_m and d_k are the unknown coefficients. The basis functions $E_{xm}(x)$ and $E_{zk}(x)$ describe the x- and z-electric field distributions on the slots and are nonzero only on these slots. In the spectral domain, the unknown electric fields are

$$\tilde{E}_x(\alpha_n) \cong \sum_{m=0}^{M} c_m \tilde{E}_{xm}(\alpha_n) \tag{7.130}$$

$$\tilde{E}_z(\alpha_n) \cong \sum_{k=1}^{K} d_k \tilde{E}_{zk}(\alpha_n) \tag{7.131}$$

Substituting Eqs. (7.130) and (7.131) into Eqs. (7.118) and (7.119) gives

$$\sum_{m=0}^{M} \tilde{G}_{11}(\alpha_n, \beta)\tilde{E}_{xm}(\alpha_n)c_m + \sum_{k=1}^{K} \tilde{G}_{12}(\alpha_n, \beta)\tilde{E}_{zk}(\alpha_n)d_k = \tilde{J}_x(\alpha_n) \tag{7.132}$$

$$\sum_{m=0}^{M} \tilde{G}_{21}(\alpha_n, \beta)\tilde{E}_{xm}(\alpha_n)c_m + \sum_{k=1}^{K} \tilde{G}_{22}(\alpha_n, \beta)\tilde{E}_{zk}(\alpha_n)d_k = \tilde{J}_z(\alpha_n) \tag{7.133}$$

We now take the *inner product* of Eqs. (7.132) and (7.133) with $\tilde{E}_{xi}(\alpha_n)$, $i = 0, 1, \ldots, M$, and $\tilde{E}_{zj}(\alpha_n)$, $j = 0, 1, \ldots, K$, respectively. That is, multiply Eqs. (7.132) and (7.133) with $\tilde{E}_{xi}(\alpha_n)$, $i = 0, 1, \ldots, M$, and $\tilde{E}_{zj}(\alpha_n)$, $j = 0, 1, \ldots, K$, respectively, and sum over α_n. This process results in the following system of coupled linear algebraic equations, referred to as the Galerkin or Rayleigh–Ritz equations:

$$\sum_{m=0}^{M} P_{11}^{im}(\beta)c_m + \sum_{k=1}^{K} P_{12}^{ik}(\beta)d_k = 0; \quad i = 0, 1, 2, \ldots, M \tag{7.134}$$

$$\sum_{m=0}^{M} P_{21}^{jm}(\beta)c_m + \sum_{k=1}^{K} P_{22}^{jk}(\beta)d_k = 0; \quad j = 0, 1, 2, \ldots, K \tag{7.135}$$

where

$$\begin{aligned} P_{11}^{im} &= \left\langle \tilde{E}_{xi}(\alpha_n), \tilde{G}_{11}(\alpha_n, \beta)\tilde{E}_{xm}(\alpha_n) \right\rangle \\ &= \sum_{n=1}^{\infty} \tilde{E}_{xi}(\alpha_n)\tilde{G}_{11}(\alpha_n, \beta)\tilde{E}_{xm}(\alpha_n) \end{aligned} \tag{7.136}$$

$$\begin{aligned} P_{12}^{ik} &= \left\langle \tilde{E}_{xi}(\alpha_n)\tilde{G}_{12}(\alpha_n, \beta)\tilde{E}_{zk}(\alpha_n) \right\rangle \\ &= \sum_{n=1}^{\infty} \tilde{E}_{xi}(\alpha_n)\tilde{G}_{12}(\alpha_n, \beta)\tilde{E}_{zk}(\alpha_n) \end{aligned} \tag{7.137}$$

$$P_{21}^{jm} = \left\langle \tilde{E}_{zj}(\alpha_n)\tilde{G}_{21}(\alpha_n, \beta)\tilde{E}_{xm}(\alpha_n) \right\rangle$$
$$= \sum_{n=1}^{\infty} \tilde{E}_{zj}(\alpha_n)\tilde{G}_{21}(\alpha_n, \beta)\tilde{E}_{xm}(\alpha_n) \qquad (7.138)$$

$$P_{22}^{jk} = \left\langle \tilde{E}_{zj}(\alpha_n)\tilde{G}_{22}(\alpha_n, \beta)\tilde{E}_{zk}(\alpha_n) \right\rangle$$
$$= \sum_{n=1}^{\infty} \tilde{E}_{zj}(\alpha_n)\tilde{G}_{22}(\alpha_n, \beta)\tilde{E}_{zk}(\alpha_n) \qquad (7.139)$$

The notation $\langle\cdot\rangle$ again indicates an inner product. The use of Galerkin's technique enables us to eliminate the unknown current densities $\tilde{J}_{x,z}(\alpha_n)$ by applying Parseval's theorem to the right-hand sides of Eqs. (7.134) and (7.135). Specifically, Parseval's theorem eliminates the summations occurring on the right-hand sides, due to the fact that the currents and electric fields are nonzero in complementary regions in the space domain.

Equations (7.134) and (7.135) form a system of homogeneous equations and, hence, their nontrivial solutions exist only if the determinant of the coefficient matrix vanishes. Setting this determinant to zero, we obtain a characteristic equation of the propagation constant, β, which can then be solved for β. The roots for β are values of the propagation constants associated with (discrete) eigenmodes of the considered CPW. The effective dielectric constant of the transmission line for a particular mode is obtained as $\varepsilon_{\text{eff}} = (\beta/k_0)^2$, where $k_0 = \omega\sqrt{\varepsilon_0\mu_0}$. Upon using the computed value for β, we can solve Eqs. (7.134) and (7.135) for c_m and d_k, from which E_x and E_z can be determined. We can then calculate all the other field components and the (dynamic) characteristic impedance as a function of frequency. As discussed in Chapter 4, because of the non-TEM nature of the considered transmission line, its characteristic impedance is not unique; that is, there exists various definitions for it. Using the definition based on power and voltage, we can express the characteristic impedance as

$$Z_o = \frac{|V_o|^2}{2P_{\text{avg}}} \qquad (7.140)$$

where

$$V_o = \int_W^{W+S} E_x(x, h_2 + h_3)\, dx \qquad (7.141)$$

is the voltage across the slot, and

$$P_{\text{avg}} = \frac{1}{2}\,\text{Re} \int_{-a}^{a}\int_0^b \left(E_x H_y^* - E_y H_x^*\right) dy\, dx \qquad (7.142)$$

represents the transmitted power across the transmission line's cross section.

To obtain numerical results for β and consequently for ε_{eff} and Z_c^0, suitable basis functions for the electric field components, E_{xm} and E_{zk} with $m = 0, 1, \ldots, M$ and $k = 1, 2, \ldots, K$, are needed. These functions affect substantially the numerical efficiency of the solution process and the accuracy of the solutions. As for the choice of the charge basis functions discussed in Section 7.1 for the quasi-static SDA, the selected basis functions should resemble the expected physical behaviors of the electric fields over the slots. In order to be able to increase the solution accuracy by increasing the number of basis functions, these functions must also belong to complete sets. Moreover, the basis functions should have closed-form Fourier transforms to reduce computing time. Lastly, they should be twice continuous differentiable to eliminate spurious solutions. In choosing the basis function to satisfy the actual field behavior, particular attention should be paid to the singularities of the fields in the vicinity of sharp edges. Incorporation of these singularities in the solution process leads to some favorable features, including enhancement of the numerical solution accuracy, the possibility of avoiding an undesirable effect called the relative convergence (i.e., converging to nonphysical solutions) [10,11], and the achievement of high computing efficiency. The principal concern of satisfying the edge requirement is to safeguard against the convergence toward a nonphysical solution. However, other benefits including enhancement of the rate of convergence and improvement of the matrix condition, if applicable, usually follow. It is shown in Mirshekar-Syahkal [3] that for CPW with strips of infinitely thin perfect conductors, the transverse electric and magnetic field components are unbounded at the strip edges and approach infinity with $r^{1/2}$. The longitudinal field components are bounded and behave as $r^{1/2}$ near the edges. With regard to these criteria, we employ the following basis functions according to Uwano and Itoh [12]:

$$E_{xm}(x) = \begin{cases} \dfrac{\cos\left(m\pi\dfrac{(x+d)}{S}\right)}{\sqrt{1-[2(x+d)/S]^2}} - \dfrac{\cos\left(m\pi\dfrac{(x-d)}{S}\right)}{\sqrt{1-[2(x-d)/S]^2}}; & m = 0, 2, \ldots \\[2ex] \dfrac{\sin\left(m\pi\dfrac{(x+d)}{S}\right)}{\sqrt{1-[2(x+d)/S]^2}} + \dfrac{\sin\left(m\pi\dfrac{(x-d)}{S}\right)}{\sqrt{1-[2(x-d)/S]^2}}; & m = 1, 3, \ldots \end{cases} \tag{7.143}$$

$$E_{zk}(x) = \begin{cases} \dfrac{\cos\left(k\pi\dfrac{(x+d)}{S}\right)}{\sqrt{1-[2(x+d)/S]^2}} + \dfrac{\cos\left(k\pi\dfrac{(x-d)}{S}\right)}{\sqrt{1-[2(x-d)/S]^2}}; & k = 1, 3, \ldots \\[2ex] \dfrac{\sin\left(k\pi\dfrac{(x+d)}{S}\right)}{\sqrt{1-[2(x+d)/S]^2}} - \dfrac{\sin\left(k\pi\dfrac{(x-d)}{S}\right)}{\sqrt{1-[2(x-d)/S]^2}}; & k = 2, 4, \ldots \end{cases} \tag{7.144}$$

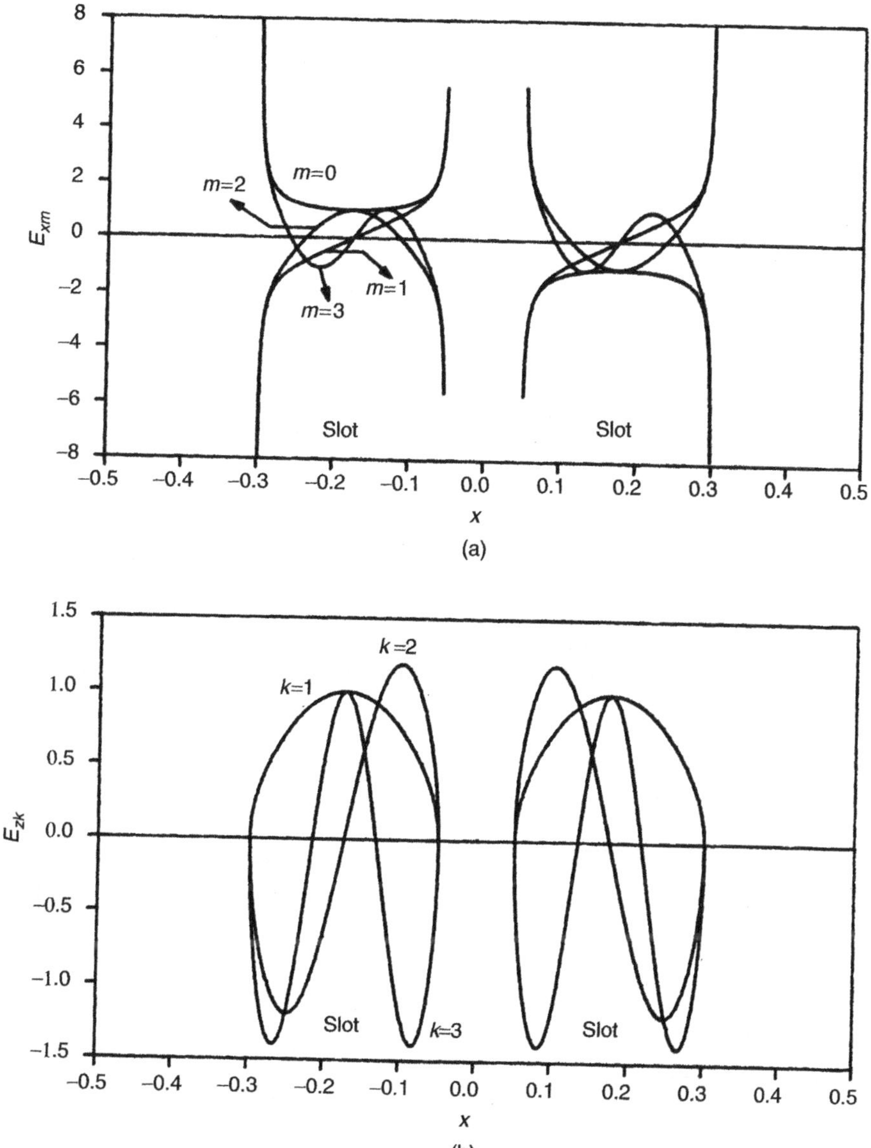

Figure 7.5 Plots of basis functions for the (a) x and (b) z electric fields.

which are defined only over the slots, where $d = W + S/2$. These functions satisfy all the constraints discussed earlier. Figure 7.5 illustrates the shapes of the first few basis functions. Using Eq. (7.66), their Fourier transforms are found to be

$$\tilde{E}_{xm}(\alpha_n) = \begin{cases} -j\dfrac{\pi S}{2}\sin\alpha_n d\left[J_0\left(\left|\dfrac{\alpha_n S + m\pi}{2}\right|\right) + J_0\left(\left|\dfrac{\alpha_n S - m\pi}{2}\right|\right)\right]; \\ m = 0, 2, \ldots \\ -j\dfrac{\pi S}{2}\cos\alpha_n d\left[J_0\left(\left|\dfrac{\alpha_n S + m\pi}{2}\right|\right) - J_0\left(\left|\dfrac{\alpha_n S - m\pi}{2}\right|\right)\right]; \\ m = 1, 3, \ldots \end{cases} \tag{7.145}$$

$$\tilde{E}_{zk}(\alpha_n) = \begin{cases} \dfrac{\pi S}{2}\cos\alpha_n d\left[J_0\left(\left|\dfrac{\alpha_n S + k\pi}{2}\right|\right) + J_0\left(\left|\dfrac{\alpha_n S - k\pi}{2}\right|\right)\right]; \\ k = 1, 3, \ldots \\ -\dfrac{\pi S}{2}\sin\alpha_n d\left[J_0\left(\left|\dfrac{\alpha_n S + k\pi}{2}\right|\right) - J_0\left(\left|\dfrac{\alpha_n S - k\pi}{2}\right|\right)\right]; \\ k = 2, 4, \ldots \end{cases} \tag{7.146}$$

where J_0 denotes the zero-order Bessel function of the first kind.

An example of computed results from the dynamic SDA is shown in Fig. 7.6, in which the propagation constants of the first four modes for a CPW are plotted

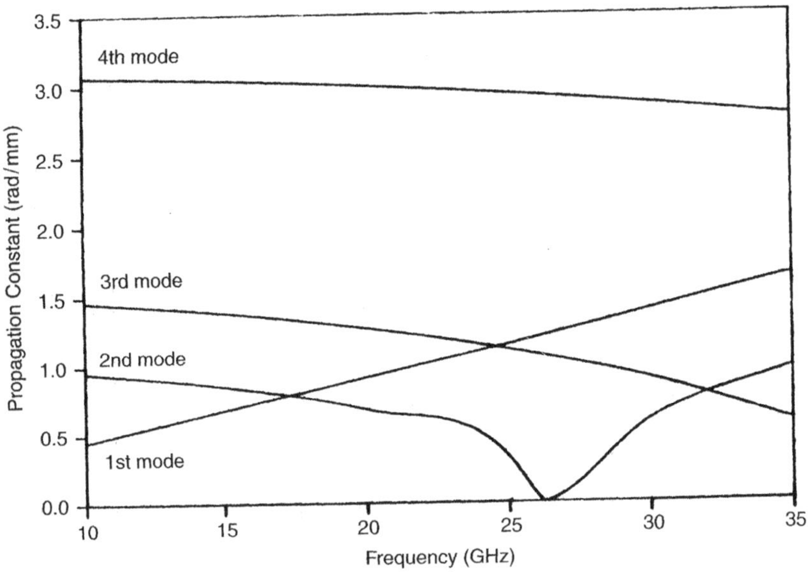

Figure 7.6 Propagation constants of the first four modes for the CPW calculated using the dynamic SDA. $a = 1.524$ mm, $h_1 = 0.762$ mm, $h_2 = 0.254$ mm, $h_3 = 0.508$ mm, $2W = 0.5$ mm, $S = 6$ mm, $\varepsilon_{r1} = 1$, $\varepsilon_{r2} = 9.6$, and $\varepsilon_{r3} = 13$.

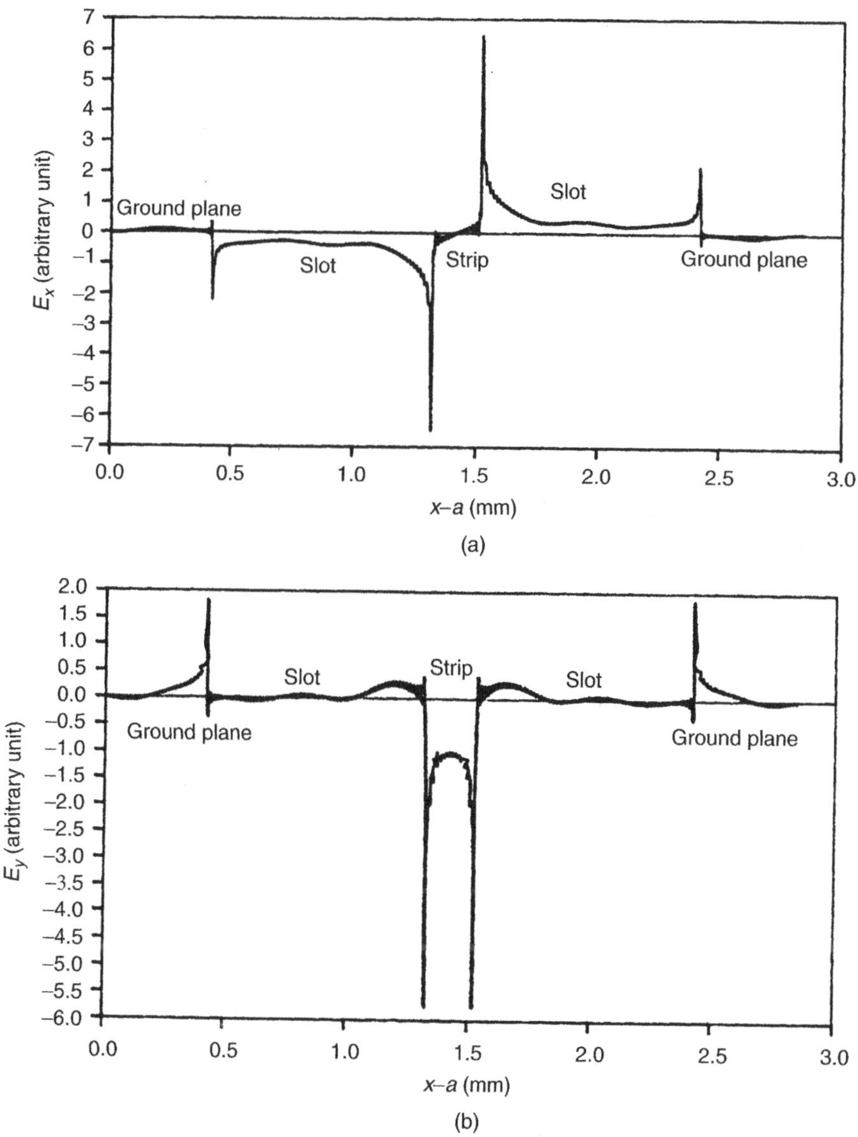

Figure 7.7 (a) x component and (b) y component of the dominant mode's electric field at the interface $y = h_2 + h_3$. $a = 1.422$ mm, $h_1 = 4.55$ mm, $h_2 = 0.127$ mm, $h_3 = 1.016$ mm, $S = 0.889$ mm, $2W = 0.102$ mm, $\varepsilon_{r1} = 3.7$, $\varepsilon_{r2} = 10$, and $\varepsilon_{r3} = 2.2$.

against frequency. The first mode, which is the dominant mode, is found to be propagating up to about 26.5 GHz. As a check for the validity of the numerically calculated eigenmodes, we calculate the electric fields at the interface $y = h_2 + h_3$ for the dominant and first higher-order evanescent modes. They are shown in Figs. 7.7 and 7.8, respectively. The expected asymmetry of the x-directed fields

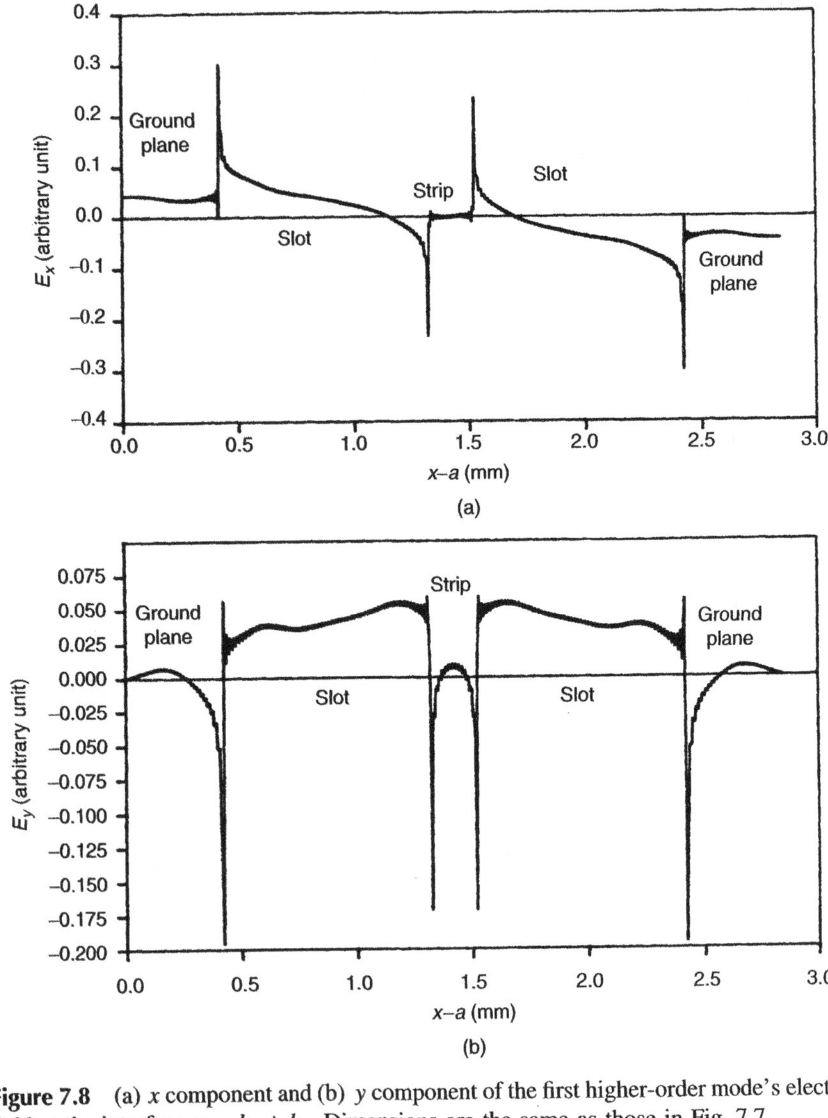

Figure 7.8 (a) x component and (b) y component of the first higher-order mode's electric field at the interface $y = h_2 + h_3$. Dimensions are the same as those in Fig. 7.7.

and the edge conditions are clearly visible in these plots. In addition, the tangential fields on the conducting strips are very close to zero.

REFERENCES

1. T. Itoh and R. Mittra, "Spectral-Domain Approach for Calculating the Dispersion Characteristic of Microstrip Line," *IEEE Trans. Microwave Theory Tech.*, Vol. MTT-2, pp. 498–499, July 1973.

2. T. Itoh and A. S. Hebert, "A Generalized Spectral Domain Analysis for Coupled Suspended Microstrip Lines with Tuning Septums," *IEEE Trans. Microwave Theory Tech.*, Vol. MTT-26, pp. 820–826, Oct. 1978.
3. D. Mirshekar-Syahkal, *Spectral Domain Method for Microwave Integrated Circuits*, Research Studies Press Ltd., Somerset, England, 1990.
4. T. Itoh, "Analysis of Microstrip Resonators," *IEEE Trans. Microwave Theory Tech.*, Vol. MTT-22, pp. 946–952, Nov. 1974.
5. Q. Zhang and T. Itoh, "Spectral-Domain Analysis of Scattering from *E*-Plane Circuit Elements," *IEEE Trans. Microwave Theory Tech.*, Vol. MTT-35, pp. 138–150, Feb. 1987.
6. C. Scott, *The Spectral Domain Method in Electromagnetics*, Artech House, Norwood, MA, 1989.
7. K. M. Rahman and C. Nguyen, "Frequency Dependent Analysis of Shielded Asymmetric Coplanar Waveguide Step Discontinuity," *IEEE MTT-S Int. Microwave Symp. Dig.*, pp. 1013–1015, 1994.
8. C. Nguyen, Spectral-Domain Analysis, in *Wiley Encyclopedia of Electrical and Electronics Engineering*, Vol. 20, J. G. Webster, Ed., John Wiley & Sons, New York, 1998, pp. 105–111.
9. K. M. Rahman and C. Nguyen, "On the Computation of Complex Modes in Lossless Shielded Asymmetric Coplanar Waveguides," *IEEE Trans. Microwave Theory Tech.*, Vol. MTT-43, pp. 2713–2716, Dec. 1995.
10. R. Mittra, "Relative Convergence of the Solution of a Doubly Infinite Set of Equations," *J. Res. NBS*, Vol. 670, pp. 245–254, 1963.
11. S. W. Lee, W. R. Jones, and J. J. Campbell, "Convergence of Numerical Solutions of Iris-Type Discontinuity Problems," *IEEE Trans. Microwave Theory Tech.*, Vol. MTT-19, pp. 528–536, 1971.
12. T. Uwano and T. Itoh, Spectral-Domain Approach, in *Numerical Techniques for Microwave and Millimeter-Wave Structures*, T. Itoh, Ed., John Wiley & Sons, New York, 1989, pp. 334–380.
13. R. F. Harrington, *Field Computation by Moment Methods*, Robert E. Krieger Publishing, Melbourne, FL, 1983, Chap. 1.

PROBLEMS

7.1 Derive Eqs. (7.50) and (7.51).

7.2 Derive Eq. (7.53).

7.3 Derive Eq. (7.69).

7.4 Derive Eq. (7.70).

7.5 Derive Eqs. (7.145) and (7.146).

7.6 Consider a microstrip line as shown in Fig. 4.2. It is assumed that the strip is a perfect conductor and has zero thickness. Also, the ground plane is assumed to be infinitely wide and a perfect conductor. The dielectric is assumed to be perfect. Assume the charge distribution on the strip is given as

$$\rho_s(x) = \begin{cases} 1, & |x| \leq W/2 \\ 0, & \text{otherwise} \end{cases}$$

(a) Formulate the quasi-static SDA for determining the capacitance per unit length, characteristic impedance, Z_o, and effective dielectric constant, ε_{eff}.

(b) Write a computer program to calculate Z_o and ε_{eff} using the formulation in part (a). Compute and plot Z_o and ε_{eff} versus W/h from 0.05 to 5 for $\varepsilon_r = 2.2$ and 10.5 and $h = 0.635$ and 1.27 mm.

7.7 Repeat Problem 7.6 using the following charge distribution on the strip:

$$\rho_s(x) = \begin{cases} \dfrac{1}{\sqrt{1 - [|x|/(W/2)]^2}} & |x| \leq \dfrac{W}{2} \\ 0, & \text{otherwise} \end{cases}$$

Compare the numerical results of Z_o and ε_{eff} to those obtained in Problem 7.6, and provide an assessment of the charge distributions used.

7.8 Repeat Problem 7.6 using the following basis functions for charge distribution on the strip:

$$\rho_{si}(x) = \frac{\cos\left((i-1)\pi \dfrac{(x + W/2)}{W}\right)}{\sqrt{1 - \left(\dfrac{2(x + W/2) - W}{W}\right)^2}}$$

Compare the numerical results of Z_o and ε_{eff} to those obtained in Problems 7.6 and 7.7, and provide an assessment of the charge distributions used. Also, compare results using one to five basis functions and 50 to 400 spectral terms.

7.9 Consider the CPW shown in Fig. 7.1. Use the quasi-static SDA formulation for this transmission line, presented in Section 7.1, to write a computer program to calculate its characteristic impedance, Z_o, and effective dielectric constant, ε_{eff}. Compute and plot Z_o and ε_{eff} versus $S_1 = S_2$ from 0.0254 to 0.762 mm, for $a = 2.54$ mm, $h_1 = h_3 = 0$ and 508 mm, $h_2 = 0.127$ mm, $\varepsilon_{r1} = 1$, $\varepsilon_{r2} = 2.2$, $\varepsilon_{r3} = 2.2$, $W = 0.0.508$ mm, and $G_1 = G_2$.

7.10 Consider a CPW as shown in Fig. 4.3(a). The ground and strip are assumed to be perfect conductors and to have zero thickness, and the dielectric substrate is assumed lossless.

(a) Formulate the quasi-static SDA for determining the capacitance per unit length, characteristic impedance, Z_o, and effective dielectric constant, ε_{eff}. Choose appropriate charge distributions for the strip and grounds.

(b) Write a computer program to calculate Z_o and ε_{eff}. Compute and plot Z_o and ε_{eff} versus the dimension ratio a/b from 0.1 to 0.9 for $\varepsilon_r = 2.2$ and 10.5 and $h/b = 0.1$, 0.5, 4, and 10. Compare results to those obtained in Problems 4.10 and 6.18.

7.11 Use the program developed in Problem 7.9 to calculate and plot Z_o and ε_{eff} of the CPW shown in Fig. 4.3(a) versus the dimension ratio a/b from 0.1 to 0.9 for $\varepsilon_r = 2.2$ and 10.5 and $h/b = 0.1$, 0.5, 4, and 10. Compare results to those obtained in Problems 4.10, 6.18, and 7.10.

7.12 Consider a CPS as shown in Fig. 4.4. The strips are assumed to be perfect conductors with negligible thickness, and the dielectric substrate is assumed lossless. Assume that the basis functions for the charge distributions on the strips are given by

$$\rho_{s1i}(x) = \begin{cases} \dfrac{\cos\left((i-1)\pi\dfrac{(x+b)}{b-a}\right)}{\sqrt{1-\left(\dfrac{2(x+b)-(b-a)}{b-a}\right)^2}}, & -b \le x \le -a \\ 0, & \text{otherwise} \end{cases}$$

$$\rho_{s2i}(x) = \begin{cases} \dfrac{\cos\left((i-1)\pi\dfrac{(x-a)}{b-a}\right)}{\sqrt{1-\left(\dfrac{2(x-a)-(b-a)}{b-a}\right)^2}}, & a \le x \le b \\ 0, & \text{otherwise} \end{cases}$$

(a) Formulate the quasi-static SDA for determining the characteristic impedance, Z_o, and effective dielectric constant, ε_{eff}.

(b) Write a computer program to calculate Z_o and ε_{eff}. Compute and plot Z_o and ε_{eff} versus *a/b* from 0.1 to 0.9, for a relative dielectric constant, ε_r, of 2.2 and 10.5, and a normalized substrate thickness, *h/b*, of 0.1, 0.5, 4, and 10. Compare results to those obtained in Problem 6.20.

7.13 Consider a broadside-coupled CPW as shown in Fig. P7.1. The enclosure is assumed to be a perfect conductor, and all the ground and central strips are assumed to be perfectly conducting with zero thickness. The dielectric substrates are assumed to be lossless. This structure supports two dominant propagating modes: the even and odd modes corresponding to an open circuit or magnetic wall (MW) and a short-circuit or electric wall (EW) at the center of the central dielectric substrate ($y = 0$), respectively. These modes are balanced modes corresponding to a MW at the center of the strip ($x = 0$). Note that the structure corresponding to the odd mode is the same as the shielded conductor-backed CPW. Assume the charge distributions on the central strips, ρ_{s1i}^U and ρ_{s1i}^L, and ground strips, ρ_{s2i}^U, ρ_{s2i}^L and ρ_{s3i}^U,

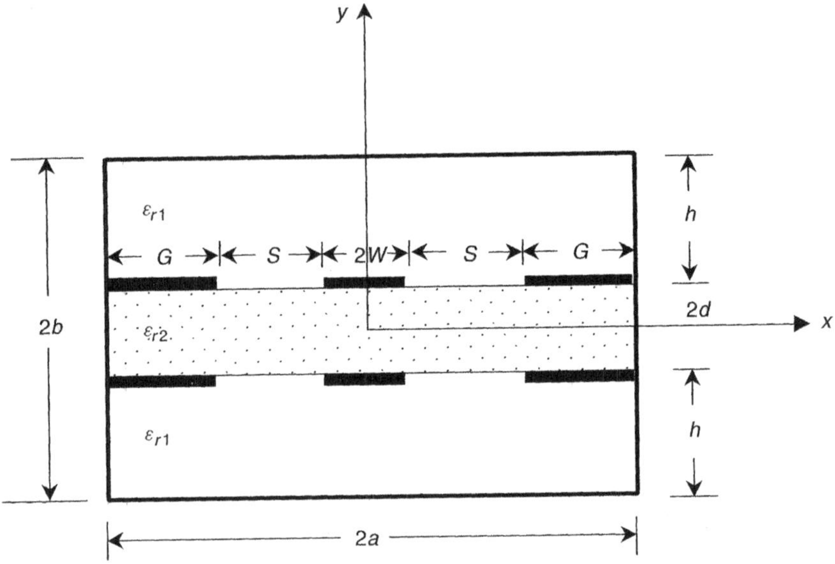

Figure P7.1 Cross section of a broadside-coupled CPW.

ρ^L_{s3i}, are given as

$$\rho^U_{s1i}(x) = \rho^L_{s1i}(x) = \frac{\cos\left((i-1)\pi\dfrac{(x+W)}{2W}\right)}{\sqrt{1-\left(\dfrac{2(x+W)-2W}{2W}\right)^2}}$$

$$\rho^U_{s2i}(x) = \rho^L_{s2i}(x) = \frac{\cos\left(\left(i-\dfrac{1}{2}\right)\pi\dfrac{(x+W+S)}{G}\right)}{\sqrt{1-\left(\dfrac{x+W+S+G}{G}\right)^2}}$$

$$\rho^U_{s3i}(x) = \rho^L_{s3i}(x) = \frac{\cos\left(\left(i-\dfrac{1}{2}\right)\pi\dfrac{(x-W-S)}{G}\right)}{\sqrt{1-\left(\dfrac{x-W-S-G}{G}\right)^2}}$$

where the superscripts U and L refer to the upper and lower strips, respectively. The subscripts 1, 2, and 3 denote the central strip and left and right ground strips, respectively.

(a) Formulate the quasi-static SDA analysis for determining the characteristic impedance, Z_{oe}, and effective dielectric constant, $\varepsilon_{\text{eff},e}$, of the even mode.

(b) Formulate the quasi-static SDA analysis for determining the characteristic impedance, Z_{oo}, and effective dielectric constant, $\varepsilon_{\text{eff},o}$, of the odd mode.

(c) Write a computer program to calculate the Z_{oe}, Z_{oo}, and $\varepsilon_{\text{eff},e}$, $\varepsilon_{\text{eff},o}$. Compute and plot these parameters versus $2W$ from 0.0254 to 1.016 mm for $2a = 2.54$ mm, $h = 0.508$ mm, $2d = 0.127$ mm, $\varepsilon_{r1} = 1$, $\varepsilon_{r2} = 10.5$, $\varepsilon_{r3} = 2.2$, and $S = 0.508$ mm.

7.14 Consider the broadside-coupled CPW in Problem 7.13.

(a) Formulate an analysis of the effective dielectric constant, $\varepsilon_{\text{eff},e}$, of the dominant even mode using the dynamic SDA.

(b) Formulate an analysis of the effective dielectric constant, $\varepsilon_{\text{eff},o}$, of the dominant odd mode using the dynamic SDA.

(c) Write a computer program to calculate $\varepsilon_{\text{eff},e}$ and $\varepsilon_{\text{eff},o}$ as a function of frequency, using the electric field's basis functions given in Eqs. (7.143) and (7.144). Compute and plot these parameters versus frequency from 1 to 50 GHz for $2a = 2.54$ mm, $h = 0.508$ mm, $2d = 0.127$ mm, $\varepsilon_{r1} = 1$, $\varepsilon_{r2} = 10.5$, $2W = 0.508$ mm, and $S = 0.0254$, 0.254, and 0.762 mm. Compare results using one to five basis functions and 50 to 400 spectral terms.

7.15 Consider again the broadside-coupled CPW in Problem 7.13, and define its characteristic impedance as

$$Z_o = \frac{V_o^2}{2P_{\text{avg}}}$$

where V_o is the voltage at the slot, and P_{avg} denotes the average power transported across the transmission line's cross section.

(a) Derive P_{avg} in terms of the field components in the Fourier-transform domain.

(b) Formulate an analysis of the characteristic impedance for the dominant even mode, Z_{oe}, using the dynamic SDA.

(c) Formulate an analysis of the characteristic impedance for the dominant odd mode, Z_{oo}, using the dynamic SDA.

(d) Write a computer program to calculate Z_{oe} and Z_{oo} as a function of frequency. The electric field's basis functions are given in Eqs. (7.143) and (7.144). Compute and plot these characteristic impedances versus frequency from 1 to 50 GHz for $2a = 3.556$ mm, $h = 2.54$ mm, $2d = 0.635$ mm, $\varepsilon_{r1} = 1$, $\varepsilon_{r2} = 10$, $S = 0.635$, 0.254, and 0.787 mm. Compare results using one to five basis functions and 50 to 400 spectral terms.

7.16 Consider again the broadside-coupled CPW in Problem 7.13. At high frequencies, due the hybrid nature of the propagation, this structure can generate higher-order modes, classified as the even and odd eigenmodes corresponding to a MW and an EW at $y = 0$, respectively.

(a) Formulate an analysis of the even eigenmodes using the dynamic SDA.

(b) Formulate an analysis of the odd eigenmodes using the dynamic SDA.

(c) Write a computer program to calculate the effective dielectric constants of the even and odd eigenmodes as a function of frequency. The electric field's basis functions are given in Eqs. (7.143) and (7.144). Compute and plot the effective dielectric constants of the dominant and first higher-order even and odd modes versus frequency from 1 to 50 GHz for $2a = 3.556$ mm, $h = 2.54$ mm, $2d = 0.635$ mm, $2W = 0.787$ mm, $S = 0.635$ mm, $\varepsilon_{r1} = 1$, and $\varepsilon_{r2} = 2.2$ and 10. These results can be used to determine the operating frequency range of the broadside-coupled CPW (i.e., the range of dominant-mode operation).

7.17 Use the formulation derived in Problem 7.16 to write a computer program to calculate the characteristic impedances of the even and odd eigenmodes as a function of frequency. The electric field's basis functions are still given in Eqs. (7.143) and (7.144). Compute and plot the characteristic impedances of the dominant and first higher-order even and odd modes versus frequency from 1 to 50 GHz for $2a = 3.556$ mm, $h = 2.54$ mm, $2d = 0.635$ mm, $2W = 0.787$ mm, and $S = 0.635$ mm, $\varepsilon_{r1} = 1$, and $\varepsilon_{r2} = 2.2$ and 10.

7.18 Consider the CPW shown in Fig. 7.4. Use the dynamic SDA formulation for this transmission line, presented in Section 7.2, to write a computer program to calculate its effective dielectric constants, ε_{eff}, for the eigenmodes. Compute and plot ε_{eff} for the first four modes versus frequency from 10 to 35 GHz for $2a = 1.524$ mm, $h_1 = 0762$ mm, $h_2 = 0.254$ mm, $h_3 = 0.508$ mm, $2W = 0.5$ mm, $S = 0.6$ mm, $\varepsilon_{r1} = 1$, $\varepsilon_{r2} = 9.6$, and $\varepsilon_{r3} = 13$.

7.19 Consider the CPW shown in Fig. 7.4. Use the dynamic SDA formulation for this transmission line, presented in Section 7.2, to write a computer program to calculate its characteristic impedance, Z_o, for the dominant (CPW) mode. Compute and plot Z_o versus frequency from 1 to 25 GHz for $2a = 1.524$ mm, $h_1 = 0762$ mm, $h_2 = 0.254$ mm, $h_3 = 0.508$ mm, $2W = 0.5$ mm, $S = 0.6$ mm, $\varepsilon_{r1} = 1$, $\varepsilon_{r2} = 9.6$, and $\varepsilon_{r3} = 13$.

7.20 Consider a slot line as shown in Fig. 4.6. The ground planes are assumed to be perfect conductors and to have zero thickness. The dielectric substrate is assumed to be lossless. The x- and z-directed electric fields are approximated by

$$E_x(x) \cong \sum_{m=1}^{M} c_m E_{xm}(x)$$

$$E_z(x) \cong \sum_{k=1}^{K} d_k E_{zk}(x)$$

where c_m and d_k are the unknown coefficients. The basis functions $E_{xm}(x)$ and $E_{zk}(x)$ are assumed as

$$E_{xm}(x) = \begin{cases} \dfrac{W[(W/2)^2 - x^2]^{m-1}}{2\sqrt{(W/2)^2 - x^2}}, & |x| \leq \dfrac{W}{2} \\ 0, & \text{otherwise} \end{cases}$$

$$E_{zk}(x) = \begin{cases} -j\dfrac{2x}{W}\sqrt{\left(\dfrac{W}{2}\right)^2 - x^2}\left[\left(\dfrac{W}{2}\right)^2 - x^2\right]^{k-1}, & |x| \leq \dfrac{W}{2} \\ 0, & \text{otherwise} \end{cases}$$

The Fourier transform of a function $f(x, y)$ is defined as

$$\tilde{f}(\alpha, y) = \int_{-\infty}^{\infty} f(x, y)e^{j\alpha x}$$

(a) Plot the first four basis functions for E_{xm} and E_{zk} versus x.

(b) Formulate an analysis for the effective dielectric constant, ε_{eff}, of the dominant mode using the dynamic SDA. The Fourier transforms of E_{xm} and E_{zk} would involve the Bessel functions of order $(m-1)$ and $(k+1)$, respectively.

(c) Write a computer program to calculate ε_{eff}. Compute and plot ε_{eff} as a function of the normalized slot width, W/h, from 0.1 to 2 for ε_r of 10.5 and frequency of 1, 10, and 20 GHz. Compare results to those obtained in Problem 4.22.

7.21 Repeat Problem 7.20, parts (b) and (c), for the characteristic impedance, Z_o, of the dominant mode, defined as

$$Z_o = \frac{V_o^2}{2P_{\text{avg}}}$$

where V_o is the voltage at the slot, and P_{avg} denotes the average power transported across the slot line's cross section.

7.22 Consider a three-layer shielded microstrip line as shown in Fig. 6.6. The enclosure and conducting strip are assumed to be perfect conductors. The strip thickness is assumed negligible, and the dielectric substrates are assumed to be lossless. The basis functions for the currents on the strip are assumed as

$$J_{xn}(x) = \frac{\sin\left(2n\pi\dfrac{(x-S-W/2)}{W}\right)}{\sqrt{1-\left(\dfrac{2(x-S-W/2)}{W}\right)^2}}$$

$$J_{zn}(x) = \frac{\cos\left(2\,(n-1)\,\pi\dfrac{(x-S-W/2)}{W}\right)}{\sqrt{1-\left(\dfrac{2(x-S-W/2)}{W}\right)^2}}$$

(a) Plot the first four basis functions for J_{xn} and J_{zn} versus x.

(b) Formulate the dynamic SDA for determining the effective dielectric constant, ε_{eff}, of the dominant mode.

(c) Write a computer program to calculate ε_{eff}. Calculate and plot ε_{eff} versus W/h_2 from 0.1 to 5, for $\varepsilon_{r2} = 2.2$, $h_1 = h_3 = 0.66$ mm, $h_2 = 0.254$ mm, $a = 2.54$ mm, $S = 0.25(a-W)$ and $0.5(a-W)$, and a frequency of 1, 10, and 100 GHz. Compare results to the quasi-static results obtained in Problem 6.7.

7.23 Repeat Problem 7.22, parts (b) and (c), for the characteristic impedance Z_o of the dominant mode. Z_o is defined as

$$Z_o = \frac{P_{\text{avg}}}{I_o^2}$$

where I_o is the current flow on the strip along the z direction, and P_{avg} represents the time-average power transmitted across the microstrip line's cross section.

7.24 Consider a multi layer slot line as shown in Fig. P7.2. The ground planes are assumed to be perfect conductors and to have zero thickness. The dielectric substrates are assumed to be lossless. The electric fields' basis functions are the same as those given in Problem 7.20.

(a) Formulate an analysis for the effective dielectric constant, ε_{eff}, of the dominant mode using the dynamic SDA. The Fourier transforms of E_{xm} and E_{zk} would involve the Bessel functions of order $(m-1)$ and $(k+1)$, respectively.

(b) Write a computer program to calculate ε_{eff}. Compute and plot ε_{eff} as a function of the normalized slot width, W/h, from 0.1 to 2 for $\varepsilon_{r1} = \varepsilon_{r2} = \varepsilon_{r4} = 1$, $\varepsilon_{r3} = 10.5$, and a frequency of 1, 10, and 20 GHz. Compare results to those obtained in Problems 4.22 and 7.20.

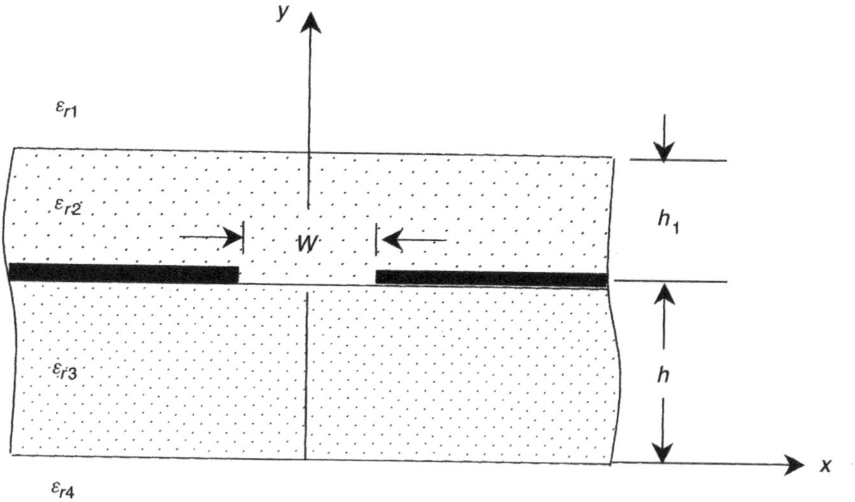

Figure P7.2 Cross section of a multilayer slot line.

(c) Compute and plot ε_{eff} as a function of frequency from 1 to 30 GHz for $W/h = 0.05, 0.1, 0.8, 1$, and 1.5 and $\varepsilon_{r1} = \varepsilon_{r4} = 1$, $\varepsilon_{r2} = 2.2$, and $\varepsilon_{r3} = 10.5$.

7.25 Consider the slot line in Problem 7.24, and define its characteristic impedance as

$$Z_o = \frac{V_o^2}{2P_{\text{avg}}}$$

where V_o is the voltage at the slot, and P_{avg} denotes the average power transported across the slot line's cross section.

(a) Formulate an analysis for the characteristic impedance, Z_o, of the dominant mode using the dynamic SDA.

(b) Write a computer program to calculate Z_o. Compute and plot Z_o as a function of the normalized slot width, W/h, from 0.1 to 2 for $\varepsilon_{r1} = \varepsilon_{r2} = \varepsilon_{r4} = 1$, $\varepsilon_{r3} = 10.5$, and a frequency of 1, 10, and 20 GHz. Compare results to those obtained in Problems 4.22 and 7.20.

(c) Compute and plot Z_o as a function of frequency from 1 to 30 GHz for $W/h = 0.05, 0.1, 0.8, 1$, and 1.5 and $\varepsilon_{r1} = \varepsilon_{r4} = 1$, $\varepsilon_{r2} = 2.2$, and $\varepsilon_{r3} = 10.5$.

7.26 Consider a microstrip resonator as shown in Fig. P7.3. Define the following two-dimensional Fourier transform of $f(x, y, z)$:

$$\tilde{f}(\alpha_n, y, \beta) = \int_{-\infty}^{\infty} \int_{-a}^{a} f(x, y, z) e^{j(\alpha_n x + \beta z)} \, dx \, dz$$

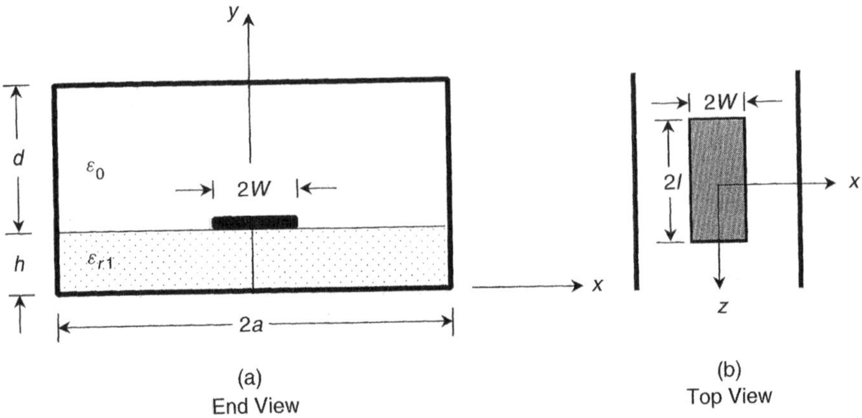

Figure P7.3 (a) End view and (b) top view of a microstrip resonator.

where α_n and β are the Fourier transform variables. Formulate the process to determine the resonant frequency using the SDA.

APPENDIX A: FOURIER TRANSFORM AND PARSEVAL'S THEOREM

Open Transmission Line Structures

For open structures, such as the open microstrip line shown in Fig. 4.2, the Fourier transform of a function $f(x, y)$ with respect to x is a continuous transform and can be defined as

$$\tilde{f}(\alpha, y) = \int_{-\infty}^{\infty} f(x, y) e^{j\alpha x}\, dx \tag{A7.1}$$

where α is the Fourier transform variable, signifying the number of cycles per unit of x. The inverse Fourier transform or Fourier integral is

$$f(x, y) = \frac{1}{2\pi} \int_{-\infty}^{\infty} \tilde{f}(\alpha, y) e^{-j\alpha x}\, d\alpha \tag{A7.2}$$

This Fourier transform is the same as the sine transform

$$\tilde{f}(\alpha, y) = 2 \int_{0}^{\infty} f(x, y) \sin \alpha x\, dx \tag{A7.3}$$

when $f(x, y)$ is odd with respect to x, that is, $f(x, y) = -f(-x, y)$, and the cosine transform

$$\tilde{f}(\alpha, y) = 2 \int_{0}^{\infty} f(x, y) \cos \alpha x\, dx \tag{A7.4}$$

if $f(x, y)$ is even with respect to x, that is, $f(x, y) = f(-x, y)$. Note that $f(x, y)$ is defined only over a certain part of the domain of x, whereas $\tilde{f}(\alpha, y)$ is defined over the entire domain of α. Other definitions for the Fourier transform may also be used. For example, the cosine transform may be written as

$$\tilde{f}(\alpha, y) = \int_0^\infty f(x, y) \cos \alpha x \, dx \tag{A7.5}$$

Parseval's theorem associated with the continuous Fourier transform has the form

$$\int_{-\infty}^{\infty} f(x, y) g^*(x, y) \, dx = \frac{1}{2\pi} \int_{-\infty}^{\infty} \tilde{f}(\alpha, y) \tilde{g}^*(\alpha, y) \, d\alpha \tag{A7.6}$$

where the superscript * denotes the complex conjugate. This Parseval's formula indicates that if the two functions $f(x, y)$ and $g(x, y)$ are nonzero in complementary regions in the space domain, the integral of their Fourier transforms' product is zero with respect to the Fourier transform variable α. This result is useful for spectral-domain analysis.

Closed Transmission Line Structures

For closed structures, such as the shielded microstrip line shown in Fig. 6.6, the Fourier transform of a function $f(x, y)$ with respect to x is a discrete or finite transform and can be defined as

$$\tilde{f}(\alpha_n, y) = \frac{1}{a} \int_{-a}^{a} f(x, y) e^{j\alpha_n x} \, dx \tag{A7.7}$$

where α_n is the Fourier transform variable and has discrete values. This Fourier transform is equivalent to the sine transform,

$$\tilde{f}(\alpha_n, y) = \frac{2}{a} \int_0^a f(x, y) \sin \alpha_n x \, dx \tag{A7.8}$$

when $f(x, y)$ is odd with respect to x, that is, $f(x, y) = -f(-x, y)$, and the cosine transform

$$\tilde{f}(\alpha_n, y) = \frac{2}{a} \int_0^a f(x, y) \cos \alpha_n x \, dx \tag{A7.9}$$

when $f(x, y)$ is even with respect to x; that is, $f(x, y) = f(-x, y)$.

The inverse Fourier transforms of Eqs. (A7.8) and (A7.9) are just the Fourier series of $f(x, y)$:

$$f(x, y) = \sum_{n=1}^{\infty} \tilde{f}(\alpha_n, y) \cos \alpha_n x \tag{A7.10}$$

$$f(x, y) = \sum_{n=1}^{\infty} \tilde{f}(\alpha_n, y) \sin \alpha_n x \tag{A7.11}$$

respectively. Other versions of the discrete Fourier transform can also be used.

Parseval's theorem associated with these transforms is written as

$$\int_0^a f(x, y) g^*(x, y)\, dx = \frac{a}{2} \sum_{n=1}^{\infty} \tilde{f}(\alpha_n, y) \tilde{g}^*(\alpha_n, y) \tag{A7.12}$$

which implies that if the two functions $f(x, y)$ and $g(x, y)$ are nonzero in complementary regions in the space domain, then the sum of their Fourier transforms' product, over the Fourier transform variable α_n, is zero. This theorem is useful in implementing spectral-domain analysis.

APPENDIX B: GALERKIN'S METHOD

In this appendix, we will present *Galerkin's method* [13]. Galerkin's method is basically a *method of moment*, which can be used for solving a linear system of equations. To this end, we consider the following inhomogeneous equation:

$$L(f) = g \tag{B7.1}$$

where L is a linear operator, f is an unknown function, and g is a known function. This kind of equation is often encountered in the analysis of microwave structures, in which L represents the system, f represents the response, and g denotes the source. Note that Eq. (B7.1) has a unique solution only if the boundary conditions on f are specified. For example, in Poisson's equation

$$\nabla^2 V(x, y, z) = -\frac{\rho(x, y, z)}{\varepsilon} \tag{B7.2}$$

$L = \nabla^2$, which is the Laplacian operator; $f = V$, representing the unknown voltage; and $g = \rho$ is the known charge distribution. Solution of Eq. (B7.1) using Galerkin's method is now described as follows.

We begin by expanding the unknown function f in a series of known basis functions, f_n, which exist in the domain of L as

$$f = \sum_{n=1}^{N} a_n f_n \tag{B7.3}$$

where a_n are the unknown coefficients. Note that if N approaches infinity and $\{f_n\}$ forms a complete set, for example, a Fourier series, then Eq. (B7.3) represents an exact solution for f. Substituting Eq. (B7.3) into (B7.1) gives

$$\sum_{n=1}^{N} a_n L(f_n) = g \tag{B7.4}$$

Now we define a suitable *inner product* between two functions f and g, $\langle f, g\rangle$, which satisfies the following properties:

$$\langle f, g\rangle = \langle g, f\rangle \tag{B7.5}$$

$$\langle af + bg, h\rangle = a\langle f, h\rangle + b\langle g, h\rangle \tag{B7.6}$$

$$\left\langle f^*, f\right\rangle > 0, \quad \text{if } f > 0 \tag{B7.7}$$

$$\left\langle f^*, f\right\rangle = 0, \quad \text{if } f = 0 \tag{B7.8}$$

where h is another function, a and b are scalars, and the superscript* indicates complex conjugate. For example, we can define

$$\langle f, g\rangle = \int w(x) f(x) g(x)\, dx \tag{B7.9}$$

where $w(x)$ is referred to as the weighting or testing function.

For the considered problem, we use N different weighting functions, $w_m(m = 1, 2, \ldots, N)$, in the range of L. Taking the inner product of Eq. (B7.4) with each of these functions, we obtain the following system of linear equations:

$$\sum_{n=1}^{N} a_n \langle w_m, L(f_n)\rangle = \langle w_m, g\rangle \tag{B7.10}$$

which can be rewritten as

$$[l_{mn}][a_n] = [g_m] \tag{B7.11}$$

where

$$[l_{mn}] = \begin{bmatrix} \langle w_1, L(f_1)\rangle & \langle w_1, L(f_2)\rangle & \cdots & \langle w_1, L(f_N)\rangle \\ \langle w_2, L(f_1)\rangle & \langle w_2, L(f_2)\rangle & \cdots & \langle w_2, L(f_N)\rangle \\ \vdots & \vdots & \vdots & \vdots \\ \langle w_N, L(f_1)\rangle & \langle w_N, L(f_2)\rangle & \cdots & \langle w_N, L(f_N)\rangle \end{bmatrix} \tag{B7.12}$$

$$[a_n] = [a_1 \quad a_2 \quad \cdots \quad a_N]^T \tag{B7.13}$$

$$\begin{aligned}[g_n] &= [g_1 \quad g_2 \quad \cdots \quad g_N]^T \\ &= [\langle w_1, g\rangle \quad \langle w_2, g\rangle \quad \cdots \quad \langle w_N, g\rangle]^T \end{aligned} \tag{B7.14}$$

with T indicating a transpose matrix. Equation (B7.11) can be solved to determine the unknown coefficients a_n. If $[l_{mn}]$ is nonsingular, then we can obtain a_n directly as

$$[a_n] = [l_{mn}]^{-1}\left[g_m\right] \tag{B7.15}$$

Substituting Eq. (B7.15) into (B7.3) yields the final solution for the unknown function f as

$$f = \left[f_n\right][l_{mn}]^{-1}\left[g_m\right] \tag{B7.16}$$

where $[f_n] = [f_1 \quad f_2 \quad \cdots \quad f_N]$. Various choices for f_n and w_n can be used, and they are critical for the numerical efficiency of the solution process and the accuracy of the solution. f_n should be linearly independent and form a complete set so the solution accuracy can be enhanced by increasing the number of f_n. They should be chosen such that $\sum_{n=1}^{N} a_n f_n$ satisfies the boundary conditions of f; individual f_n, however, do not have to satisfy these boundary conditions closely. In practice, f_n are normally chosen to closely describe f. Moreover, f_n should be twice continuously differentiable to avoid spurious solutions. w_n should also be linearly independent and form a complete set. Very often in microwave problems, w_n and f_n are chosen to be equal, and the method is referred to as Galerkin's method. Thus, the spectral-domain method is basically a method of moment implemented using Galerkin's technique in the Fourier-transform domain.

CHAPTER EIGHT

Mode-Matching Method

The mode-matching method [1] is a useful technique for formulation of boundary-value problems, especially for structures consisting of two or more separate regions. It is based on matching the fields at the boundaries of different regions and hence lends itself naturally to the analysis of microwave boundary-value problems. The mode-matching method has widely been used for scattering and transmission problems, as well as transmission line analysis. Scattering problems include discontinuities in waveguides and transmission lines, such as the microstrip line and the coplanar waveguide (CPW), and obstacles in a medium. Transmission problems include analysis of filters, such as fin line bandpass filters, impedance transformers, and power dividers and transitions such as waveguide-to-microstrip line. Transmission line analyses include determination of the transmission line's propagation constant and characteristic impedance, such as those of a microstrip line, a CPW, and coplanar strips (CPSs).

In this chapter, we will describe the mode-matching method for analyzing planar transmission lines. Coplanar strips with finite strip metallization thickness will be used to illustrate the formulation process [2]. We will also describe the mode-matching method for the analysis of planar transmission line discontinuities. A CPW will be used to serve this purpose [3].

8.1 MODE-MATCHING ANALYSIS OF PLANAR TRANSMISSION LINES

Figure 8.1 shows a cross section of the coplanar strips (CPSs) used to illustrate the formulation process. The structure is assumed to be uniform and infinitely long. The conducting strips are assumed to be perfect, and the dielectric substrate is considered lossless. This structure supports, in general, hybrid modes consisting of both TE and TM. The dominant mode however, is, quasi-TEM. Modes in the structure are odd modes corresponding to an electric wall (EW) at the plane of symmetry of the two strips.

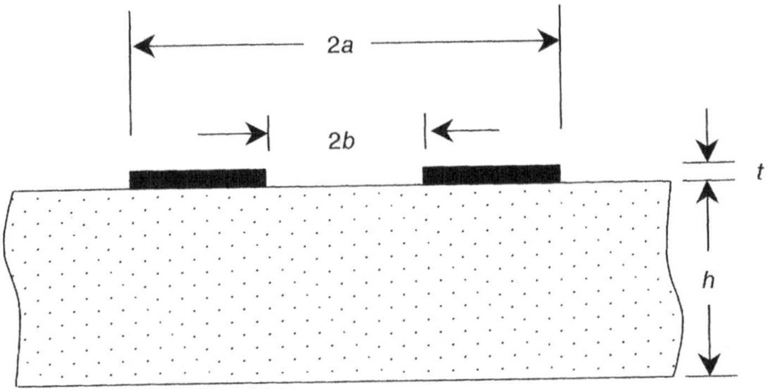

Figure 8.1 Cross section of a CPS.

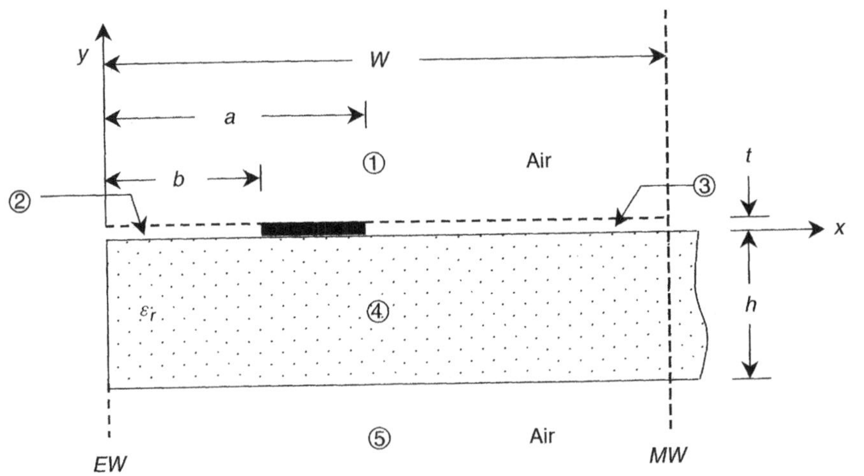

Figure 8.2 Half-structure of the CPS used for analysis. EW and MW stand for electric wall and magnetic wall, respectively.

In view of the symmetry, only one-half of the structure ($x \geq 0$) shown in Fig. 8.2 needs to be considered. To facilitate the solution process, a longitudinal magnetic wall (MW) is inserted perpendicular to the structure, sufficiently apart from the strips, as seen in Fig. 8.2. It should be noted here that a longitudinal electric wall (EW) can also be used in place of the MW and will yield the same results. Of course, appropriate boundary conditions must be used corresponding to the EW. These walls are needed for Fourier series expansions of the fields. The fields are normally confined between the two strips; hence the effect of this hypothetical MW can be neglected. This step is avoided for closed transmission line structures such as a shielded microstrip line. The considered half structure is now divided into five subregions. In mode-matching analysis, hybrid fields in all the subregions are determined first. These fields are then matched along the transverse plane at the boundaries of different regions to yield a set of equations, from which parameters such as the propagation constants and characteristic impedances are determined.

8.1.1 Electric and Magnetic Field Expressions

The hybrid fields in each of the subregions i, $i = 1, 2, \ldots, 5$, can be expressed in terms of the scalar electric, ψ^e , and magnetic, ψ^h, potentials using Eqs. (2.51a), (2.51b), (2.53a), and (2.53c) given in Chapter 2:

$$E_z = k_c^2 \psi^e e^{-\gamma z} \tag{8.1}$$

$$\mathbf{E}_t = -\gamma e^{-\gamma z} \nabla_t \psi^e + j\omega\mu e^{-\gamma z} \mathbf{a}_z \times \nabla_t \psi^h \tag{8.2}$$

$$H_z = k_c^2 \psi^h e^{-\gamma z} \tag{8.3}$$

$$\mathbf{H}_t = -\gamma e^{-\gamma z} \nabla_t \psi^h - j\omega\varepsilon e^{-\gamma z} \mathbf{a}_z \times \nabla_t \psi^e \tag{8.4}$$

where k_c is the cutoff wave number and γ is the propagation constant. We assume that the wave propagates in the $+z$ direction. The subscript t denotes the transverse (x and y) components. Note that ψ^e and ψ^h equal zero for TE and TM modes, respectively. ψ^e and ψ^h are solutions of the wave equations (2.46d) and (2.46b),

$$\nabla_t^2 \psi^e(x, y) + k_c^2 \psi^e(x, y) = 0 \tag{8.5a}$$

$$\nabla_t^2 \psi^h(x, y) + k_c^2 \psi^h(x, y) = 0 \tag{8.5b}$$

and can be expanded in a series of eigenfunctions with respect to the y direction as follows:

$$\psi_1^e(x, y) = \sum_{n=1}^{N} A_n \sin(\beta_n x) \exp[-\alpha_{1n}(y - t)] \tag{8.6}$$

$$\psi_1^h(x, y) = \sum_{n=1}^{N} B_n \cos(\beta_n x) \exp[-\alpha_{1n}(y - t)] \tag{8.7}$$

$$\psi_2^e(x, y) = \sum_{n=1}^{P} \sin(\beta_{pn} x)[C_{1n} \sin(\alpha_{2n} y) + C_{2n} \cos(\alpha_{2n} y)] \tag{8.8}$$

$$\psi_2^h(x, y) = \sum_{n=0}^{P} \cos(\beta_{pn} x)[D_{1n} \sin(\alpha_{2n} y) + D_{2n} \cos(\alpha_{2n} y)] \tag{8.9}$$

$$\psi_3^e(x, y) = \sum_{n=1}^{Q} \sin[\beta_{qn}(x - a)][P_{1n} \sin(\alpha_{3n} y) + P_{2n} \cos(\alpha_{3n} y)] \tag{8.10}$$

$$\psi_3^h(x, y) = \sum_{n=1}^{Q} \cos[\beta_{qn}(x - a)][Q_{1n} \sin(\alpha_{3n} y) + Q_{2n} \cos(\alpha_{3n} y)] \tag{8.11}$$

$$\psi_4^e(x, y) = \sum_{n=1}^{N} \sin(\beta_n x)[R_{1n} \sin(\alpha_{4n} y) + R_{2n} \cos(\alpha_{4n} y)] \tag{8.12}$$

$$\psi_4^h(x, y) = \sum_{n=1}^{N} \cos(\beta_n x)[S_{1n} \sin(\alpha_{4n} y) + S_{2n} \cos(\alpha_{4n} y)] \tag{8.13}$$

$$\psi_5^e(x, y) = \sum_{n=1}^{N} K_n \sin(\beta_n x) \exp[\alpha_{1n}(y + h)] \tag{8.14}$$

$$\psi_5^h(x, y) = \sum_{n=1}^{N} L_n \sin(\beta_n x) \exp[\alpha_{1n}(y + h)] \tag{8.15}$$

where

$$\beta_n = \frac{\left(n - \frac{1}{2}\right)\pi}{W}, \qquad n = 1, 2, 3, \ldots \tag{8.16}$$

$$\beta_{pn} = \frac{n\pi}{b}, \qquad n = 0, 1, 2, \ldots \tag{8.17}$$

$$\beta_{qn} = \frac{\left(n - \frac{1}{2}\right)\pi}{W - a}, \qquad n = 1, 2, 3, \ldots \tag{8.18}$$

$$\alpha_{1n}^2 = \beta_n^2 - \gamma^2 - \omega^2 \varepsilon_0 \mu_0 \tag{8.19}$$

$$\alpha_{2n}^2 = \omega^2 \varepsilon_0 \mu_0 + \gamma^2 - \beta_{pn}^2 \tag{8.20}$$

$$\alpha_{3n}^2 = \omega^2 \varepsilon_0 \mu_0 + \gamma^2 - \beta_{qn}^2 \tag{8.21}$$

$$\alpha_{4n}^2 = \omega^2 \varepsilon_0 \varepsilon_r \mu_0 + \gamma^2 - \beta_n^2 \tag{8.22}$$

$A_n, B_n, \ldots, L_n$ are the unknown Fourier coefficients; P, Q, and N represent the numbers of Fourier terms in regions 2, 3, and all the other regions, respectively. The scalar potentials, expressed in Eqs. (8.6)–(8.15), already satisfy the boundary conditions at the magnetic wall at $x = W$ and the electric walls at $x = 0$, b, and a.

Substituting Eqs. (8.6)–(8.15) into (8.1)–(8.4) then yields the components of the electric and magnetic fields in all the regions as

$$E_{x1} = \sum_{n=1}^{N} \left(-\gamma B_n - \frac{A_n \alpha_{1n} \beta_n}{j\omega\varepsilon_0} \right) \cos(\beta_n x) \exp[-\alpha_{1n}(y - t)] \tag{8.23}$$

$$E_{x2} = \left(\sum_{n=0}^{P} -\gamma [D_{1n} \sin(\alpha_{2n} y) + D_{2n} \cos(\alpha_{2n} y)] + \sum_{n=1}^{P} \frac{\alpha_{2n} \beta_{pn}}{j\omega\varepsilon_0} [C_{1n} \cos(\alpha_{2n} y) - C_{2n} \sin(\alpha_{2n} y)] \right) \cos(\beta_{pn} x) \tag{8.24}$$

$$E_{x3} = \sum_{n=1}^{Q} \left(-\gamma[Q_{1n} \sin(\alpha_{3n} y) + Q_{2n} \cos(\alpha_{3n} y)] + \frac{\alpha_{3n}\beta_{qn}}{j\omega\varepsilon_0}[P_{1n} \cos(\alpha_{3n} y) - P_{2n} \sin(\alpha_{3n} y)] \right) \cos(\beta_{qn}(x-a)) \quad (8.25)$$

$$E_{x4} = \sum_{n=1}^{N} \left(-\gamma[S_{1n} \sin(\alpha_{4n} y) + S_{2n} \cos(\alpha_{4n} y)] + \frac{\alpha_{4n}\beta_n}{j\omega\varepsilon_0\varepsilon_r}[R_{1n} \cos(\alpha_{4n} y) - R_{2n} \sin(\alpha_{4n} y)] \right) \cos(P_n x) \quad (8.26)$$

$$E_{x5} = \sum_{n=1}^{N} \left(-\gamma L_n + \frac{K_n \alpha_{1n}\beta_n}{j\omega\varepsilon_0} \right) \cos(\beta_n x) \exp[\alpha_{1n}(y+h)] \quad (8.27)$$

$$E_{y1} = \sum_{n=1}^{N} \left(j\omega\mu_0 + \frac{\alpha_{1n}^2}{j\omega\varepsilon_0} \right) A_n \sin(\beta_n x) \exp[-\alpha_{1n}(y-t)] \quad (8.28)$$

$$E_{y2} = \sum_{n=1}^{P} \left[\left(j\omega\mu_0 - \frac{\alpha_{2n}^2}{j\omega\varepsilon_0} \right) [C_{1n} \sin(\alpha_{2n} y) + C_{2n} \cos(\alpha_{2n} y)] \right] \sin(\beta_{pn} x) \quad (8.29)$$

$$E_{y3} = \sum_{n=1}^{Q} \left[\left(-j\omega\mu_0 - \frac{\alpha_{3n}^2}{j\omega\varepsilon_0} \right) [P_{1n} \sin(\alpha_{3n} y) + P_{2n} \cos(\alpha_{3n} y)] \right] \sin(\beta_{qn} x) \quad (8.30)$$

$$E_{y4} = \sum_{n=1}^{N} \left[\left(-j\omega\mu_0 - \frac{\alpha_{4n}^2}{j\omega\varepsilon_0\varepsilon_r} \right) [R_{1n} \sin(\alpha_{4n} y) + R_{2n} \cos(\alpha_{4n} y)] \right] \sin(\beta_n x) \quad (8.31)$$

$$E_{y5} = \sum_{n=1}^{N} \left(-j\omega\mu_0 K_n + \frac{\alpha_{1n}^2}{j\omega\varepsilon_0} K_n \right) \sin(\beta_n x) \exp[\alpha_{1n}(y+h)] \quad (8.32)$$

$$E_{z1} = \sum_{n=1}^{N} \left(B_n\beta_n + \frac{\alpha_{1n}\gamma A_n}{j\omega\varepsilon_0} \right) \sin(\beta_n x) \exp[-\alpha_{1n}(y-t)] \quad (8.33)$$

$$E_{z2} = \sum_{n=1}^{P} \left(\beta_{pn}[D_{1n} \sin(\alpha_{2n} y) + D_{2n} \cos(\alpha_{2n} y)] - \frac{\gamma\alpha_{2n}}{j\omega\varepsilon_0}[C_{1n} \cos(\alpha_{2n} y) - C_{2n} \sin(\alpha_{2n} y)] \right) \sin(\beta_{pn}x) \tag{8.34}$$

$$E_{z3} = \sum_{n=1}^{Q} \left(\beta_{pn}[Q_{1n} \sin(\alpha_{3n} y) + Q_{2n} \cos(\alpha_{3n} y)] - \frac{\gamma\alpha_{3n}}{j\omega\varepsilon_0}[P_{1n} \cos(\alpha_{3n} y) + P_{2n} \sin(\alpha_{3n} y)] \right) \sin[\beta_{qn}(x-a)] \tag{8.35}$$

$$E_{z4} = \sum_{n=1}^{N} \left(\beta_{n}[S_{1n} \sin(\alpha_{4n} y) + S_{2n} \cos(\alpha_{4n} y)] - \frac{\gamma\alpha_{4n}}{j\omega\varepsilon_0\varepsilon_r}[R_{1n} \cos(\alpha_{4n} y) - R_{2n} \sin(\alpha_{4n} y)] \right) \sin(P_n x) \tag{8.36}$$

$$E_{z5} = \sum_{n=1}^{N} \left(L_n\beta_n - \frac{\gamma\alpha_{1n}K_n}{j\omega\varepsilon_0} \right) \sin(\beta_n x) \exp[\alpha_{1n}(y+h)] \tag{8.37}$$

$$H_{x1} = \sum_{n=1}^{N} \left(\gamma A_n + \frac{\beta_n B_n \alpha_{1n}}{j\omega\mu_0} \right) \sin(\beta_n x) \exp[-\alpha_{1n}(y-t)] \tag{8.38}$$

$$H_{x2} = \sum_{n=1}^{P} \left(\gamma[C_{1n} \sin(\alpha_{2n} y) + C_{2n} \cos(\alpha_{2n} y)] - \frac{\alpha_{2n}\beta_{pn}}{j\omega\mu_0}[D_{1n} \cos(\alpha_{2n} y) - D_{2n} \sin(\alpha_{2n} y)] \right) \sin(\beta_{pn}x) \tag{8.39}$$

$$H_{x3} = \sum_{n=1}^{Q} \left(\gamma[P_{1n} \sin(\alpha_{3n} y) + P_{2n} \cos(\alpha_{3n} y)] - \frac{\alpha_{3n}\beta_{qn}}{j\omega\mu_0}[Q_{1n} \cos(\alpha_{3n} y) - Q_{2n} \sin(\alpha_{3n} y)] \right) \sin[\beta_{qn}(x-a)] \tag{8.40}$$

$$H_{x4} = \sum_{n=1}^{N} \left(\gamma[R_{1n} \sin(\alpha_{4n} y) + R_{2n} \cos(\alpha_{4n} y)] - \frac{\beta_n\alpha_{4n}}{j\omega\mu_0}[S_{1n} \cos(\alpha_{4n} y) - S_{2n} \sin(\alpha_{4n} y)] \right) \sin(\beta_n x) \tag{8.41}$$

$$H_{x5} = \sum_{n=1}^{N} \left(\gamma K_n - \frac{\beta_n L_n \alpha_{1n}}{j\omega\mu_0} \right) \sin(\beta_n x) \exp[\alpha_{1n}(y+h)] \tag{8.42}$$

$$H_{y1} = \sum_{n=0}^{N} \left(-j\omega\varepsilon_0 B_n + \frac{\alpha_{1n}^2 B_n}{j\omega\mu_0} \right) \cos(\beta_n x) \exp[-\alpha_{1n}(y-t)] \tag{8.43}$$

$$H_{y2} = \sum_{n=0}^{P} \left[\left(-j\omega\varepsilon_0 - \frac{\alpha_{2n}^2}{j\omega\varepsilon_0} \right) [D_{1n} \sin(\alpha_{2n} y) + D_{2n} \cos(\alpha_{2n} y)] \right] \cos(\beta_{pn} x) \tag{8.44}$$

$$H_{y3} = \sum_{n=0}^{Q} \left[\left(-j\omega\varepsilon_0 - \frac{\alpha_{3n}^2}{j\omega\varepsilon_0} \right) [Q_{1n} \sin(\alpha_{3n} y) + Q_{2n} \cos(\alpha_{3n} y)] \right] \cos[\beta_{qn}(x-a)] \tag{8.45}$$

$$H_{y4} = \sum_{n=0}^{N} \left(-j\omega\varepsilon_0\varepsilon_r - \frac{\alpha_{4n}^2}{j\omega\mu_0} \right) [S_{1n} \sin(\alpha_{4n} y) + S_{2n} \cos(\alpha_{4n} y)] \tag{8.46}$$

$$H_{y5} = \sum_{n=0}^{N} \left(-j\omega\varepsilon_0 L_n + \frac{\alpha_{1n}^2 K_n}{j\omega\mu_0} \right) \cos(\beta_n x) \exp[\alpha_{1n}(y+h)] \tag{8.47}$$

$$H_{z1} = \sum_{n=1}^{N} \left(\beta_n A_n + \frac{\gamma B_n \alpha_{1n}}{j\omega\mu_0} \right) \cos(\beta_n x) \exp[-\alpha_{1n}(y-t)] \tag{8.48}$$

$$H_{z2} = \left(\sum_{n=1}^{P} \beta_{pn} [C_{1n} \sin(\alpha_{2n} y) + C_{2n} \cos(\alpha_{2n} y)] - \sum_{n=0}^{P} \frac{\gamma\alpha_{2n}}{j\omega\mu_0} [D_{1n} \cos(\alpha_{2n} y) - D_{2n} \sin(\alpha_{2n} y)] \right) \cos(\beta_{pn} x) \tag{8.49}$$

$$H_{z3} = \sum_{n=1}^{Q} \left(\beta_{qn} [P_{1n} \sin(\alpha_{3n} y) + P_{2n} \cos(\alpha_{3n} y)] - \frac{\gamma\alpha_{3n}}{j\omega\mu_0} [Q_{1n} \cos(\alpha_{3n} y) - Q_{2n} \cos(\alpha_{3n} y)] \right) \cos[\beta_{qn}(x-a)] \tag{8.50}$$

$$H_{z4} = \sum_{n=1}^{N} \left(\beta_n [R_{1n} \sin(\alpha_{4n} y) + R_{2n} \cos(\alpha_{4n} y)] - \frac{\gamma\alpha_{4n}}{j\omega\mu_0} [S_{1n} \cos(\alpha_{4n} y) - S_{2n} \sin(\alpha_{4n} y)] \right) \cos(\beta_n x) \tag{8.51}$$

$$H_{z5} = \sum_{n=1}^{N} \left(K_n \beta_n - \frac{\alpha_{1n} \gamma L_n}{j\omega\mu_0} \right) \cos(\beta_n x) \exp[\alpha_{1n}(y+h)] \tag{8.52}$$

8.1.2 Mode-Matching Equations

We now impose the following remaining boundary conditions along each of the interfaces at $y = t$, 0, and $-h$:

At $y = t$,

$$E_{x1}(x, t) = E_{x2}(x, t), \quad x \in [0, b] \tag{8.53}$$

$$E_{z1}(x, t) = E_{z2}(x, t), \quad x \in [0, b] \tag{8.54}$$

$$E_{x1}(x, t) = E_{x3}(x, t), \quad x \in [a, W] \tag{8.55}$$

At $y = 0$,

$$E_{x4}(x, 0) = E_{x2}(x, t), \quad x \in [0, b] \tag{8.56}$$

$$E_{z4}(x, 0) = E_{z2}(x, t), \quad x \in [0, b] \tag{8.57}$$

$$E_{x4}(x, 0) = E_{x3}(x, t), \quad x \in [a, W] \tag{8.58}$$

$$E_{z4}(x, 0) = E_{z3}(x, t), \quad x \in [a, W] \tag{8.59}$$

At $y = -h$,

$$E_{x4}(x, -h) = E_{x5}(x, -h), \quad x \in [0, W] \tag{8.60}$$

$$E_{z4}(x, -h) = E_{z5}(x, -h), \quad x \in [0, W] \tag{8.61}$$

we utilize the orthogonality properties of the eigenfunctions to obtain the following system of homogeneous equations:

$$\sum_{n=1}^{P} (H_{1mn} C_{1n} + H_{2mn} C_{2n}) + \sum_{n=0}^{P} (H_{3mn} D_{1n} + H_{4mn} D_{2n})$$
$$+ \sum_{n=1}^{Q} (H_{5mn} P_{1n} + H_{6mn} P_{2n}) + \sum_{n=1}^{Q} (H_{7mn} Q_{1n} + H_{8mn} Q_{2n}) = 0,$$
$$m = 1, 2, \ldots, P \tag{8.62a}$$

$$\sum_{n=1}^{P} (I_{1mn} C_{1n} + I_{2mn} C_{2n}) + \sum_{n=0}^{P} (I_{3mn} D_{1n} + I_{4mn} D_{2n})$$
$$+ \sum_{n=1}^{Q} (I_{5mn} P_{1n} + I_{6mn} P_{2n}) + \sum_{n=1}^{Q} (I_{7mn} Q_{1n} + I_{8mn} Q_{2n}) = 0,$$
$$m = 0, 2, \ldots, P \tag{8.62b}$$

$$\sum_{n=1}^{P}(J_{1mn}C_{1n}+J_{2mn}C_{2n})+\sum_{n=0}^{P}(J_{3mn}D_{1n}+J_{4mn}D_{2n})$$
$$+\sum_{n=1}^{Q}(J_{5mn}P_{1n}+J_{6mn}P_{2n})+\sum_{n=1}^{Q}(J_{7mn}Q_{1n}+J_{8mn}Q_{2n})=0,$$
$$m=1,2,\ldots,Q \tag{8.62c}$$

$$\sum_{n=1}^{P}(K_{1mn}C_{1n}+K_{2mn}C_{2n})+\sum_{n=0}^{P}(K_{3mn}D_{1n}+K_{4mn}D_{2n})$$
$$+\sum_{n=1}^{Q}(K_{5mn}P_{1n}+K_{6mn}P_{2n})+\sum_{n=1}^{Q}(K_{7mn}Q_{1n}+K_{8mn}Q_{2n})=0,$$
$$m=1,2,\ldots,Q \tag{8.62d}$$

$$\sum_{n=1}^{P}(L_{1mn}C_{1n}+L_{2mn}C_{2n})+\sum_{n=0}^{P}(L_{3mn}D_{1n}+L_{4mn}D_{2n})$$
$$+\sum_{n=1}^{Q}(L_{5mn}P_{1n}+L_{6mn}P_{2n})+\sum_{n=1}^{Q}(L_{7mn}Q_{1n}+L_{8mn}Q_{2n})=0,$$
$$m=1,2,\ldots,P \tag{8.62e}$$

$$\sum_{n=1}^{P}(M_{1mn}C_{1n}+M_{2mn}C_{2n})+\sum_{n=0}^{P}(M_{3mn}D_{1n}+M_{4mn}D_{2n})$$
$$+\sum_{n=1}^{Q}(M_{5mn}P_{1n}+M_{6mn}P_{2n})+\sum_{n=1}^{Q}(M_{7mn}Q_{1n}+M_{8mn}Q_{2n})=0,$$
$$m=0,1,2,\ldots,P \tag{8.62f}$$

$$\sum_{n=1}^{P}(N_{1mn}C_{1n}+N_{2mn}C_{2n})+\sum_{n=0}^{P}(N_{3mn}D_{1n}+N_{4mn}D_{2n})$$
$$+\sum_{n=1}^{Q}(N_{5mn}P_{1n}+N_{6mn}P_{2n})+\sum_{n=1}^{Q}(N_{7mn}Q_{1n}+N_{8mn}Q_{2n})=0,$$
$$m=1,2,\ldots,Q \tag{8.62g}$$

$$\sum_{n=1}^{P}(T_{1mn}C_{1n}+T_{2mn}C_{2n})+\sum_{n=0}^{P}(T_{3mn}D_{1n}+T_{4mn}D_{2n})$$
$$+\sum_{n=1}^{Q}(T_{5mn}P_{1n}+T_{6mn}P_{2n})+\sum_{n=1}^{Q}(T_{7mn}Q_{1n}+T_{8mn}Q_{2n})=0,$$
$$m=1,2,\ldots,Q \tag{8.62h}$$

where the coefficients H, I, J, K, L, M, N, and T are given in Appendix A at the end of this chapter. These equations are referred to as the *mode-matching equations*. For nontrivial solutions, the determinant of the coefficient matrix of Eq. (8.62) must vanish, from which the propagation constant γ can be solved. The effective dielectric constant is then obtained as

$$\varepsilon_{\text{eff}} = \frac{\beta^2}{\omega^2 \varepsilon_0 \mu_0} \tag{8.63}$$

where β is the phase constant.

The characteristic impedance of a CPS, due to its hybrid nature of propagation, cannot be uniquely defined as discussed in Chapter 4. Here we use the following definition:

$$Z_0 = \frac{|V_o|^2}{2P_{\text{avg}}} \tag{8.64}$$

where V_o represents the voltage between the strips and P_{avg} is the average power flowing through the cross section of the structure. V_o is given by

$$V_o = 2\int_0^b E_x(x, 0)\, dx \tag{8.65}$$

P_{avg} is obtained as

$$P_{\text{avg}} = 2\int_{-\infty}^{\infty}\int_0^W \mathbf{E} \times \mathbf{H}^* \cdot dx\, dy\, \mathbf{a}_z \tag{8.66}$$

and equals the summation of individual powers contributed by all five regions.

Figure 8.3 shows an example of the numerical results obtained from the mode-matching analysis of a CPS. The effective dielectric constant and characteristic impedance are plotted as a function of frequency with the strip thickness as a parameter. An illustration of the numerical convergence of the mode-matching analysis is given in Fig. 8.4, which shows the effective dielectric constant and characteristic impedance of the CPS as a function of the numbers of eigenmodes in regions 2, P, and 3, Q. The abscissa is the total number of modes in the two regions. For each curve, P is kept constant while Q is varied. It is readily seen that the convergence is approached as the number of modes is increased in both regions. All numerical results in Fig. 8.3 were computed by using $P = Q = 10$, which ensures the convergence of the solutions for the dimensions used.

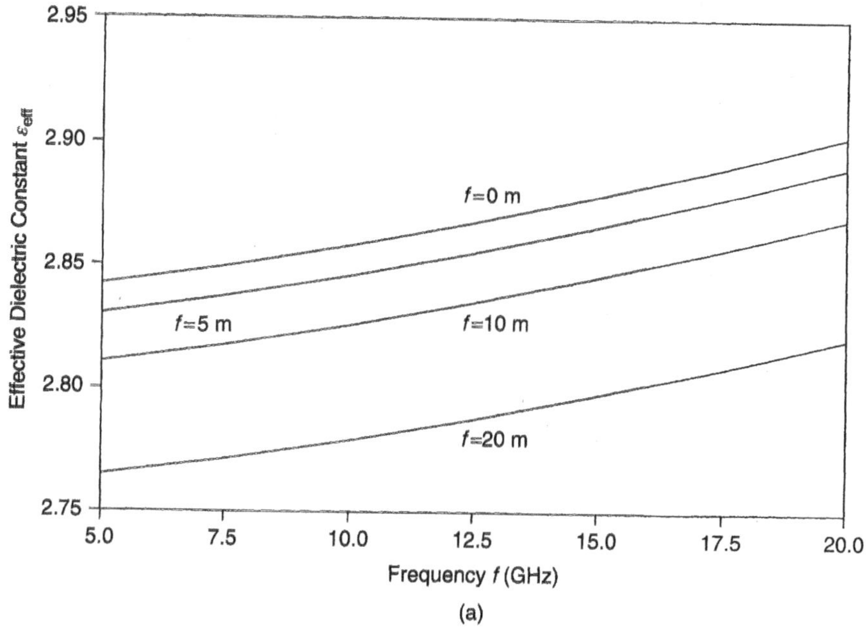

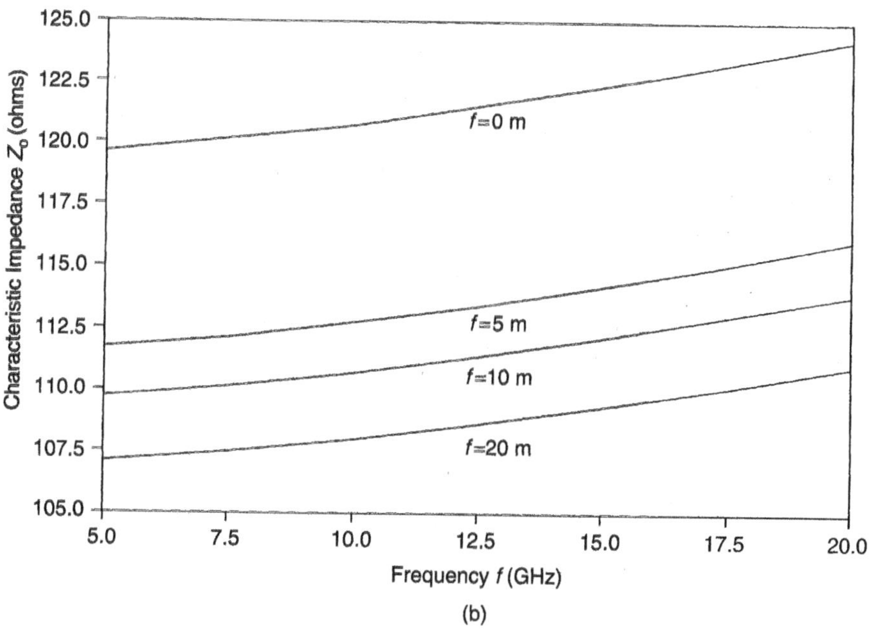

Figure 8.3 (a) Effective dielectric constant and (b) characteristic impedance of a CPS versus frequency for different values of the metallization thickness t. $a = 0.5$ mm, $b = 0.2$ mm, $h = 0.127$ mm, and $\varepsilon_r = 9.6$.

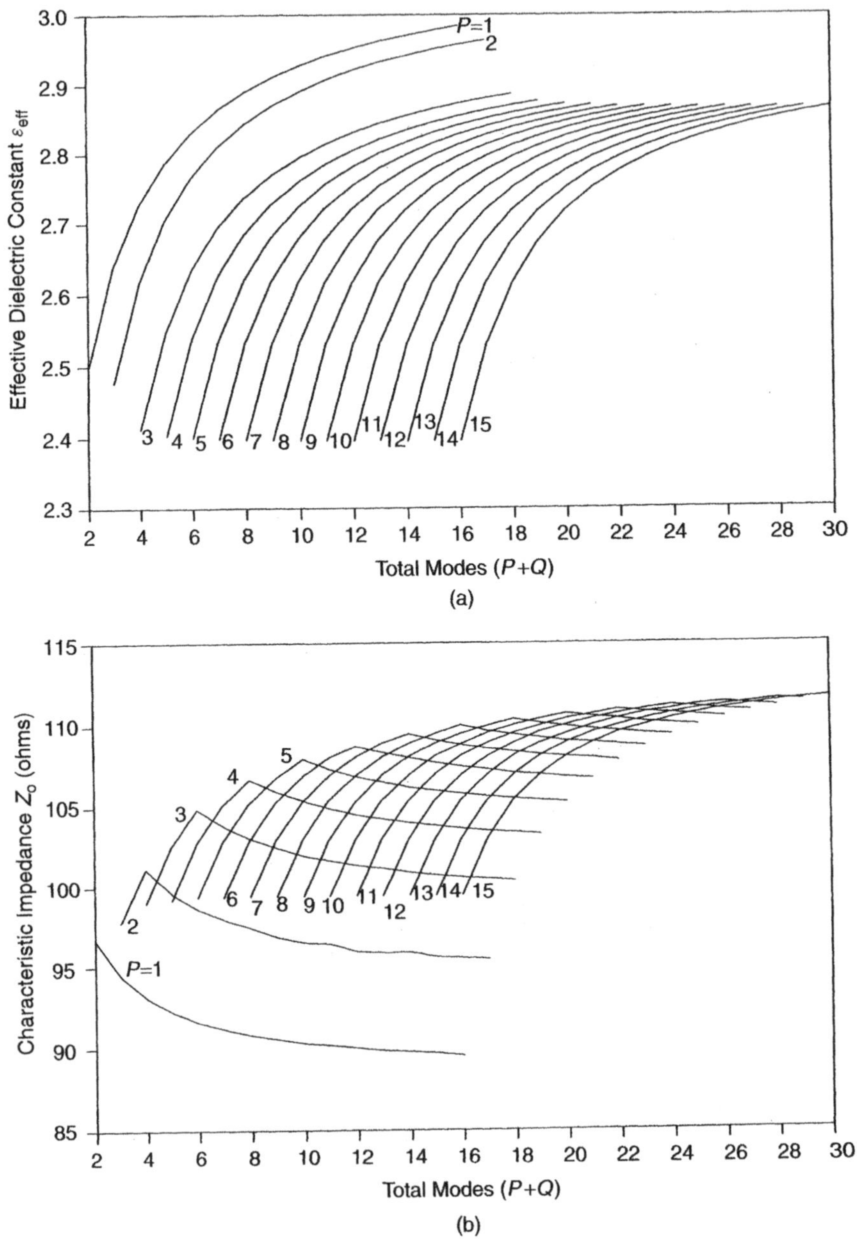

Figure 8.4 Relative convergence of (a) the effective dielectric constant and (b) characteristic impedance as a function of the total number of modes. P = number of modes in region 2; Q = number of modes in region 3. $a = 0.5$ mm, $b = 0.2$ mm, $h = 0.127$ mm, $t = 10$ μm, $\varepsilon_r = 9.6$, and $f = 10$ GHz.

8.2 MODE-MATCHING ANALYSIS OF PLANAR TRANSMISSION LINE DISCONTINUITIES

Figures 8.5 and 8.6 show a cross section of the shielded three-layer CPW and step discontinuities used to demonstrate the mode-matching analysis. The enclosure, ground planes, and central strip are assumed to be perfect conductors, and the dielectric substrates are assumed to be lossless. Also, the thickness of the ground planes and strip is neglected. The discontinuities are classified as single-step [Fig. 8.6(a)], double-step [Fig. 8.6(b)], and multiple-step [Fig. 8.6(c)]. They are general in that the widths of the strip and ground planes can be changed separately or simultaneously. Both symmetrical and asymmetrical changes can also be accommodated. In order to implement the mode matching method, all the hybrid modes, including the propagating and evanescent modes, of the shielded CPW must first be determined.

8.2.1 Electric and Magnetic Field Expressions

The electric and magnetic fields in all regions i, $i = 1, 2, 3$, of the CPW shown in Fig. 8.5 can be expressed in Fourier series, which satisfy the wave equations and the enclosure's boundary conditions at $y = 0, 2a$ and $x = 0, 2b$ as follows.

$$E_x^{(1)} = \sum_{n=1}^{\infty} A_n^{(1)} \cos\left[\alpha_n^{(1)}(x - 2b)\right] \sin(\alpha_n y) \tag{8.67}$$

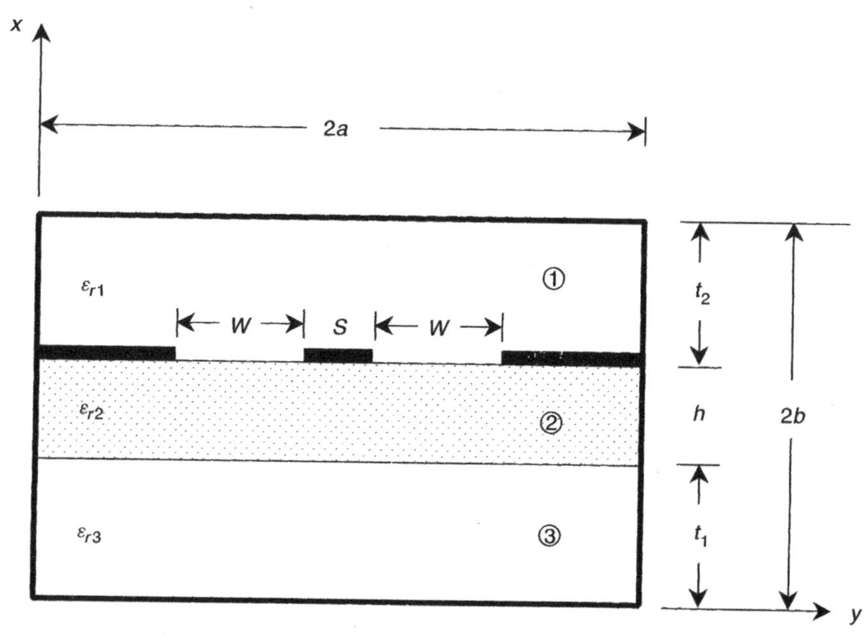

Figure 8.5 Cross section of a shielded three-layer CPW.

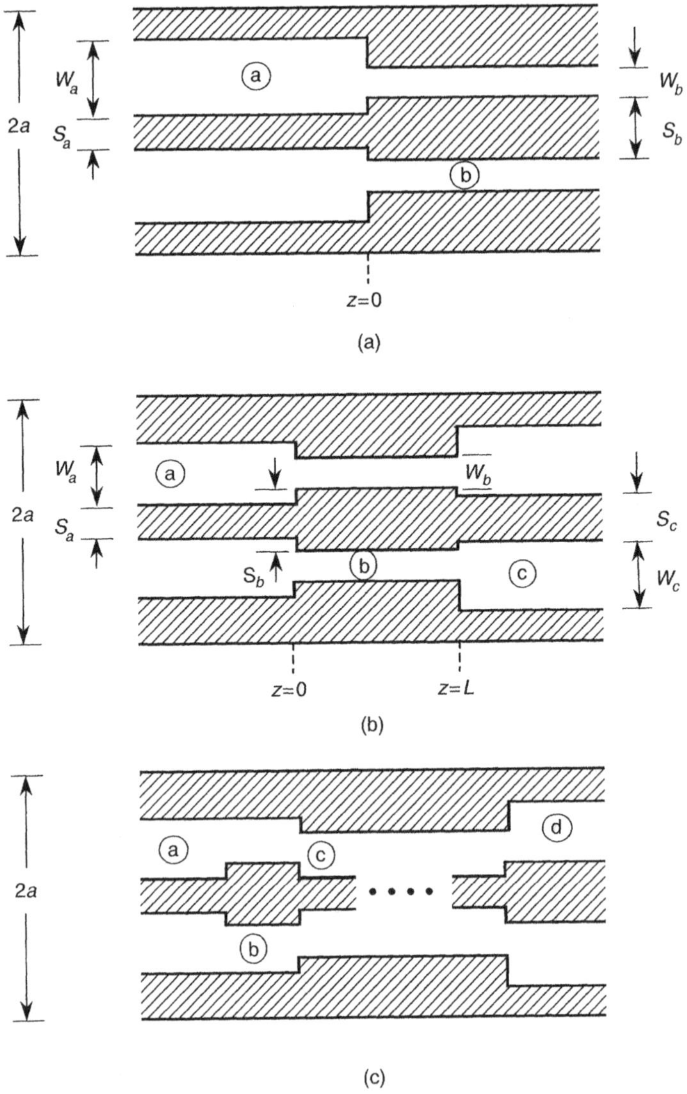

Figure 8.6 (a) Single-step, (b) double-step, and (c) multiple-step CPW discontinuities.

$$E_y^{(1)} = \sum_{n=1}^{\infty} \frac{1}{\alpha_n^2 + \beta^2} \left(-A_n^{(1)}\alpha_n^{(1)}\alpha_n - j\beta k_0 B_n^{(1)}\right) \times \sin\left[\alpha_n^{(1)}(x-2b)\right]\cos(\alpha_n y) \tag{8.68}$$

$$E_z^{(1)} = \sum_{n=1}^{\infty} \frac{1}{\alpha_n^2 + \beta^2} \left(j\beta\alpha_n^{(1)}A_n^{(1)} + k_0\alpha_n B_n^{(1)}\right) \sin\left[\alpha_n^{(1)}(x-2b)\right]\sin(\alpha_n y) \tag{8.69}$$

$$H_x^{(1)} = \sum_{n=1}^{\infty} B_n^{(1)} \sin\left[\alpha_n^{(1)}(x-2b)\right]\cos(\alpha_n y) \tag{8.70}$$

$$H_y^{(1)} = \sum_{n=1}^{\infty} \frac{1}{\alpha_n^2+\beta^2}\left(-j\beta k_0\varepsilon_{r1}A_n^{(1)} - \alpha_n^{(1)}\alpha_n B_n^{(1)}\right) \times \cos\left[\alpha_n^{(1)}(x-2b)\right]\sin(\alpha_n y) \tag{8.71}$$

$$H_z^{(1)} = \sum_{n=1}^{\infty} \frac{1}{\alpha_n^2+\beta^2}\left(-k_0\varepsilon_{r1}\alpha_n A_n^{(1)} - j\beta\alpha_n^{(1)}B_n^{(1)}\right) \times \cos\left[\alpha_n^{(1)}(x-2b)\right]\cos(\alpha_n y) \tag{8.72}$$

$$E_x^{(2)} = \sum_{n=1}^{\infty} A_n^{(2)} \sin[\alpha_n^{(2)}(x-t_1)]\sin(\alpha_n y) + \sum_{n=1}^{\infty} P_n^{(2)} \times \cos[\alpha_n^{(2)}(x-t_1)]\sin(\alpha_n y) \tag{8.73}$$

$$E_y^{(2)} = \sum_{n=1}^{\infty} \frac{1}{\alpha_n^2+\beta^2}\left\{\left(\alpha_n\alpha_n^{(2)}A_n^{(2)} - j\beta k_0 B_n^{(2)}\right)\cos\left[\alpha_n^{(2)}(x-t_1)\right]\cos(\alpha_n y) + \left(-\alpha_n\alpha_n^{(2)}P_n^{(2)} - j\beta k_0 Q_n^{(2)}\right)\sin\left[\alpha_n^{(2)}(x-t_1)\right]\cos(\alpha_n y)\right\} \tag{8.74}$$

$$E_z^{(2)} = \sum_{n=1}^{\infty} \frac{1}{\alpha_n^2+\beta^2}\left\{\left(-j\beta\alpha_n^{(2)}A_n^{(2)} + \alpha_n k_0 B_n^{(2)}\right)\cos\left[\alpha_n^{(2)}(x-t_1)\right]\sin(\alpha_n y) + \left(j\beta\alpha_n^{(2)}P_n^{(2)} + \alpha_n k_0 Q_n^{(2)}\right)\sin\left[\alpha_n^{(2)}(x-t_1)\right]\sin(\alpha_n y)\right\} \tag{8.75}$$

$$H_x^{(2)} = \sum_{n=1}^{\infty} B_n^{(2)} \cos\left[\alpha_n^{(2)}(x-t_1)\right]\cos(\alpha_n y) + \sum_{n=1}^{\infty} Q_n^{(2)} \times \sin\left[\alpha_n^{(2)}(x-t_1)\right]\cos(\alpha_n y) \tag{8.76}$$

$$H_y^{(2)} = \sum_{n=1}^{\infty} \frac{1}{\alpha_n^2+\beta^2}\left\{\left(-j\beta k_0\varepsilon_{r2}A_n^{(2)} + \alpha_n^{(2)}\alpha_n B_n^{(2)}\right)\sin\left[\alpha_n^{(2)}(x-t_1)\right]\sin(\alpha_n y) + \left(-j\beta k_0\varepsilon_{r2}P_n^{(2)} - \alpha_n^{(2)}\alpha_n Q_n^{(2)}\right)\cos\left[\alpha_n^{(2)}(x-t_1)\right]\cos(\alpha_n y)\right\} \tag{8.77}$$

$$H_z^{(2)} = \sum_{n=1}^{\infty} \frac{1}{\alpha_n^2+\beta^2}\left\{\left(-k_0\varepsilon_{r2}\alpha_n A_n^{(2)} + j\beta\alpha_n^{(2)}B_n^{(2)}\right)\sin\left[\alpha_n^{(2)}(x-t_1)\right]\cos(\alpha_n y) + \left(-k_0\varepsilon_{r2}\alpha_n P_n^{(2)} + j\beta\alpha_n^{(2)}Q_n^{(2)}\right)\cos\left[\alpha_n^{(2)}(x-t_1)\right]\cos(\alpha_n y)\right\} \tag{8.78}$$

$$E_x^{(3)} = \sum_{n=1}^{\infty} A_n^{(3)} \cos\left(\alpha_n^{(3)}x\right)\sin(\alpha_n y) \tag{8.79}$$

$$E_y^{(3)} = \sum_{n=1}^{\infty} \frac{1}{\alpha_n^2+\beta^2}\left(-\alpha_n^{(3)}\alpha_n A_n^{(3)} - j\beta k_0 B_n^{(3)}\right)\sin\left(\alpha_n^{(3)}x\right)\cos(\alpha_n y) \tag{8.80}$$

$$E_z^{(3)} = \sum_{n=1}^{\infty} \frac{1}{\alpha_n^2 + \beta^2} \left(j\beta\alpha_n^{(3)} A_n^{(3)} + \alpha_n k_0 B_n^{(3)} \right) \sin\left(\alpha_n^{(3)} x\right) \sin(\alpha_n y) \tag{8.81}$$

$$H_x^{(3)} = \sum_{n=1}^{\infty} B_n^{(3)} \sin\left(\alpha_n^{(3)} x\right) \cos(\alpha_n y) \tag{8.82}$$

$$H_y^{(3)} = \sum_{n=1}^{\infty} \frac{1}{\alpha_n^2 + \beta^2} \left(-j\beta k_0 \varepsilon_{r3} A_n^{(3)} - \alpha_n^{(3)} \alpha_n B_n^{(3)} \right) \cos\left(\alpha_n^{(3)} x\right) \sin(\alpha_n y) \tag{8.83}$$

$$H_z^{(3)} = \sum_{n=1}^{\infty} \frac{1}{\alpha_n^2 + \beta^2} \left(-k_0 \varepsilon_{r3} \alpha_n A_n^{(3)} - j\beta\alpha_n^{(3)} B_n^{(3)} \right) \cos\left(\alpha_n^{(3)} x\right) \cos(\alpha_n y) \tag{8.84}$$

where

$$\alpha_n^{(1)} = \sqrt{\varepsilon_{r1} k_0^2 - \beta^2 - \alpha_n^2} \tag{8.85}$$

$$\alpha_n^{(2)} = \sqrt{\varepsilon_{r2} k_0^2 - \beta^2 - \alpha_n^2} \tag{8.86}$$

$$\alpha_n^{(3)} = \sqrt{\varepsilon_{r3} k_0^2 - \beta^2 - \alpha_n^2} \tag{8.87}$$

$$\alpha_n = \frac{n\pi}{2a} \tag{8.88}$$

k_0 is the free-space wave number, and β represents the propagation constants of all the eigenmodes. The superscript (1, 2, or 3) indicates the corresponding region. The expansion coefficients, the A's, B's, P's and Q's, can be determined by matching the tangential electric and magnetic fields across the interface at $x = t_1$ and the tangential electric fields at $x = t_1 + h$ as

$$A_n^{(2)} = \frac{j\beta \tilde{E}_{zn}^{(2)}(\alpha_n) + \alpha_n \tilde{E}_{yn}^{(2)}(\alpha_n)}{a\alpha_n^{(2)} \left(\cos(\alpha_n^{(2)} h) + \dfrac{\alpha_n^{(2)} \varepsilon_{r3} \sin\left(\alpha_n^{(2)} h\right)}{\alpha_n^{(3)} \varepsilon_{r2} \tan\left(\alpha_n^{(3)} t_1\right)} \right)} \tag{8.89}$$

$$B_n^{(2)} = \frac{\alpha_n \tilde{E}_{zn}^{(2)}(\alpha_n) + j\beta \tilde{E}_{yn}^{(2)}(\alpha_n)}{k_0 a \left(\cos(\alpha_n^{(2)} h) + \dfrac{\alpha_n^{(3)} \sin\left(\alpha_n^{(2)} h\right)}{\alpha_n^{(2)} \tan\left(\alpha_n^{(3)} t_1\right)} \right)} \tag{8.90}$$

$$A_n^{(1)} = \frac{\alpha_n^{(2)}}{\alpha_n^{(1)} \sin\left(\alpha_n^{(1)} t_2\right)} \left(\cos\left(\alpha_n^{(2)} h\right) + \frac{\alpha_n^{(2)} \varepsilon_{r3} \sin\left(\alpha_n^{(2)} h\right)}{\alpha_n^{(3)} \varepsilon_{r2} \tan\left(\alpha_n^{(3)} t_1\right)} \right) A_n^{(2)} \tag{8.91}$$

$$B_n^{(1)} = -\frac{1}{\sin\left(\alpha_n^{(1)} t_2\right)} \left(\cos\left(\alpha_n^{(2)} h\right) + \frac{\alpha_n^{(3)} \sin\left(\alpha_n^{(2)} h\right)}{\alpha_n^{(2)} \tan\left(\alpha_n^{(3)} t_1\right)} \right) B_n^{(2)} \tag{8.92}$$

$$A_n^{(3)} = \frac{\alpha_n^{(2)}}{\alpha_n^{(3)} \sin\left(\alpha_n^{(3)} t_1\right)} A_n^{(2)} \tag{8.93}$$

$$B_n^{(3)} = \frac{B_n^{(2)}}{\sin\left(\alpha_n^{(3)} t_1\right)} \tag{8.94}$$

$$P_n^{(2)} = -\frac{\alpha_n^{(2)} \varepsilon_{r3}}{\alpha_n^{(3)} \varepsilon_{r2} \tan\left(\alpha_n^{(3)} t_1\right)} A_n^{(2)} \tag{8.95}$$

$$Q_n^{(2)} = \frac{\alpha_n^{(3)}}{\alpha_n^{(2)} \tan\left(\alpha_n^{(3)} t_1\right)} B_n^{(2)} \tag{8.96}$$

where $\tilde{E}_{yn}^{(2)}$ and $\tilde{E}_{zn}^{(2)}$ are the Fourier transforms of the unknown y and z components of the slot field at $x = t_1 + h$, obtained using the spectral-domain technique described in Chapter 7. The electric and magnetic fields given in Eqs. (8.67)–(8.84) will be used in the following analysis of the single-, double-, and multiple-step discontinuities using the mode-matching method.

8.2.2 Single-Step Discontinuity

The single-step discontinuity, as shown in Fig. 8.6(a), is divided into two regions, a and b. Region b is assumed to extend to infinity. The hybrid modes on the left-hand side (region a) and the right-hand side (region b) of the plane of the discontinuity ($z = 0$) are first determined. For the mode-matching analysis, it is sufficient to impose only the continuity condition of the normal transverse fields at the discontinuity. The boundary conditions for the longitudinal fields will automatically be satisfied through Maxwell's equations. The transverse electric and magnetic fields for a mode i in region a can be expressed for each of the three regions, 1, 2, and 3, shown in Fig. 8.5 as follows:

Region 1:

$$\mathbf{E}_{ai}^{(1)} = \sum_{n=1}^{\infty} A_{nai}^{(1)} \cos\left[\alpha_{nai}^{(1)}(x - 2b)\right] \sin(\alpha_n y)\mathbf{a}_x + \sum_{n=1}^{\infty} M_{nai}^{(1)} \sin\left[\alpha_{nai}^{(1)}(x - 2b)\right] \cos(\alpha_n y)\mathbf{a}_y \tag{8.97}$$

$$\mathbf{H}_{ai}^{(1)} = \sum_{n=1}^{\infty} B_{nai}^{(1)} \sin\left[\alpha_{nai}^{(1)}(x - 2b)\right] \cos(\alpha_n y)\mathbf{a}_x + \sum_{n=1}^{\infty} S_{nai}^{(1)} \cos\left[\alpha_{nai}^{(1)}(x - 2b)\right] \sin(\alpha_n y)\mathbf{a}_y \tag{8.98}$$

where

$$M_{nai}^{(1)} = \frac{1}{\alpha_n^2 + \beta_i^2} \left(-A_{nai}^{(1)} \alpha_{nai}^{(1)} \alpha_n - j\beta_i k_0 B_{nai}^{(1)} \right) \tag{8.99}$$

$$S_{nai}^{(1)} = \frac{1}{\alpha_n^2 + \beta_i^2} \left(-j\beta_i k_0 \varepsilon_{r1} A_{nai}^{(1)} - \alpha_{nai}^{(1)} \alpha_n B_{nai}^{(1)} \right) \tag{8.100}$$

Region 2:

$$\begin{aligned} \mathbf{E}_{ai}^{(2)} &= \sum_{n=1}^{\infty} \left\{ A_{nai}^{(2)} \sin\left[\alpha_{nai}^{(2)}(x - t_1)\right] + P_{nai}^{(2)} \cos\left[\alpha_{nai}^{(2)}(x - t_1)\right] \right\} \sin(\alpha_n y)\mathbf{a}_x \\ &+ \sum_{n=1}^{\infty} \left\{ M_{nai}^{(2)} \sin\left[\alpha_{nai}^{(2)}(x - t_1)\right] + R_{nai}^{(2)} \cos\left[\alpha_{nai}^{(2)}(x - t_1)\right] \right\} \cos(\alpha_n y)\mathbf{a}_y \end{aligned} \tag{8.101}$$

$$\begin{aligned} \mathbf{H}_{ai}^{(2)} &= \sum_{n=1}^{\infty} \left\{ B_{nai}^{(2)} \cos\left[\alpha_{nai}^{(2)}(x - t_1)\right] + Q_{nai}^{(2)} \sin\left[\alpha_{nai}^{(2)}(x - t_1)\right] \right\} \cos(\alpha_n y)\mathbf{a}_x \\ &+ \sum_{n=1}^{\infty} \left\{ S_{nai}^{(2)} \cos\left[\alpha_{nai}^{(2)}(x - t_1)\right] + T_{nai}^{(2)} \sin\left[\alpha_{nai}^{(2)}(x - t_1)\right] \right\} \sin(\alpha_n y) a_y \end{aligned} \tag{8.102}$$

where

$$R_{nai}^{(2)} = \frac{1}{\alpha_n^2 + \beta_i^2} \left(\alpha_n \alpha_{nai}^{(2)} A_{nai}^{(2)} - j\beta_i k_0 B_{nai}^{(2)} \right) \tag{8.103}$$

$$M_{nai}^{(2)} = \frac{1}{\alpha_n^2 + \beta_i^2} \left(-\alpha_n \alpha_{nai}^{(2)} P_{nai}^{(2)} - j\beta_i k_0 Q_{nai}^{(2)} \right) \tag{8.104}$$

$$S_{nai}^{(2)} = \frac{1}{\alpha_n^2 + \beta_i^2} \left(-j\beta_i k_0 \varepsilon_{r2} P_{nai}^{(2)} - \alpha_{nai}^{(2)} \alpha_n Q_{nai}^{(2)} \right) \tag{8.105}$$

$$T_{nai}^{(2)} = \frac{1}{\alpha_n^2 + \beta_i^2} \left(-j\beta_i k_0 \varepsilon_{r2} A_{nai}^{(2)} - \alpha_{nai}^{(2)} \alpha_n B_{nai}^{(2)} \right) \tag{8.106}$$

Region 3:

$$\begin{aligned} \mathbf{E}_{ai}^{(3)} &= \sum_{n=1}^{\infty} A_{nai}^{(3)} \cos\left(\alpha_{nai}^{(3)} x\right) \sin(\alpha_n y)\mathbf{a}_x \\ &+ \sum_{n=1}^{\infty} M_{nai}^{(3)} \sin\left(\alpha_{nai}^{(3)} x\right) \cos(\alpha_n y)\mathbf{a}_y \end{aligned} \tag{8.107}$$

$$\mathbf{H}_{ai}^{(3)} = \sum_{n=1}^{\infty} B_{nai}^{(3)} \sin\left(\alpha_{nai}^{(3)} x\right) \cos(\alpha_n y)\mathbf{a}_x + \sum_{n=1}^{\infty} S_{nai}^{(3)} \cos\left(\alpha_{nai}^{(3)} x\right) \sin(\alpha_n y)\mathbf{a}_y \tag{8.108}$$

where

$$M_{nai}^{(3)} = \frac{1}{\alpha_n^2 + \beta_i^2}\left(-A_{nai}^{(3)}\alpha_{nai}^{(3)}\alpha_n - j\beta_i k_0 B_{nai}^{(3)}\right) \tag{8.109}$$

$$S_{nai}^{(3)} = \frac{1}{\alpha_n^2 + \beta_i^2}\left(-j\beta_i k_0 \varepsilon_{r1} A_{nai}^{(3)} - \alpha_{nai}^{(3)}\alpha_n B_{nai}^{(3)}\right) \tag{8.110}$$

$\alpha_{nai}^{(1)}$, $\alpha_{nai}^{(2)}$, $\alpha_{nai}^{(3)}$, α_n, $A_{nai}^{(1)}$, $A_{nai}^{(2)}$, A_{nai}^{3}, $B_{nai}^{(1)}$, $B_{nai}^{(2)}$, and $B_{nai}^{(3)}$ are given in Eqs. (8.85)–(8.94). Note that the subscripts a and i have been added to signify region a and mode i, respectively. The fields in region b can also be obtained similarly.

The transverse electric, $\mathbf{e}_i(x, y, z)$, and magnetic, $\mathbf{h}_i(x, y, z)$, fields for mode i are written as:

$$\mathbf{e}_i(x, y, z) = a_i \mathbf{e}_i(x, y) e^{\pm\gamma_i z} \tag{8.111}$$

$$\mathbf{h}_i(x, y, z) = a_i \mathbf{h}_i(x, y) e^{\pm\gamma_i z} \tag{8.112}$$

where $\mathbf{e}_i(x, y)$ and $\mathbf{h}_i(x, y)$ are given in Eqs. (8.97)–(8.110) for region a. a_i is the amplitude for mode i. γ_i denotes the propagation constant of mode i and is imaginary and real for the propagating and evanescent modes, respectively.

We now assume that a wave, corresponding to $i = 1$, with a unit amplitude ($a_1 = 1$) is incident upon the junction at $z = 0$ from region a. To satisfy the boundary conditions, this wave will excite all the other modes, including propagating and evanescent modes, at the junction. Although the evanescent modes die down with distance away from the junction, their effects would be significant near the junction and thus should be considered. The propagating and evanescent modes, generated at the discontinuity, are then reflected into region a and transmitted into region b. The total transverse electric and magnetic fields in region a, just to the left of the plane of the discontinuity, can be expressed in terms of the hybrid modes in this region as

$$\mathbf{E}_a = (1 + \rho)\mathbf{e}_1^a + \sum_{i=2}^{\infty} a_i \mathbf{e}_i^a \tag{8.113}$$

$$\mathbf{H}_a = (1 - \rho)\mathbf{h}_1^a - \sum_{i=2}^{\infty} a_i \mathbf{h}_i^a \tag{8.114}$$

where $\mathbf{e}_i^a$ and $\mathbf{h}_i^a$ are the normalized electric and magnetic fields of mode i ($i = 1, 2, 3, \ldots$) in region a, respectively. ρ is the reflection coefficient of the incident mode characterized by $\mathbf{e}_1^a$ and $\mathbf{h}_1^a$. a_i is the complex amplitude of mode i scattered into region a from the incident mode at the discontinuity. Expanding the total transverse fields, just to the right of the discontinuity, in region b in terms of the transmitted modes from region a, we get

$$\mathbf{E}_b = \sum_{j=1}^{\infty} b_j \mathbf{e}_j^b \tag{8.115}$$

$$\mathbf{H}_b = \sum_{j=1}^{\infty} b_j \mathbf{h}_j^b \tag{8.116}$$

where $\mathbf{e}_j^b$ and $\mathbf{h}_j^b$ are the normalized electric and magnetic fields of mode j ($j = 1, 2, 3, \ldots$) in region b, respectively. b_j, $j = 1, 2, 3, \ldots$, is the complex amplitude of mode j transmitted into region b, which is generated by the incident mode at the discontinuity. We assume that region b is continuous in the z direction, so there are no back-scattered modes in this region.

Imposing the continuity condition of the transverse electric and magnetic fields at the discontinuity yields the following mode-matching equations:

$$(1+\rho)\mathbf{e}_1^a + \sum_{i=2}^{\infty} a_i \mathbf{e}_i^a = \sum_{j=1}^{\infty} b_j \mathbf{e}_j^b \tag{8.117}$$

$$(1-\rho)\mathbf{h}_1^a - \sum_{i=2}^{\infty} a_i \mathbf{h}_i^a = \sum_{j=1}^{\infty} b_j \mathbf{h}_j^b \tag{8.118}$$

In order to solve Eqs. (8.117)–(8.118), we introduce an inner product defined by

$$I_{ij}^{\gamma\lambda} = \left\langle \mathbf{e}_i^{\gamma}(x, y), \mathbf{h}_j^{\lambda}(x, y) \right\rangle = \int_S \mathbf{e}_i^{\gamma}(x, y) \times \mathbf{h}_j^{\lambda}(x, y) \cdot d\mathbf{S} \tag{8.119}$$

where $\gamma = a$ or b, $\lambda = a$ or b, and S is the cross section of the considered CPW. Here, the inner product is performed in the spectral domain to keep the computation consistent with the mode solution of the CPW obtained using the spectral-domain approach. The inner product in the spectral domain is

$$\left\langle \tilde{e}_i^{\gamma}, \tilde{h}_j^{\lambda} \right\rangle = \frac{1}{4a} \sum_{n=1}^{\infty} \int_0^{2b} \left[\tilde{e}_{ix}^{\gamma}(x, \alpha_n) \tilde{h}_{jy}^{\lambda}(x, \alpha_n) - \tilde{e}_{iy}^{\gamma}(x, \alpha_n) \tilde{h}_{jx}^{\lambda}(x, \alpha_n) \right] dx \tag{8.120}$$

Using the terminology of Wexler [1], waveguide step discontinuities can be classified into two types: boundary reduction and boundary enlargement. A boundary reduction occurs when the waveguide cross section for the incident wave is larger than that for the transmitted wave. Otherwise, it is referred to as the boundary enlargement. The difference in the formulation for the two problems is in the definition of the inner product, which optimizes the convergence of the solution as a function of the number of modes used. For many waveguide discontinuities, such as those in circular and rectangular waveguides, the boundary enlargement and boundary reduction are well defined. This is, however, not the case for the CPW, as the enclosure has the same dimension on both sides of the discontinuity. Therefore, the definition of boundary enlargement or reduction must be studied carefully for CPW step discontinuities to obtain an optimally convergent solution with a relatively small number of modes. When boundary enlargement or reduction is clearly defined, the most rapid convergence is obtained when the inner product is defined such as when the electric and magnetic fields are taken from the smaller and larger waveguides, respectively, because of the effective enforcement of the discontinuity's boundary conditions through the inner products. This enables the boundary condition at the discontinuity to be satisfied fairly well with a minimum number of modes.

For a CPW step discontinuity having $W_a < W_b$, which can be referred to as the boundary enlargement or step-up problem, we take the inner products of Eqs. (8.117) and (8.118) with $\mathbf{h}_m^b$ and $\mathbf{e}_n^a$, respectively. By truncating the infinite series to N_a and N_b, which signify the numbers of eigenmodes in regions a and b, respectively, we obtain the following system of $(N_a + N_b)$ linear equations:

$$(1+\rho)I_{1m}^{ab} + \sum_{i=2}^{N_a} a_i I_{im}^{ab} - \sum_{j=1}^{N_b} b_j I_{jm}^{bb} = 0, \quad m = 1, \ldots, N_b \quad (8.121a)$$

$$(1-\rho)I_{n1}^{aa} - \sum_{i=2}^{N_a} a_i I_{ni}^{aa} - \sum_{j=1}^{N_b} b_j I_{nj}^{ab} = 0, \quad n = 1, \ldots, N_a \quad ((8.121b)$$

where the inner product I's are given in Appendix B at the end of this chapter. For the step-down junction with $W_b > W_a$ (boundary reduction problem), we have

$$(1+\rho)I_{1m}^{aa} + \sum_{i=2}^{N_a} a_i I_{im}^{aa} - \sum_{j=1}^{N_b} b_j I_{jm}^{ba} = 0, \quad m = 1, \ldots, N_a \quad (8.122)$$

$$(1-\rho)I_{n1}^{ba} - \sum_{i=2}^{N_a} a_i I_{ni}^{ba} - \sum_{j=1}^{N_b} b_j I_{nj}^{bb} = 0, \quad n = 1, \ldots, N_b \quad (8.123)$$

The above equations can immediately be solved to obtain the solutions for the $(N_a + N_b)$ unknowns, ρ, a_i $(i = 2, \ldots, N_a)$, and b_j $(j = 1, 2, \ldots, N_b)$, which characterize the scattering (S) parameters of the discontinuity.

8.2.3 Double-Step Discontinuity

The double-step discontinuity is shown in Fig. 8.6(b). It contains three different regions: $a(z \leq 0)$, $b(0 \leq z \leq L)$, and $c(z \geq L)$. Regions a and c are assumed to be semi-infinitely long. The hybrid modes in regions a, b, and c are first determined, similar to the case of a single-step discontinuity. For instance, the transverse fields in region a are given by Eqs. (8.97)–(8.112). We assume that a wave of unit amplitude ($a_1 = 1$) is incident upon the first junction ($z = 0$) from region a. This incident wave excites all the propagating and evanescent modes at this junction. These generated modes are then reflected and transmitted into regions a and b, respectively. The total transverse electric and magnetic fields, just to the left of the plane of the discontinuity at $z = 0$, in region a can thus be described in terms of the hybrid modes in this region as given in Eqs. (8.113) and (8.114), respectively.

The modes in region b consist of transmitted modes from region a (through the first junction at $z = 0$) and the reflected modes from the second junction at $z = L$ (due to the modes incident upon this junction from region b). The transverse electric field of mode j ($j = 1, 2, 3, \ldots$) transmitted into region b from region a is given as $b_j \mathbf{e}_j^b$, where $\mathbf{e}_j^b$ is the normalized transverse electric field of mode j in region b, and b_j is the complex amplitude of mode j transmitted into region b, which is generated by the incident mode at the first junction. Each transmitted mode j (propagating or evanescent), upon reaching the second junction, excites all the other modes at that junction, which are then reflected and transmitted into regions b and c, respectively. The transverse electric field of a mode k reflected from the second junction is contributed from an infinite number of modes j incident at that junction and can thus be written as

$$r_k \mathbf{e}_k^b = \sum_{j=1}^{\infty} b_j \mathbf{e}_k^b e^{-j\beta_j^b L} \tag{8.124}$$

where r_k is the overall reflection amplitude of mode k at the second junction. The total transverse electric and magnetic fields in region b can now be given as

$$\mathbf{E}_b = \sum_{j=1}^{\infty} b_j \mathbf{e}_j^b e^{-j\beta_1^b z} + \sum_{k=1}^{\infty} r_k \mathbf{e}_k^b e^{-j\beta_k^b (L-z)} \tag{8.125}$$

$$\mathbf{H}_b = \sum_{j=1}^{\infty} b_j \mathbf{h}_j^b e^{-j\beta_1^b z} - \sum_{k=1}^{\infty} r_k \mathbf{h}_k^b e^{-j\beta_k^b (L-z)} \tag{8.126}$$

where $\mathbf{e}_j^b$ and $\mathbf{h}_j^b$ are the normalized transverse electric and magnetic fields of mode j ($j = 1, 2, 3, \ldots$) in region b, respectively. b_j, $j = 1, 2, 3, \ldots$, is the complex amplitude of mode j at the first junction, and r_k is the complex amplitude of mode k reflected from the second junction.

In region c, there are only transmitted modes from region b since region c is assumed to be semi-infinitely long. The total transverse fields in region c can therefore be expressed as

$$\mathbf{E}_c = \sum_{r=1}^{\infty} c_r \mathbf{e}_r^c e^{-j\beta_r^c (z-L)} \tag{8.127}$$

$$\mathbf{H}_c = \sum_{r=1}^{\infty} c_r \mathbf{h}_r^c e^{-j\beta_r^c (z-L)} \tag{8.128}$$

where $\mathbf{e}_r^c$ and $\mathbf{h}_r^c$ are the normalized electric and magnetic fields of mode r ($r = 1, 2, 3, \ldots$) in region c, respectively. c_r is the complex amplitude of mode r transmitted into region c, which is generated by an incident mode at the second junction.

We now impose the boundary conditions at the two junctions. By applying the continuity of the transverse electric and magnetic fields at the $z = 0$ plane, we obtain

$$(1+\rho)\mathbf{e}_1^a + \sum_{i=2}^{\infty} a_i \mathbf{e}_i^a = \sum_{j=1}^{\infty} b_j \mathbf{e}_j^b + \sum_{k=1}^{\infty} r_k \mathbf{e}_k^b e^{-j\beta_k^b L} \tag{8.129}$$

$$(1-\rho)\mathbf{h}_1^a - \sum_{i=2}^{\infty} a_i \mathbf{h}_i^a = \sum_{j=1}^{\infty} b_j \mathbf{h}_j^b - \sum_{k=1}^{\infty} r_k \mathbf{h}_k^b e^{-j\beta_k^b L} \tag{8.130}$$

Similarly, we have at the $z = L$ plane

$$\sum_{j=1}^{\infty} b_j \mathbf{e}_j^b e^{-j\beta_1^b L} + \sum_{k=1}^{\infty} r_k \mathbf{e}_k^b = \sum_{r=1}^{\infty} c_r \mathbf{e}_r^c \tag{8.131}$$

$$\sum_{j=1}^{\infty} b_j \mathbf{h}_j^b e^{-j\beta_1^b L} - \sum_{k=1}^{\infty} r_k \mathbf{h}_k^b = \sum_{r=1}^{\infty} c_r \mathbf{h}_r^c \tag{8.132}$$

Taking the inner products of Eqs. (8.129)–(8.132) with $\mathbf{h}_m^a$, $\mathbf{e}_n^b$, $\mathbf{h}_p^c$, and $\mathbf{e}_q^b$, respectively, in accordance with the inner product defined in Eq. (8.119), with γ and λ now denoting a, b, or c, yields the following system of $(N_a + 2N_b + N_c)$ linear equations:

$$(1+\rho)I_{1m}^{aa} + \sum_{i=2}^{N_a} a_i I_{im}^{aa} - \sum_{j=1}^{N_b} b_j I_{jm}^{ba}$$

$$- \sum_{k=1}^{N_b} r_k I_{km}^{ba} \exp(-j\beta_k^b L) = 0, \quad m = 1, \ldots, N_a \tag{8.133}$$

$$(1-\rho)I_{n1}^{ba} - \sum_{i=2}^{N_a} a_i I_{ni}^{ba} - \sum_{j=1}^{N_b} b_j I_{nj}^{bb}$$

$$+ \sum_{k=1}^{N_b} r_k I_{nk}^{bb} \exp(-j\beta_k^b L) = 0, \quad n = 1, \ldots, N_b \tag{8.134}$$

$$\sum_{j=1}^{N_b} b_j I_{jp}^{bc} \exp(-j\beta_j^b L) + \sum_{k=1}^{N_b} r_k I_{kp}^{bc} - \sum_{r=1}^{N_c} c_p I_{rp}^{cc} = 0,$$

$$p = 1, \ldots, N_c \tag{8.135}$$

$$\sum_{j=1}^{N_b} b_j I_{qj}^{bb} \exp(-j\beta_j^b L) - \sum_{k=1}^{N_b} r_k I_{qk}^{bb} - \sum_{r=1}^{N_c} c_r I_{qr}^{bc} = 0,$$

$$q = 1, \ldots, N_b \tag{8.136}$$

where the inner product I's are given in Appendix B at the end of this chapter. The infinite series are truncated to N_a, N_b, and N_c that approximate the number of eigenmodes in regions a, b, and c, respectively. These equations can now be solved to determine the unknowns, ρ, a_i $(i = 2, 3, \ldots, N_a)$, b_j and r_k $(j, k = 1, 2, \ldots, N_b)$, and c_r $(r = 1, 2, \ldots, N_c)$.

8.2.4 Multiple-Step Discontinuity

Figure 8.6(c) shows the multiple-step discontinuity. Its analysis is very similar to that of the double-step discontinuity. In the multistep problem, the composite generalized S matrix is obtained, taking into account the interactions between the junctions via the fundamental as well as higher-order modes.

As examples of numerical results obtained from the mode-matching analysis of step discontinuities of a shielded three-layer CPW, we show in Figs. 8.7–8.9 the calculated S parameters of single-, double-, and triple-step discontinuities versus frequency, respectively.

A remark needs to be made at this point that the eigenmodes of the considered CPW should satisfy the following modal orthogonality condition:

$$I_{ij}^{SS} = \int_S \mathbf{e}_i^S \times \mathbf{h}_j^S \cdot d\mathbf{S} = \delta_{ij} \tag{8.137}$$

where δ_{ij} is the *Kronecker delta function* and $S = a$, b, c, or d. However, as the CPW eigenmode calculations are only approximate due to the necessary numerical truncations, deviation from the orthogonality criterion may occur. To illustrate this phenomenon, we compare the S parameters of single-step CPW discontinuities, calculated with and without the assumption of the orthogonality condition, in Figs. 8.10 and 8.11. In Fig. 8.10, the mode calculations are very accurate, and hence the results agree very well. However, there is a small variation between the results in Fig. 8.11 due to slightly inaccurate calculations of the eigenmodes.

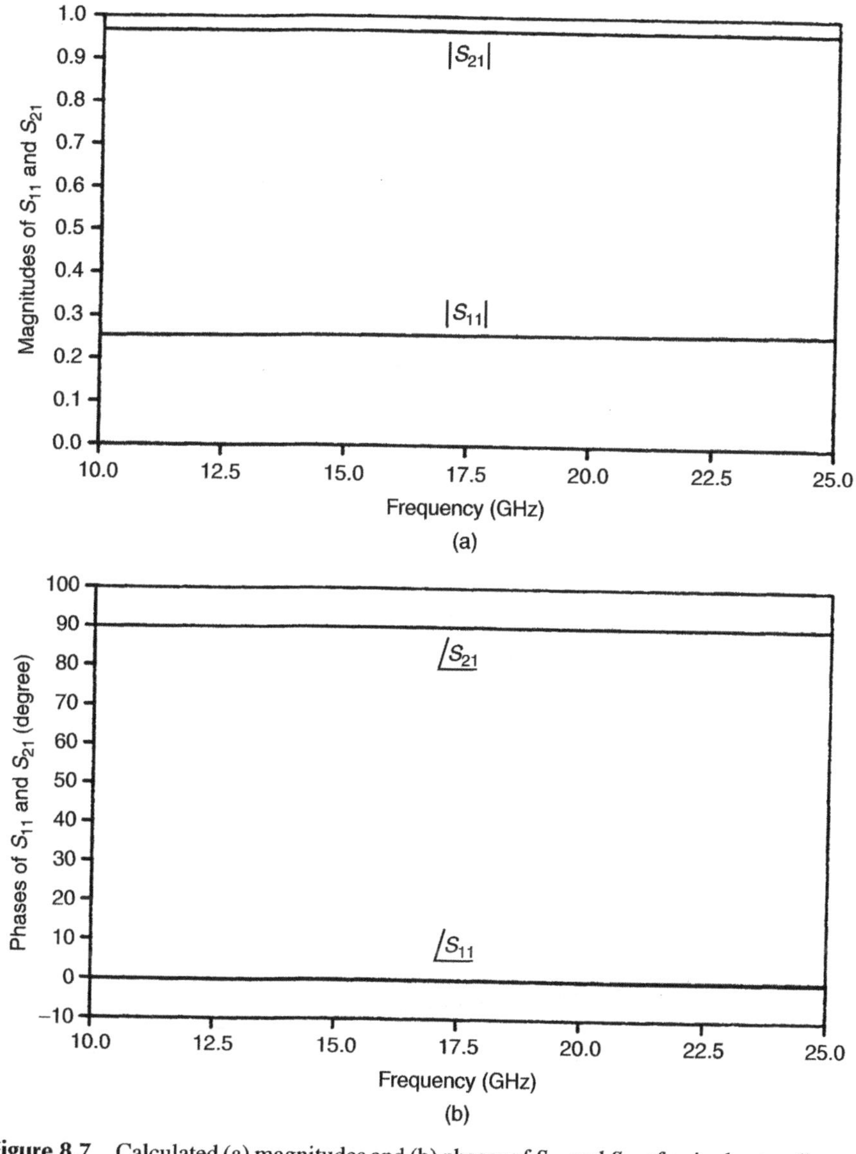

Figure 8.7 Calculated (a) magnitudes and (b) phases of S_{11} and S_{21} of a single-step discontinuity. $a = 1.4224$ mm, $t_1 = 1.016$ mm, $t_2 = 4.5466$ mm, $h = 0.127$ mm, $W_a = 0.3$ mm, $S_a = 0.527$ mm, $W_b = 0.5$ mm, $S_b = 0.127$ mm, $\varepsilon_{r1} = 3.7$, $\varepsilon_{r2} = 10$, and $\varepsilon_{r3} = 2.2$.

In order to assess the accuracy of the calculated S parameters, the validity of the power conservation ($|S_{11}|^2 + |S_{21}|^2 = 1$), the boundary condition at the $x = t_1 + h$ plane, the orthogonality condition, and the convergence of the S parameters should be verified numerically. In our calculations, the power conservation condition is satisfied to within 0.9% when the orthogonality is not assumed, but to

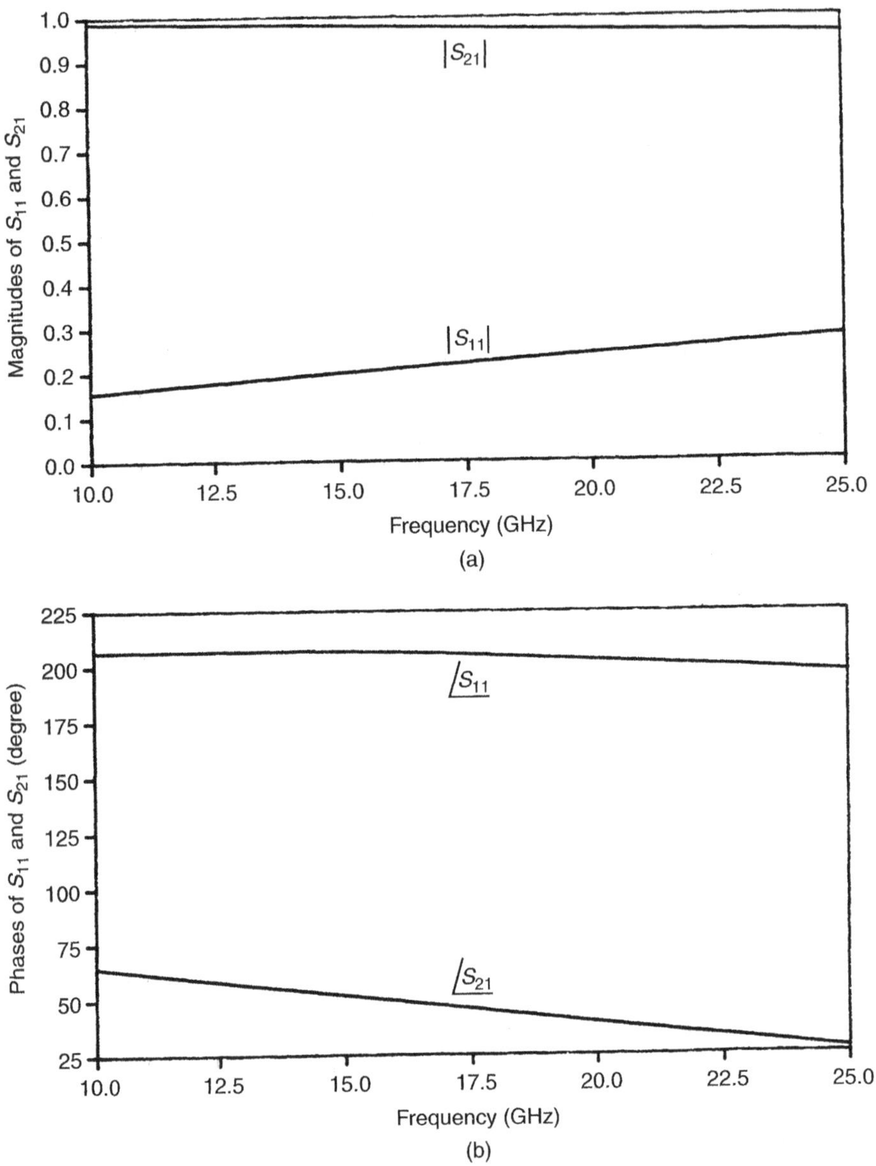

Figure 8.8 Calculated (a) magnitudes and (b) phases of S_{11} and S_{21} of a double-step discontinuity. $a = 1.4224$ mm, $t_1 = 1.016$ mm, $t_2 = 4.5466$ mm, $h = 0.127$ mm, $L = 1$ mm, $W_a = 0.9$ mm, $S_a = 0.2$ mm, $W_b = 0.6$ mm, $S_b = 0.6$ mm, $W_c = 0.8$ mm, $S_c = 0.4$ mm, $\varepsilon_{r1} = 3.7$, $\varepsilon_{r2} = 10$, and $\varepsilon_{r3} = 2.2$.

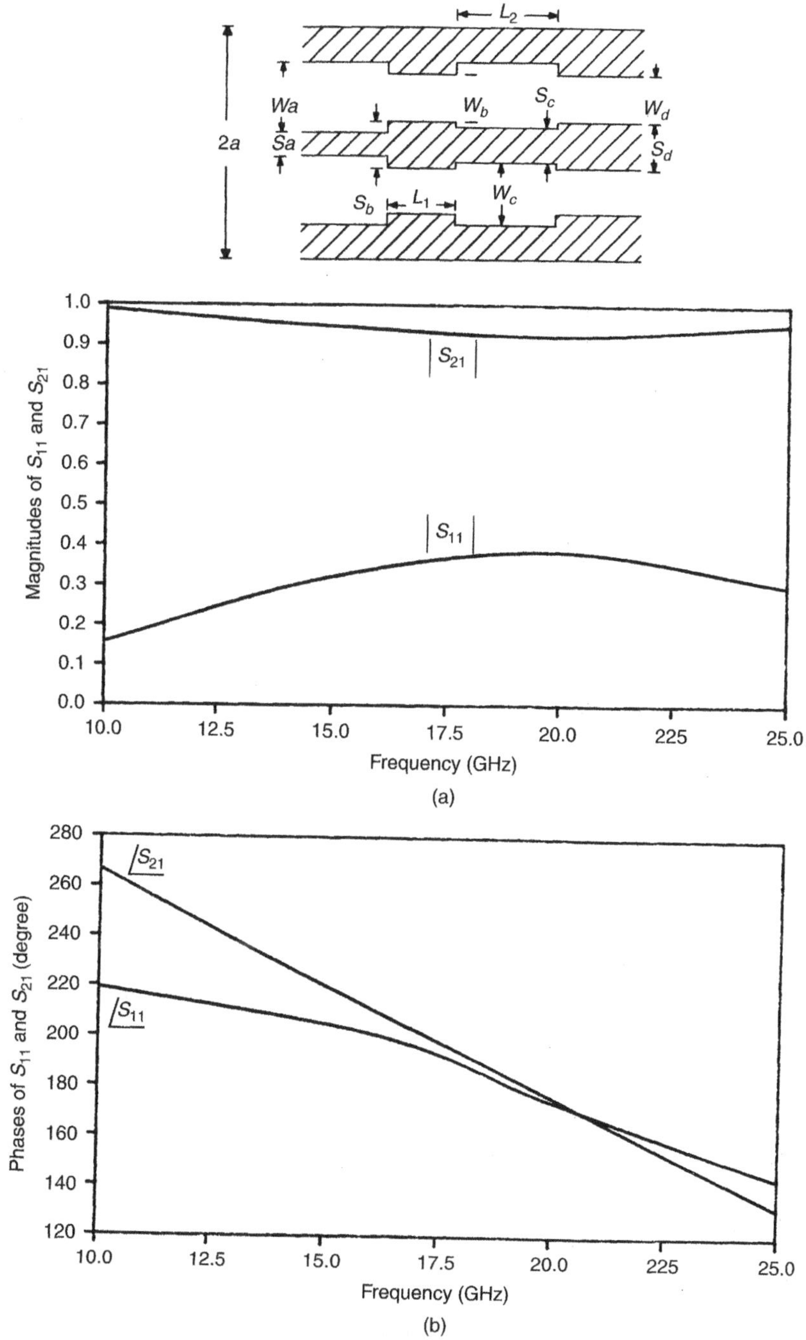

Figure 8.9 Calculated (a) magnitudes and (b) phases of S_{11} and S_{21} of a triple-step discontinuity. $a = 1.4224$ mm, $t_1 = 1.016$ mm, $t_2 = 4.5466$ mm, $h = 0.127$ mm, $L_1 = 2$ mm, $L_2 = 4.5$ mm, $W_a = 0.9$ mm, $S_a = 0.2$ mm, $W_b = 0.6$ mm, $S_b = 0.6$ mm, $W_c = 0.8$ mm, $S_c = 0.4$ mm, $W_d = 0.6$ mm, $S_d = 0.6$ mm, $\varepsilon_{r1} = 3.7$, $\varepsilon_{r2} = 10$, and $\varepsilon_{r3} = 2.2$.

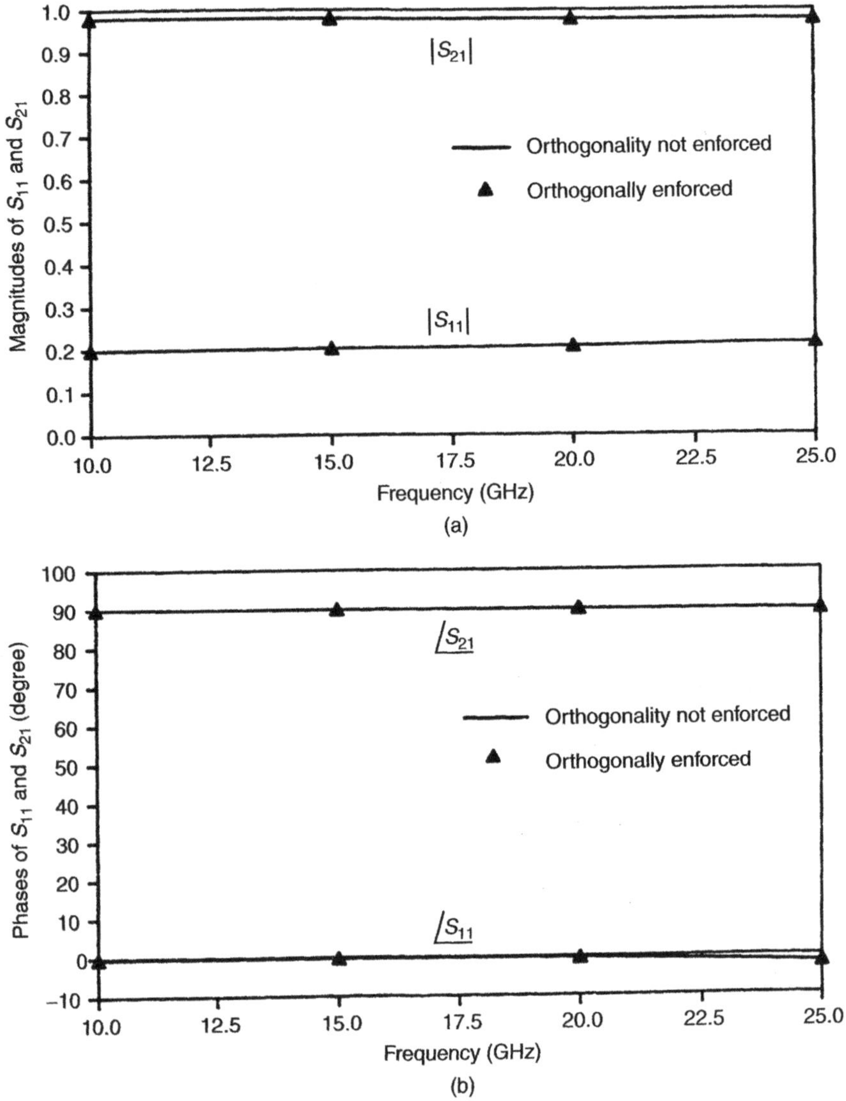

Figure 8.10 Comparison of the calculated (a) magnitudes and (b) phases of S_{11} and S_{21} of a single-step discontinuity with and without the orthogonality enforced. $a = 1.4224$ mm, $t_1 = 1.016$ mm, $t_2 = 4.5466$ mm, $h = 0.127$ mm, $W_a = 0.3$ mm, $S_a = 0.6$ mm, $W_b = 0.9$ mm, $S_b = 0.2$ mm, $\varepsilon_{r1} = 3.7$, $\varepsilon_{r2} = 10$, and $\varepsilon_{r3} = 2.2$.

within only 0.005% when it is assumed. Furthermore, a better convergence of the numerically computed S parameters is achieved if the orthogonality was assumed. Consequently, we enforced the orthogonality condition in all the calculations presented in Figs. 8.7–8.9. However, the actual orthogonality conditions for these results had been checked and verified to assure their accuracy.

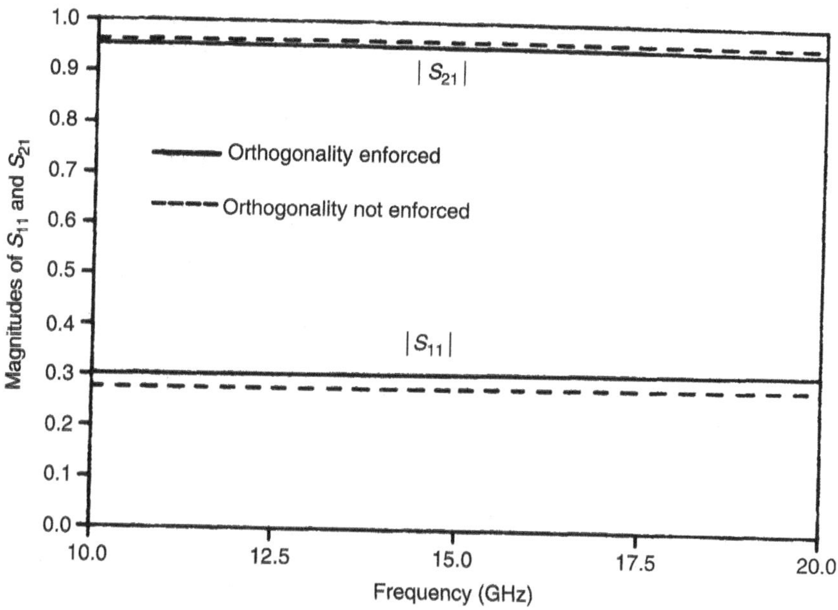

Figure 8.11 Comparison of the calculated magnitudes of S_{11} and S_{21} of another single-step discontinuity with and without the orthogonality enforced. $a = 1.4224$ mm, $t_1 = 1.016$ mm, $t_2 = 4.5466$ mm, $h = 0.127$ mm, $W_a = 0.3$ mm, $S_a = 0.6$ mm, $W_b = 0.9$ mm, $S_b = 0.2$ mm, $\varepsilon_{r1} = 3.7$, $\varepsilon_{r2} = 10$, and $\varepsilon_{r3} = 2.2$.

The numerical convergence for the S parameters of the discontinuities calculated using the mode-matching method is affected by the number of eigenmodes used. Figure 8.12 shows this convergence behavior of the magnitude and phase of S_{11} for a single-step CPW discontinuity. For each of the curves in Fig. 8.12, the number of modes, P, to the left of the discontinuity (region a) is kept constant, while the number of modes, Q, to the right of the discontinuity (region b) is varied. The convergence error is very small. For instance, a good convergence is obtained for P and Q greater than 9. The data presented for single steps in Figs. 8.7, 8.10, and 8.11 were all calculated with six eigenmodes in each region, which, as can be seen from the conversion analysis presented in Fig. 8.12, gives fairly accurate results. Similar behavior was also observed for S_{21}. Figure 8.13 shows the numerical convergence for the magnitude and phase of S_{11} of a double-step CPW discontinuity as a function of the number of eigenmodes. For each curve, the numbers of modes, P and Q, in the outer regions, a and c, respectively, are kept equal and constant, whereas the number of modes in the middle region, R, is varied. It is readily seen that the numbers of modes in regions a and c have little effect on the convergence of the S parameters. On the other hand, the number of modes in region b significantly affects the convergence, which is expected, because there are more interactions between modes in the midregion than in the outer regions. The interactions are especially strong for structures with small midregion lengths since evanescent modes created at one junction are able

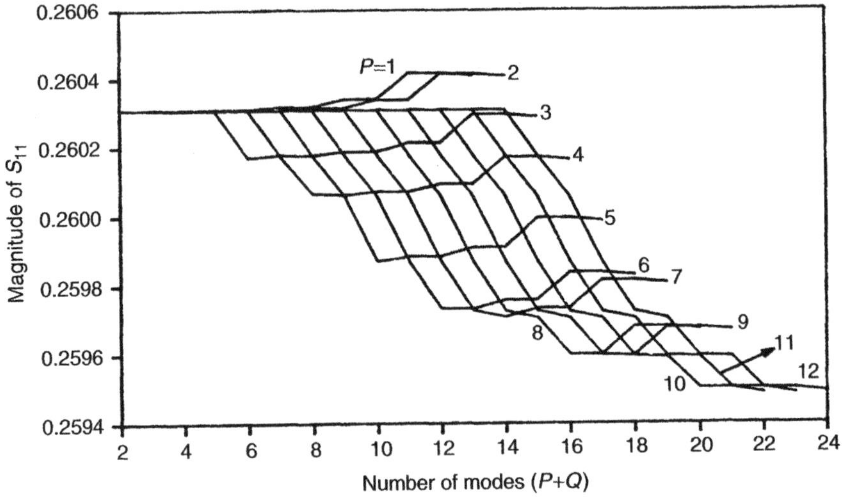

(a)

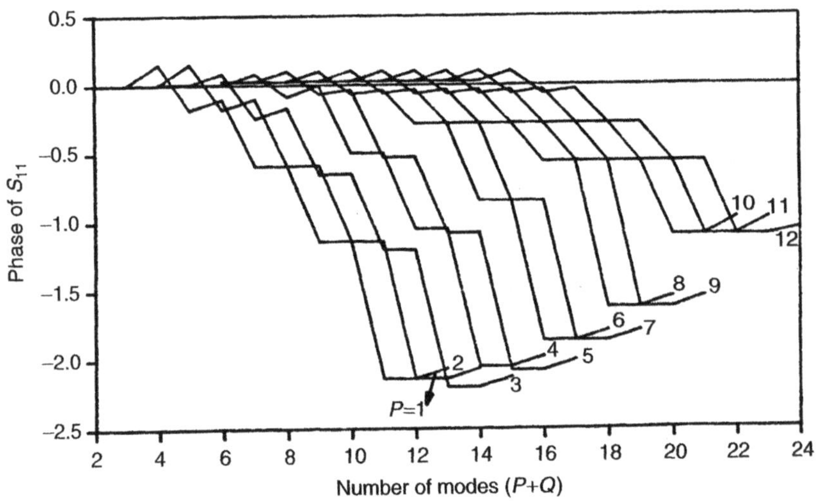

(b)

Figure 8.12 Relative convergence of the (a) magnitude and (b) phase of S_{11} of a single-step discontinuity. P = number of modes in region a and Q = number of modes in region b. $f = 20$ GHz, $a = 1.4224$ mm, $t_1 = 1.016$ mm, $t_2 = 4.5466$ mm, $h = 0.127$ mm, $W_a = 0.3$ mm, $S_a = 0.527$ mm, $W_b = 0.5$ mm, $S_b = 0.127$ mm, $\varepsilon_{r1} = 3.7$, $\varepsilon_{r2} = 10$, and $\varepsilon_{r3} = 2.2$.

to reach other junctions before decaying completely. Consequently, more modes are generally needed in the midregions for double-step as well as multiple-step discontinuities in order for the convergence to occur. Our double- and triple-step S-parameter data in Figs. 8.8 and 8.9 were obtained with six modes in the outer regions and twelve modes in the midregions.

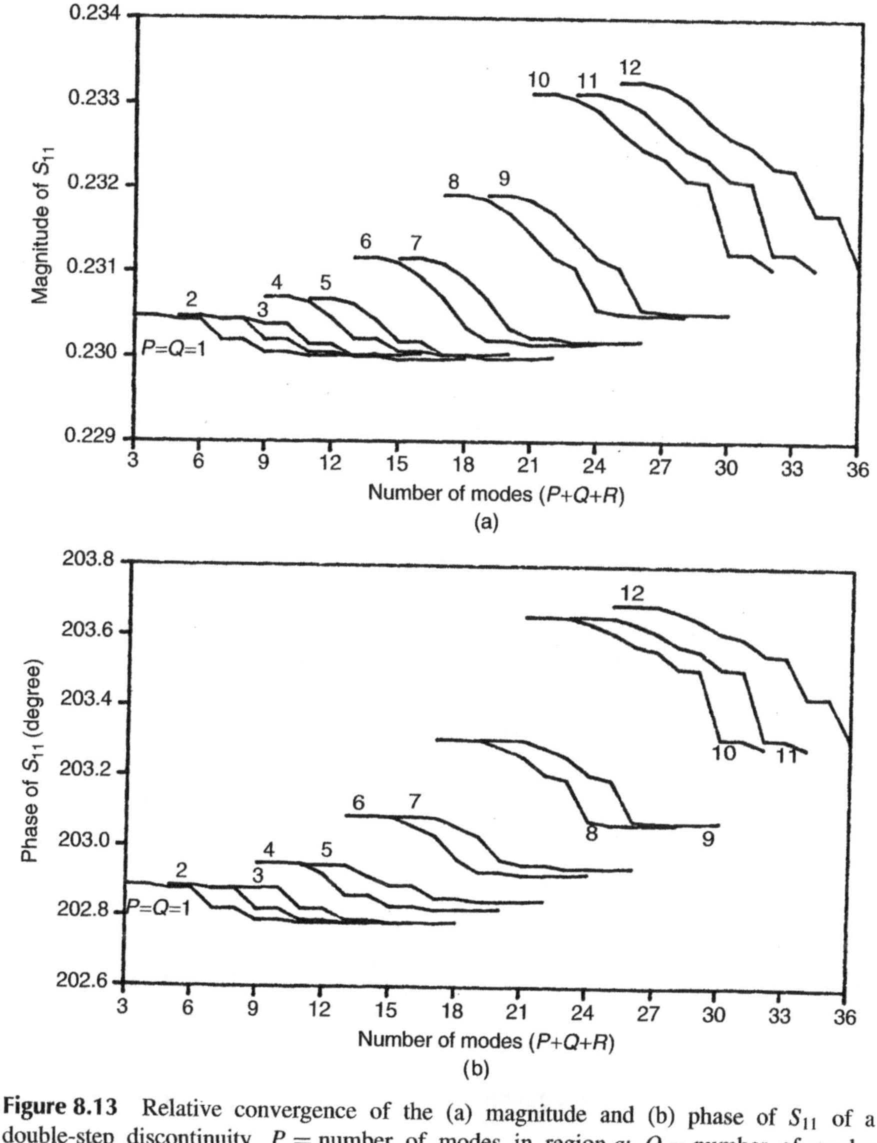

Figure 8.13 Relative convergence of the (a) magnitude and (b) phase of S_{11} of a double-step discontinuity. P = number of modes in region a; Q = number of modes in region c; and R = number of modes in region b. $f = 10$ GHz, $a = 1.4224$ mm, $t_1 = 1.016$ mm, $t_2 = 4.5466$ mm, $h = 0.127$ mm, $L = 2$ mm, $W_a = 0.9$ mm, $S_a = 0.2$ mm, $W_b = 0.6$ mm, $S_b = 0.6$ mm, $W_c = 0.8$ mm, $S_c = 0.4$ mm, $\varepsilon_{r1} = 3.7$, $\varepsilon_{r2} = 10$, and $\varepsilon_{r3} = 2.2$.

REFERENCES

1. A. Wexler, "Solution of Waveguide Discontinuities by Modal Analysis," *IEEE Trans. Microwave Theory Tech.*, Vol. MTT-15, pp. 508–517, Sept. 1967.

2. K. M. Rahman and C. Nguyen, "Full-Wave Analysis of Coplanar Strips Considering the Finite Metallization Thickness," *IEEE Trans. Microwave Theory Tech.*, Vol. MTT-42, pp. 2177–2179, Nov. 1994.

3 K. M. Rahman and C. Nguyen, "On the Analysis of Single and Multiple Step Discontinuities for a Shielded Three-Layer Coplanar Waveguide," *IEEE Trans. Microwave Theory Tech.*, Vol. MTT-41, pp. 1484–1488, Sept. 1993.

PROBLEMS

8.1 Consider the CPS as shown in Fig. 8.1. Use the formulation presented in Section 8.1 to write a computer program for calculating the effective dielectric constant, ε_{eff}, and characteristic impedance, Z_o, for the quasi-TEM mode. Compute and plot ε_{eff} and Z_o versus h/λ from 0.01 to 1 for $2b/h = 1$, $(a-b)/h = 1.5$, $\varepsilon_r = 9$, and $t = 0$, 5, 10, and 20 μm. λ is the free-space wavelength.

8.2 Consider a microstrip line as shown in Fig. P8.1. The enclosure and strip are assumed to be perfect conductors. The dielectric substrates are assumed to be lossless. By placing a magnetic wall at the center of the strip and dividing the structure into subregions as shown, formulate the mode-matching process for determining the effective dielectric constant, ε_{eff}, and characteristic impedance, Z_o, for the eigenmodes. Z_o is defined as in Eq. (4.9).

8.3 Use the formulation derived in Problem 8.2 to write a computer program for calculating ε_{eff} and Z_o for the quasi-TEM mode. Compute and plot these parameters versus frequency from 1 to 40 GHz for $\varepsilon_{r1} = 10.5$,

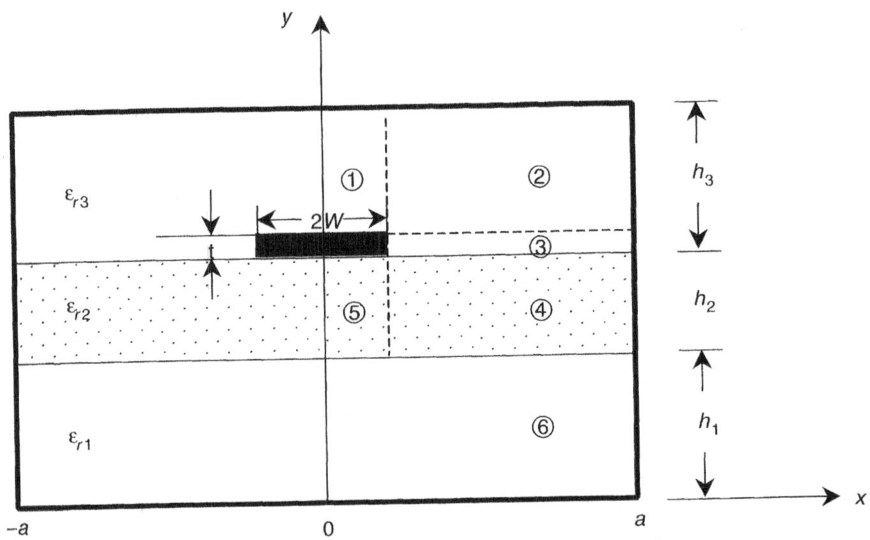

Figure P8.1 Cross section of a shielded three-layer microstrip line.

$\varepsilon_{r2} = 2.2$, $\varepsilon_{r3} = 1$, $a = 1.27$ mm, $h_1 = h_2 = 0.254$ mm, $h_3 = 0.635$ mm, $W = 0.254$ mm, and $t/(h_1 + h_2)$ of 0, 0.01, 0.02, 0.05, and 0.1.

8.4 Repeat Problem 8.3 for ε_{eff} of the first three higher-order modes. The frequency is from 1 to 60 GHz.

8.5 Use the computer program developed in Problem 8.3 to compute and plot Z_o and ε_{eff} for the quasi-TEM mode versus W/h_2 from 0.05 to 5 for ε_{r2} of 2.2 and 10.5, t/h_2 of 0, 0.01, 0.02, 0.05, and 0.1, $h_1 = 0$, h_3 and a very large, and a frequency of 0.1, 10, and 20 GHz. This structure is basically an open microstrip line considered in Problem 4.7. Compare results at 0.1 GHz to the quasi-static results obtained in Problem 4.7.

8.6 Consider a two-layer conductor-backed CPW as shown in Fig. P8. 2. The central and ground strips are assumed to be perfect conductors, and the dielectrics are assumed lossless.

(a) Formulate the mode-matching method for determining the characteristic impedance, Z_o, and effective dielectric constant, ε_{eff}, for the quasi-TEM mode. Z_o is defined as in Eq. (4.10).

(b) Write a computer program to calculate Z_o and ε_{eff}. Compute and plot Z_o and ε_{eff} versus a/b from 0.1 to 0.9 for $\varepsilon_{r1} = 1$, $\varepsilon_{r2} = 10.5$, $h_2/b = 0.1, 0.5, 4$, and 10, $t/a = 0$, 0.01, 0.02, 0.05, and 0.1, h_1 approaching infinity, and $f = 0.1$, 10, and 20 GHz. This structure is the conventional CPW as shown in Fig. 4.3(a). Compare results at 0.1 GHz to the quasi-static results obtained in Problem 4.12.

(c) Compute and plot Z_o and ε_{eff} versus a/b from 0.1 to 0.9 for $\varepsilon_{r1} = 12.9$, $\varepsilon_{r2} = 3.9$, $h_2/b = 0.1, 0.5, 4$, and 10, $h_1/h_2 = 1, 5, 10$, and 15, $t/a = 0$, 0.01, 0.02, 0.05, and 0.1, and $f = 0.1$, 10, and 20 GHz.

8.7 Derive Eqs. (8.23)–(8.52).

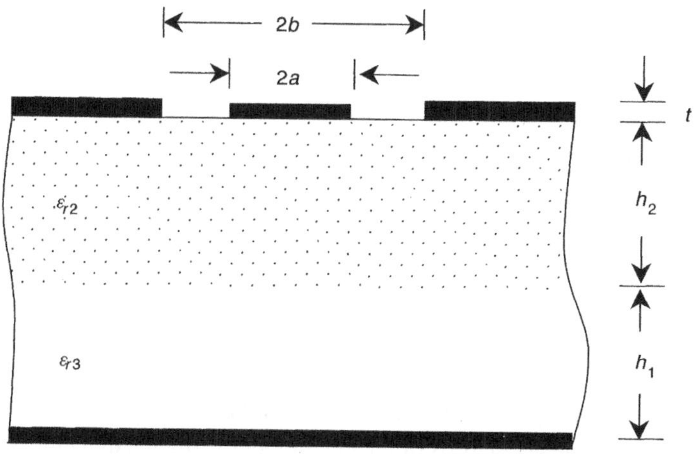

Figure P8.2 Cross section of a two-layer CPW.

8.8 Derive Eq. (8.62) and the coefficients given in Appendix A at the end of this chapter.

8.9 Derive Eqs. (8.67)–(8.88).

8.10 Derive Eqs. (8.89)–(8.96).

8.11 Derive Eqs. (8.97)–(8.110).

8.12 Derive Eq. (8.120).

8.13 Derive Eqs. (8.121)–(8.123) and the inner products given in Appendix B at the end of this chapter.

8.14 Derive Eqs. (8.133)–(8.136).

8.15 Consider the CPW shown in Fig. 8.5. Use the formulation presented in Section 8.2.2 to write a computer program to calculate S_{11} and S_{21} of a single-step discontinuity. Compute and plot S_{11} and S_{21} versus frequency from 20 to 40 GHz for $a = 1.78$ mm, $t_1 = t_2 = 3.432$ mm, $h = 0.254$ mm, $W_a = 0.1$ mm, $S_a = 0.8$ mm, $W_b = 0.6$ mm, $\varepsilon_{r1} = \varepsilon_{r3} = 1$, and $\varepsilon_{r2} = 9.6$. The orthogonality condition is enforced in the calculations. Also, verify the power conservation described by $|S_{11}|^2 + |S_{21}|^2 = 1$.

8.16 Repeat Problem 8.15 without enforcement of the orthogonality condition and compare the results to those obtained in Problem 8.15.

8.17 Consider the CPW shown in Fig. 8.5. Use the formulation presented in Section 8.2.3 to write a computer program to calculate S_{11} and S_{21} of a double-step discontinuity. Compute and plot S_{11} and S_{21} versus the length of the central section for $a = 1.4224$ mm, $t_1 = 1.016$ mm, $t_2 = 4.5466$ mm, $h = 0.127$ mm, $W_a = W_c = 0.5$ mm, $S_a = S_c = 0.127$ mm, $W_b = 0.3$ mm, $S_b = 0.527$ mm, $\varepsilon_{r1} = 3.7$, $\varepsilon_{r2} = 10$, $\varepsilon_{r3} = 2.2$, and $f = 10$ GHz. The orthogonality condition is enforced. Also, verify the power conservation.

8.18 Reproduce the results in Fig. 8.12.

8.19 Consider again the CPW shown in Fig. 8.5. Use the formulation presented in Sections 8.2.3 and 8.2.4 to write a computer program to calculate S_{11} and S_{21} of a triple-step discontinuity. Reproduce the results in Fig. 8.9. Also, verify the power conservation.

8.20 Consider a shielded asymmetrical CPW as shown in Fig. 7.1. The enclosure and ground and central strips are assumed to be perfect conductors. The dielectric substrates are assumed to be lossless. Also, the strip metallization thickness is neglected. Due to the asymmetry in the structure, the CPW (π), slot-line (c), and parallel-plate modes will all be excited. Here, however, we assume an equal potential for the two ground planes, so only π and parallel-plate modes exist in the structure.

(a) Formulate an analysis of the eigenmodes using the dynamic SDA. Note that different unknown coefficients are needed for the expansions of the electric fields on different slots. The following basis functions can be used:

$$E_{xm}^{(i)}(x) = \begin{cases} \dfrac{\cos\left(\dfrac{m\pi(x - G_i - x_i)}{S_i}\right)}{\sqrt{1 - \left(\dfrac{2(x - G_i - x_i)}{S_i}\right)^2}}, & m = 0, 2, \ldots \\[2ex] \dfrac{\sin\left(\dfrac{m\pi(x - G_i - x_i)}{S_i}\right)}{\sqrt{1 - \left(\dfrac{2(x - G_i - x_i)}{S_i}\right)^2}}, & m = 1, 3, \ldots \end{cases}$$

$$E_{zm}^{(i)}(x) = \begin{cases} \dfrac{\cos\left(\dfrac{m\pi(x - G_i - x_i)}{S_i}\right)}{\sqrt{1 - \left(\dfrac{2(x - G_i - x_i)}{S_i}\right)^2}}, & m = 1, 3, \ldots \\[2ex] \dfrac{\sin\left(\dfrac{m\pi(x - G_i - x_i)}{S_i} S_i\right)}{\sqrt{1 - \left(\dfrac{2(x - G_i - x_i}{S_i}\right)^2}}, & m = 2, 4, \ldots \end{cases}$$

where $y_1 = S_1/2$ and $y_2 = W + S_1 + S_2/2$, and $i = 1, 2$ denoting the right and left slots, respectively.

(b) Formulate the mode-matching method for analyzing the single-step discontinuity shown in Fig. P8.3.

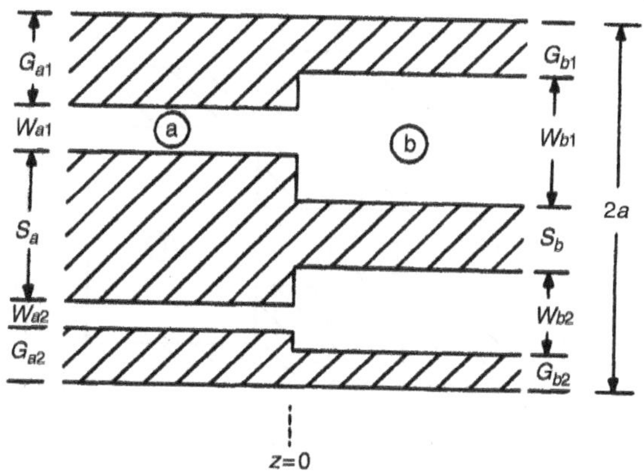

Figure P8.3 Single-step discontinuity in the CPW of Fig. 7.1.

(c) Write a computer program to calculate S_{11} and S_{21}. Compute and plot S_{11} and S_{21} versus frequency from 5 to 40 GHz for $a = 3.556$ mm, $h_1 = h_3 = 3.429$ mm, $h_2 = 0.254$ mm, $S_{a1} = 0.254$ mm, $S_{a2} = 0.381$ mm, $W_a = 0.889$ mm, $G_{a1} = 1.143$ mm, $S_{b1} = 0.762$ mm, $S_{b2} = 0.635$ mm, $W_b = 0.381$ mm, $G_{b1} = 0.8255$ mm, $\varepsilon_{r1} = \varepsilon_{r3} = 1$, and $\varepsilon_{r2} = 2.2$. Compare results with and without the assumption of modal orthogonality. Also, verify the power conservation with and without enforcing the orthogonality.

8.21 Repeat parts (b) and (c) of Problem 8.20 for the double-step discontinuity shown in Fig. P8.4. $a = 3.556$ mm, $h_1 = h_3 = 3.429$ mm, $h_2 = 0.254$ mm, $S_{a1} = 0.508$ mm, $S_{a2} = 0.762$ mm, $W_a = 0.508$ mm, $G_{a1} = 1.016$ mm, $S_{b1} = 0.254$ mm, $S_{b2} = 0.381$ mm, $W_b = 0.889$ mm, $G_{b1} = 1.143$ mm, $S_{c1} = 0.762$ mm, $S_{c2} = 0.635$ mm, $W_c = 0.381$ mm, $G_{c1} = 0.8255$ mm, $L = 1.5$ mm, $\varepsilon_{r1} = \varepsilon_{r3} = 1$, and $\varepsilon_{r2} = 2.2$.

8.22 Repeat parts (b) and (c) of Problem 8.20 for the triple-step discontinuity shown in Fig. P8.5. $a = 3.556$ mm, $h_1 = h_3 = 3.429$ mm, $h_2 = 0.254$ mm, $S_{a1} = 0.508$ mm, $S_{a2} = 0.762$ mm, $W_a = 0.508$ mm, $G_{a1} = 1.016$ mm, $S_{b1} = S_{d1} = 0.254$ mm, $S_{b2} = S_{d2} = 0.381$ mm, $W_b = W_d = 0.889$ mm, $G_{b1} = G_{d1} = 1.143$ mm, $S_{c1} = 0.762$ mm, $S_{c2} = 0.635$ mm, $W_c = 0.381$ mm, $G_{c1} = 0.8255$ mm, $L_1 = 1.5$ mm, $L_2 = 2$ mm, $\varepsilon_{r1} = \varepsilon_{r3} = 1$, and $\varepsilon_{r2} = 2.2$.

8.23 Use the computer program developed in Problem 8.21 to calculate S_{11} and S_{21} of the double-step discontinuity, as shown in Fig. P8.4, versus the length L of the middle section. $a = 3.556$ mm, $h_1 = h_3 = 3.429$ mm, $h_2 = 0.254$ mm, $S_{a1} = S_{c1} = 0.508$ mm, $S_{a2} = S_{c2} = 0.762$ mm, $W_a = W_c = 0.508$ mm, $G_{a1} = G_{c1} = 1.016$ mm, $S_{b1} = 0.254$ mm, $S_{b2} = 0.381$ mm, $W_b = 0.889$ mm, $G_{b1} = 1.143$ mm, $\varepsilon_{r1} =$

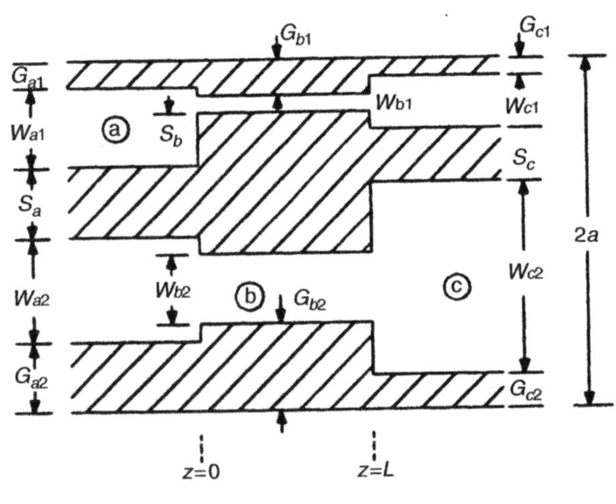

Figure P8.4 Double-step discontinuity in the CPW of Fig. 7.1.

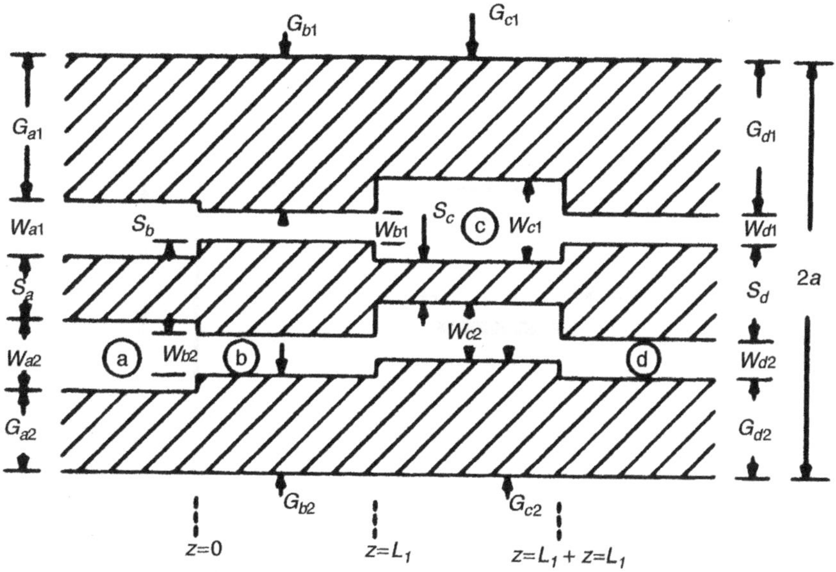

Figure P8.5 Triple-step discontinuity in the CPW of Fig. 7.1.

$\varepsilon_{r3} = 1$, and $\varepsilon_{r2} = 2.2$. Numerical results would indicate that S_{11} and S_{21} approach zero and unity, respectively, as L is reduced. Why?

8.24 Consider a suspended strip line as shown in Fig. P8.6. The enclosure and strip are assumed to be perfect conductors. The dielectric substrate is assumed to be lossless. Also, the strip metallization thickness is infinitely thin.

(a) Formulate an analysis of the eigenmodes using the dynamic SDA. The following basis functions are used for the current distributions:

$$J_{xm}(x) = \frac{\sin[2m\pi(x-a)/W]}{\sqrt{1-[2(x-a)/W]^2}}$$

$$J_{zm}(x) = \frac{\cos[2\pi(m-1)(x-a)/W]}{\sqrt{1-[2(x-a)/W]^2}}$$

(b) Formulate the mode-matching method for analyzing the single-step discontinuity shown in Fig. P8.7.

(c) Write a computer program to calculate S_{11} and S_{21}. Compute and plot S_{11} and S_{21} versus frequency from 10 to 30 GHz for $2a = h_1 = 3.556$ mm, $h_3 = 3.302$ mm, $h_2 = 0.254$ mm, $S_a = 1.7372$ mm, $S_b = 1.143$ mm, and $\varepsilon_r = 2.2$. Compare results with and without the assumption of modal orthogonality. Also, verify the power conservation with and without enforcing the orthogonality.

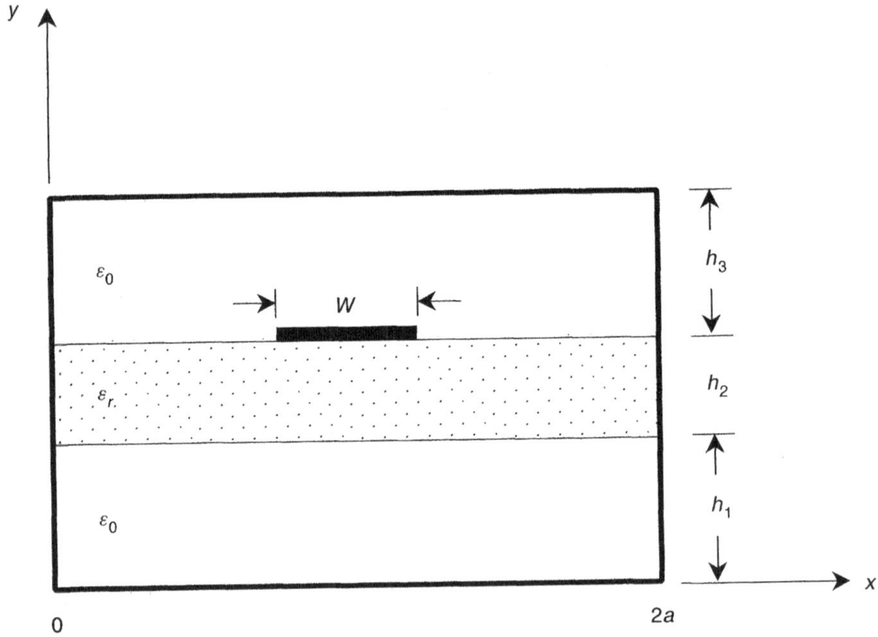

Figure P8.6 Cross section of a suspended strip line.

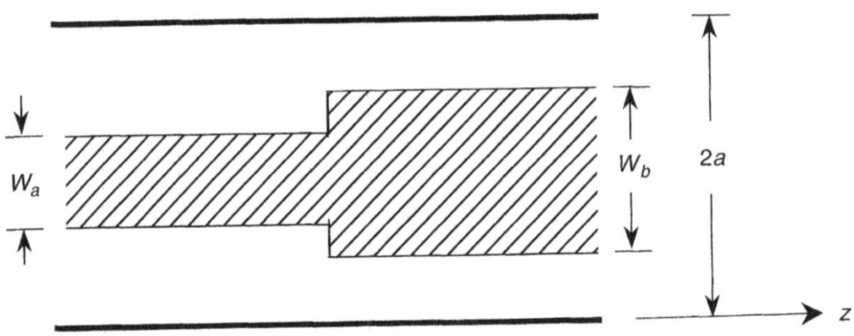

Figure P8.7 Suspended strip line's single-step discontinuity.

APPENDIX A: COEFFICIENTS IN EQS. (8.62)

The coefficients H, I, J, K, L, M, N, and T in Eqs. (8.62) are derived as follows:

$$H_{1mn} = h_{1n} A_{sm}^{s12} \cos(\alpha_{2n} t) - \delta_{mn} \gamma (Wb/4) \sin(\alpha_{2m} t)$$

$$H_{2mn} = -h_{1n} A_{sm}^{s12} \sin(\alpha_{2n} t) - \delta_{mn} \gamma (Wb/4) \cos(\alpha_{2m} t)$$

$$H_{3mn} = h_{2n} A_{sm}^{s12} \sin(\alpha_{2n} t) + \delta_{mn} (Wb/4) \alpha_{2m} \beta_{pm} (1/j\omega\mu_0) \cos(\alpha_{2m} t)$$

$$H_{4mn} = h_{2n}A_{sm}^{s12}\cos(\alpha_{2n}t) - \delta_{mn}(Wb/4)\alpha_{2m}\beta_{pm}(1/j\omega\mu_0)\sin(\alpha_{2m}t)$$

$$H_{5mn} = h_{3n}A_{sm}^{s12}\cos(\alpha_{3n}t)$$

$$H_{6mn} = -h_{3n}A_{sm}^{s12}\sin(\alpha_{3n}t)$$

$$H_{7mn} = h_{4n}A_{sm}^{s12}\sin(\alpha_{3n}t)$$

$$H_{8mn} = h_{4n}A_{sm}^{s12}\cos(\alpha_{3n}t)$$

$$I_{1mn} = f_{1n}A_{sm}^{c12}\cos(\alpha_{2n}t) - \delta_{mn}(Wb/4)\beta_{pm}\sin(\alpha_{2m}t)$$

$$I_{2mn} = -f_{1n}A_{sm}^{c12}\sin(\alpha_{2n}t) - \delta_{mn}(Wb/4)\beta_{pm}\cos(\alpha_{2m}t)$$

$$I_{3mn} = f_{2n}A_{sm}^{c12}\sin(\alpha_{2n}t) + \delta_{mn}(Wb/4)\frac{\gamma\alpha_{2m}(1+\delta_{m0})}{j\omega\mu_0}\cos(\alpha_{2m}t)$$

$$I_{4mn} = f_{2n}A_{sm}^{c12}\cos(\alpha_{2n}t) - \delta_{mn}(Wb/4)\frac{\gamma\alpha_{2m}(1+\delta_{m0})}{j\omega\mu_0}\sin(\alpha_{2m}t)$$

$$I_{5mn} = f_{3n}A_{sm}^{c12}\cos(\alpha_{3n}t)$$

$$I_{6mn} = -f_{3n}A_{sm}^{c12}\sin(\alpha_{3n}t)$$

$$I_{7mn} = f_{4n}A_{sm}^{c12}\sin(\alpha_{3n}t)$$

$$I_{8mn} = f_{4n}A_{sm}^{c12}\cos(\alpha_{3n}t)$$

$$J_{1mn} = h_{1n}A_{sm}^{s13}\cos(\alpha_{2n}t)$$

$$J_{2mn} = -h_{1n}A_{sm}^{s13}\sin(\alpha_{2n}t)$$

$$J_{3mn} = h_{2n}A_{sm}^{s13}\sin(\alpha_{2n}t)$$

$$J_{4mn} = h_{2n}A_{sm}^{s13}\cos(\alpha_{2n}t)$$

$$J_{5mn} = h_{3n}A_{sm}^{s13}\cos(\alpha_{3n}t) - \delta_{mn}[(W-a)W/4]\gamma\sin(\alpha_{3m}t)$$

$$J_{6mn} = -h_{3n}A_{sm}^{s13}\sin(\alpha_{3n}t) - \delta_{mn}\gamma[(W-a)W/4]\cos(\alpha_{2m}t)$$

$$J_{7mn} = h_{4n}A_{sm}^{s13}\sin(\alpha_{3n}t) + \delta_{mn}[(W-a)W/4]\alpha_{3m}\beta_{qm}(1/j\omega\mu_0)\cos(\alpha_{3m}t)$$

$$J_{8mn} = h_{4n}A_{sm}^{s13}\cos(\alpha_{3n}t) - \delta_{mn}[(W-a)W/4]\alpha_{3m}\beta_{qm}(1/j\omega\mu_0)\sin(\alpha_{3m}t)$$

$$K_{1mn} = f_{1n}A_{sm}^{c13}\cos(\alpha_{2n}t)$$

$$K_{2mn} = -f_{1n}A_{sm}^{c13}\sin(\alpha_{2n}t)$$

$$K_{3mn} = f_{2n}A_{sm}^{c13}\sin(\alpha_{2n}t)$$

$$K_{4mn} = f_{2n}A_{sm}^{c13}\cos(\alpha_{2n}t)$$

$$K_{5mn} = f_{3n}A_{sm}^{c13}\cos(\alpha_{3n}t) - \delta_{mn}[(W-a)W/4]\beta_{qm}\sin(\alpha_{3m}t)$$

$$K_{6mn} = -f_{3n}A_{sm}^{c13}\sin(\alpha_{3n}t) - \delta_{mn}[(W-a)W/4]\beta_{qm}\cos(\alpha_{3m}t)$$

$$K_{7mn} = f_{4n}A_{sm}^{c13}\sin(\alpha_{3n}t) + \delta_{mn}[(W-a)W/4]\frac{\gamma\alpha_{3m}}{j\omega\mu_0}\cos(\alpha_{3m}t)$$

$$K_{8mn} = f_{4n}A_{sm}^{c13}\cos(\alpha_{3n}t) - \delta_{mn}[(W-a)W/4]\frac{\gamma\alpha_{3m}}{j\omega\mu_0}\sin(\alpha_{3m}t)$$

$$L_{1mn} = \sum_{s=1}^{N}\frac{\gamma}{\beta_s^2-\gamma^2}\left(-\frac{\varepsilon_r}{\alpha_{4s}}\frac{R_{2s}}{R_{1s}}T_{sn}^1 - \frac{\beta_s\alpha_{4s}}{j\omega\varepsilon_0}\frac{S_{1s}}{S_{2s}}T_{sn}^2\right)A_{sm}^{s12}\alpha_{2n}$$

$$L_{2mn} = -\delta_{mn}\frac{Wb}{4}\gamma$$

$$L_{3mn} = \delta_{mn}\frac{Wb}{4}\frac{\alpha_{2m}\beta_{pm}}{j\omega\mu_0}$$

$$L_{4mn} = \sum_{s=1}^{N}\frac{1}{\beta_s^2-\gamma^2}\left(-\frac{j\omega\varepsilon_0\varepsilon_r}{\alpha_{4s}}\gamma^2\frac{R_{2s}}{R_{1s}}T_{sn}^3 - \beta_s\alpha_{4s}\frac{S_{1s}}{S_{2s}}T_{sn}^4\right)A_{sm}^{s12}$$

$$L_{5mn} = \sum_{s=1}^{N}\frac{\gamma}{\beta_s^2-\gamma^2}\left(-\frac{\varepsilon_r}{\alpha_{4s}}\frac{R_{2s}}{R_{1s}}T_{sn}^5 - \frac{\beta_s\alpha_{4s}}{j\omega\varepsilon_0}\frac{S_{1s}}{S_{2s}}T_{sn}^6\right)A_{sm}^{s12}\alpha_{3n}$$

$$L_{6mn} = 0$$

$$L_{7mn} = 0$$

$$L_{8mn} = \sum_{s=1}^{N}\frac{1}{\beta_s^2-\gamma^2}\left(-\frac{j\omega\varepsilon_0\varepsilon_r}{\alpha_{4s}}\gamma^2\frac{R_{2s}}{R_{1s}}T_{sn}^7 - \beta_s\alpha_{4s}\frac{S_{1s}}{S_{2s}}T_{sn}^8\right)A_{sm}^{s12}$$

$$M_{1mn} = \sum_{s=1}^{N}\frac{\gamma}{\beta_s^2-\gamma^2}\left(-\frac{\varepsilon_r\beta_s}{\alpha_{4s}}\frac{R_{2s}}{R_{1s}}T_{sn}^1 - \frac{\gamma^2\alpha_{4s}}{j\omega\varepsilon_0}\frac{S_{1s}}{S_{2s}}T_{sn}^2\right)A_{sm}^{c12}\alpha_{2n}$$

$$M_{2mn} = -\delta_{mn}\frac{Wb}{4}\beta_{pm}$$

$$M_{3mn} = \delta_{mn}\frac{Wb}{4}\frac{\alpha_{2m}\gamma}{j\omega\mu_0}(1+\delta_{m0})$$

$$M_{4mn} = \sum_{s=1}^{N}\frac{1}{\beta_s^2-\gamma^2}\left(-\frac{j\omega\varepsilon_0\varepsilon_r}{\alpha_{4s}}\gamma\beta_s\frac{R_{2s}}{R_{1s}}T_{sn}^3 - \gamma\alpha_{4s}\frac{S_{1s}}{S_{2s}}T_{sn}^4\right)A_{sm}^{c12}$$

$$M_{5mn} = \sum_{s=1}^{N}\frac{1}{\beta_s^2-\gamma^2}\left(-\frac{\varepsilon_r}{\alpha_{4s}}\beta_s\frac{R_{2s}}{R_{1s}}T_{sn}^5 - \frac{\gamma^2\alpha_{4s}}{j\omega\varepsilon_0}\frac{S_{1s}}{S_{2s}}T_{sn}^6\right)A_{sm}^{c12}\alpha_{3n}$$

$$M_{6mn} = 0$$

$$M_{7mn} = 0$$

$$M_{8mn} = \sum_{s=1}^{N} \frac{1}{\beta_s^2 - \gamma^2} \left(-\frac{j\omega\varepsilon_0\varepsilon_r}{\alpha_{4s}} \gamma\beta_s \frac{R_{2s}}{R_{1s}} T_{sn}^7 - \gamma\alpha_{4s} \frac{S_{1s}}{S_{2s}} T_{sn}^8 \right) A_{sm}^{c12}$$

$$N_{1mn} = \sum_{s=1}^{N} \frac{\gamma}{\beta_s^2 - \gamma^2} \left(-\frac{\varepsilon_r}{\alpha_{4s}} \frac{R_{2s}}{R_{1s}} T_{sn}^1 - \frac{\beta_s\alpha_{4s}}{j\omega\varepsilon_0} \frac{S_{1s}}{S_{2s}} T_{sn}^2 \right) A_{sm}^{s13} \alpha_{2n}$$

$$N_{2mn} = 0$$

$$N_{3mn} = 0$$

$$N_{4mn} = \sum_{s=1}^{N} \frac{1}{\beta_s^2 - \gamma^2} \left(-\frac{j\omega\varepsilon_0\varepsilon_r}{\alpha_{4s}} \gamma^2 \frac{R_{2s}}{R_{1s}} T_{sn}^3 - \beta_s\alpha_{4s} \frac{S_{1s}}{S_{2s}} T_{sn}^4 \right) A_{sm}^{s13}$$

$$N_{5mn} = \sum_{s=1}^{N} \frac{\gamma}{\beta_s^2 - \gamma^2} \left(-\frac{\varepsilon_r}{\alpha_{4s}} \frac{R_{2s}}{R_{1s}} T_{sn}^5 - \frac{\beta_s\alpha_{4s}}{j\omega\varepsilon_0} \frac{S_{1s}}{S_{2s}} T_{sn}^6 \right) A_{sm}^{s13} \alpha_{3n}$$

$$N_{6mn} = -\delta_{mn} \frac{W(W-a)}{4} \gamma$$

$$N_{7mn} = \delta_{mn} \frac{W(W-a)}{4} \frac{\alpha_{3m}\beta_{qm}}{j\omega\mu_0}$$

$$N_{8mn} = \sum_{s=1}^{N} \frac{1}{\beta_s^2 - \gamma^2} \left(-\frac{j\omega\varepsilon_0\varepsilon_r}{\alpha_{4s}} \gamma^2 \frac{R_{2s}}{R_{1s}} T_{sn}^7 - \beta_s\alpha_{4s} \frac{S_{1s}}{S_{2s}} T_{sn}^8 \right) A_{sm}^{s13}$$

$$T_{1mn} = \sum_{s=1}^{N} \frac{\gamma}{\beta_s^2 - \gamma^2} \left(-\frac{\varepsilon_r\beta_s}{\alpha_{4s}} \frac{R_{2s}}{R_{1s}} T_{sn}^1 - \frac{\gamma^2\alpha_{4s}}{j\omega\varepsilon_0} \frac{S_{1s}}{S_{2s}} T_{sn}^2 \right) A_{sm}^{c13} \alpha_{2n}$$

$$T_{2mn} = 0$$

$$T_{3mn} = 0$$

$$T_{4mn} = \sum_{s=1}^{N} \frac{1}{\beta_s^2 - \gamma^2} \left(-\frac{j\omega\varepsilon_0\varepsilon_r}{\alpha_{4s}} \gamma\beta_s \frac{R_{2s}}{R_{1s}} T_{sn}^3 - \gamma\alpha_{4s} \frac{S_{1s}}{S_{2s}} T_{sn}^4 \right) A_{sm}^{c13}$$

$$T_{5mn} = \sum_{s=1}^{N} \frac{\gamma}{\beta_s^2 - \gamma^2} \left(-\frac{\varepsilon_r}{\alpha_{4s}} \beta_s \frac{R_{2s}}{R_{1s}} T_{sn}^5 - \frac{\gamma^2\alpha_{4s}}{j\omega\varepsilon_0} \frac{S_{1s}}{S_{2s}} T_{sn}^6 \right) A_{sm}^{c13} \alpha_{3n}$$

$$T_{6mn} = -\delta_{mn} \frac{W(W-a)}{4} \beta_{qm}$$

$$T_{7mn} = \delta_{mn} \frac{W(W-a)}{4} \frac{\alpha_{3m}\gamma}{j\omega\mu_0} (1 + \delta_{m0})$$

$$T_{8mn} = \sum_{s=1}^{N} \frac{1}{\beta_s^2 - \gamma^2} \left(-\frac{j\omega\varepsilon_0\varepsilon_r}{\alpha_{4s}} \gamma\beta_s \frac{R_{2s}}{R_{1s}} T_{sn}^7 - \gamma\alpha_{4s} \frac{S_{1s}}{S_{2s}} T_{sn}^8 \right) A_{sm}^{c13}$$

where

$$h_{1n} = \sum_{s=1}^{N} \frac{1}{\beta_s^2 - \gamma^2} \left(\frac{1}{\alpha_{1s}} T_{sn}^1 + \frac{\alpha_{1s}\beta_s}{j\omega\varepsilon_0} T_{sn}^2 \right) \alpha_{2n}$$

$$h_{2n} = \sum_{s=1}^{N} \frac{1}{\beta_s^2 - \gamma^2} \left(\frac{j\gamma^2\omega\varepsilon_0}{\alpha_{1s}} T_{sn}^3 + \alpha_{1s}\beta_s T_{sn}^2 \right)$$

$$h_{3n} = \sum_{s=1}^{N} \frac{1}{\beta_s^2 - \gamma^2} \left(\frac{1}{\alpha_{1s}} T_{sn}^5 + \frac{\alpha_{1s}\beta_s}{j\omega\varepsilon_0} T_{sn}^6 \right) \alpha_{3n}$$

$$h_{4n} = \sum_{s=1}^{N} \frac{1}{\beta_s^2 - \gamma^2} \left(\frac{j\gamma^2\omega\varepsilon_0}{\alpha_{1s}} T_{sn}^7 + \alpha_{1s}\beta_s T_{sn}^8 \right)$$

$$f_{1n} = \sum_{s=1}^{N} \frac{1}{\beta_s^2 - \gamma^2} \left(\frac{\beta_s}{\alpha_{1s}} T_{sn}^1 + \frac{\alpha_{1s}\gamma^2}{j\omega\varepsilon_0} T_{sn}^2 \right) \alpha_{2n}$$

$$f_{2n} = \sum_{s=1}^{N} \frac{1}{\beta_s^2 - \gamma^2} \left(\frac{j\omega\varepsilon_0\gamma\beta_s}{\alpha_{1s}} T_{sn}^3 + \gamma\alpha_{1s} T_{sn}^4 \right)$$

$$f_{3n} = \sum_{s=1}^{N} \frac{1}{\beta_s^2 - \gamma^2} \left(\frac{\beta_s}{\alpha_{1s}} T_{sn}^5 + \frac{\alpha_{1s}\gamma^2}{j\omega\varepsilon_0} T_{sn}^6 \right) \alpha_{3n}$$

$$f_{4n} = \sum_{s=1}^{N} \frac{1}{\beta_s^2 - \gamma^2} \left(\frac{j\omega\varepsilon_0\gamma\beta_s}{\alpha_{1s}} T_{sn}^7 + \gamma\alpha_{1s} T_{sn}^8 \right)$$

$$T_{sn}^1 = \gamma^2 A_{sn}^{s12} - \beta_{pn}\beta_s A_{sn}^{c12}$$

$$T_{sn}^2 = \left(\frac{1}{j\omega\mu_0} \right) (\beta_{pn} A_{sn}^{c12} - \beta_s A_{sn}^{s12})$$

$$T_{sn}^3 = \beta_s A_{sn}^{c12} - \beta_{pn}\beta_s A_{sn}^{s12}$$

$$T_{sn}^4 = \left(\frac{1}{j\omega\mu_0} \right) (\beta_s\beta_{pn} A_{sn}^{s12} - \gamma^2 A_{sn}^{c12})$$

$$T_{sn}^5 = \gamma^2 A_{sn}^{s13} - \beta_{qn}\beta_s A_{sn}^{c13}$$

$$T_{sn}^6 = \left(\frac{1}{j\omega\mu_0} \right) (\beta_{qn} A_{sn}^{c13} - \beta_s A_{sn}^{s13})$$

$$T_{sn}^7 = \beta_s A_{sn}^{c13} - \beta_{qn} A_{sn}^{s13}$$

$$T_{sn}^8 = \left(\frac{1}{j\omega\mu_0} \right) (\beta_s\beta_{qn} A_{sn}^{s13} - \gamma^2 A_{sn}^{c13})$$

$$A_{sn}^{s12} = \int_0^b \sin(\beta_s x) \sin(\beta_{pn} x)\, dx$$

$$A_{sn}^{c12} = \int_0^b \cos(\beta_s x)\cos(\beta_{pn} x)\,dx$$

$$A_{sn}^{c13} = \int_a^W \cos(\beta_s x)\cos(\beta_{qn} x)\,dx$$

$$A_{sn}^{c13} = \int_a^W \cos(\beta_s x)\cos(\beta_{qn} x)\,dx$$

δ_{mn} is the *Kronecker delta function.* R_{2s}/R_{1s} and S_{1s}/S_{2s} are obtained from the boundary condition at $y = -h$.

APPENDIX B: INNER PRODUCTS IN EQS. (8.120)–(8.123)

The inner products I_{ij}^{ab} used in Eqs. (8.120)–(8.123) are given as follows:

$$\begin{aligned} I_{ij}^{ab} &= \int_s \mathbf{e}_i^a \times \mathbf{h}_j^b \cdot dx\,dy\,\mathbf{a}_z \\ &= I_{ij}^{ab(1)} + I_{ij}^{ab(2)} + I_{ij}^{ab(3)} \end{aligned} \tag{B8.1}$$

$$\begin{aligned} I_{ij}^{ab(1)} = \frac{a}{2}\sum_{n=1}^{\infty}\Bigg(& \left(A_{na}^{(1)}S_{nb}^{(1)} - B_{nb}^{(1)}M_{na}^{(1)}\right)\frac{\sin\left[\left(\alpha_{nai}^{(1)} - \alpha_{nbj}^{(1)}\right)t_2\right]}{\alpha_{nai}^{(1)} - \alpha_{nbj}^{(1)}} \\ & + \left(A_{na}^{(1)}S_{nb}^{(1)} + M_{na}^{(1)}B_{nb}^{(1)}\right)\frac{\sin\left[\left(\alpha_{nai}^{(1)} + \alpha_{nbj}^{(1)}\right)t_2\right]}{\alpha_{nai}^{(1)} + \alpha_{nbj}^{(1)}}\Bigg) \end{aligned} \tag{B8.2}$$

$$\begin{aligned} I_{ij}^{ab(2)} = \frac{a}{2}\sum_{n=1}^{\infty}\Bigg[& \left(-A_{na}^{(2)}S_{nb}^{(2)} + M_{na}^{(2)}B_{nb}^{(2)}\right)\left(\frac{\cos\left[\left(\alpha_{nai}^{(2)} + \alpha_{nbj}^{(2)}\right)h\right]}{\alpha_{nai}^{(2)} + \alpha_{nbj}^{(2)}}\right. \\ & \left. + \frac{\cos\left[\left(\alpha_{nai}^{(2)} - \alpha_{nbj}^{(2)}\right)h\right]}{\alpha_{nai}^{(2)} + \alpha_{nbj}^{(2)}} - \frac{2\alpha_{nai}^{(2)}}{\left(\alpha_{nai}^{(2)}\right)^2 - \left(\alpha_{nbj}^{(2)}\right)^2}\right) + \left(A_{na}^{(2)}T_{nb}^{(2)} - M_{na}^{(2)}Q_{nb}^{(2)}\right) \\ & \times \left(\frac{\sin\left[\left(\alpha_{nai}^{(2)} - \alpha_{nbj}^{(2)}\right)h\right]}{\alpha_{nai}^{(2)} - \alpha_{nbj}^{(2)}} - \frac{\sin\left[\left(\alpha_{nai}^{(2)} + \alpha_{nbj}^{(2)}\right)h\right]}{\alpha_{nai}^{(2)} + \alpha_{nbj}^{(2)}}\right) \\ & + \left(P_{na}^{(2)}S_{nb}^{(2)} - R_{na}^{(2)}B_{nb}^{(2)}\right) \times \left(\frac{\sin\left[\left(\alpha_{nai}^{(2)} + \alpha_{nbj}^{(2)}\right)h\right]}{\alpha_{nai}^{(2)} + \alpha_{nbj}^{(2)}} + \frac{\sin\left[\left(\alpha_{nai}^{(2)} - \alpha_{nbj}^{(2)}\right)h\right]}{\alpha_{nai}^{(2)} - \alpha_{nbj}^{(2)}}\right) \end{aligned}$$

$$+\left(-P_{na}^{(2)}T_{nb}^{(2)} \times R_{na}^{(2)}Q_{nb}^{(2)}\right) \times \left(\frac{\cos\left[\left(\alpha_{nai}^{(2)}+\alpha_{nbj}^{(2)}\right)h\right]}{\alpha_{nai}^{(2)}+\alpha_{nbj}^{(2)}}\right.$$

$$\left.-\frac{\cos\left[\left(\alpha_{nai}^{(2)}-\alpha_{nbj}^{(2)}\right)h\right]}{\alpha_{nai}^{(2)}-\alpha_{nbj}^{(2)}}+\frac{2\alpha_{nbj}^{(2)}}{\left(\alpha_{nai}^{(2)}\right)^2-\left(\alpha_{nbj}^{(2)}\right)^2}\right)\Bigg] \tag{B8.3}$$

$$I_{ij}^{ab(3)} = \frac{a}{2}\sum_{n=1}^{\infty}\left(\left(A_{na}^{(3)}S_{nb}^{(3)} - B_{nb}^{(3)}M_{na}^{(3)}\right)\frac{\sin\left[\left(\alpha_{nai}^{(3)}-\alpha_{nbj}^{(3)}\right)t_1\right]}{\alpha_{nai}^{(3)}-\alpha_{nbj}^{(3)}}\right.$$

$$\left.+\left(A_{na}^{(3)}S_{nb}^{(3)} + M_{na}^{(3)}B_{nb}^{(3)}\right)\frac{\sin\left[\left(\alpha_{nai}^{(3)}+\alpha_{nbj}^{(3)}\right)t_1\right]}{\alpha_{nai}^{(3)}+\alpha_{nbj}^{(3)}}\right) \tag{B8.4}$$

$$\begin{aligned} I_{ii}^{aa} &= \int_s \mathbf{e}_i^a \times \mathbf{h}_i^a \cdot dx\,dy\,\mathbf{a}_z \\ &= I_{ij}^{aa(1)} + I_{ij}^{aa(2)} + I_{ij}^{aa(3)} \end{aligned} \tag{B8.5}$$

$$I_{ii}^{aa(1)} = \frac{a}{2}\sum_{n=1}^{\infty}\left(\left(A_{na}^{(1)}S_{na}^{(1)} - M_{na}^{(1)}B_{na}^{(1)}\right)t_2\right.$$

$$\left.+\left(A_{na}^{(1)}S_{na}^{(1)} + M_{na}^{(1)}B_{na}^{(1)}\right)\frac{\sin\left(2\alpha_{nai}^{(1)}t_2\right)}{2\alpha_{nai}^{(1)}}\right) \tag{B8.6}$$

$$I_{ii}^{aa(2)} = \frac{a}{2}\sum_{n=1}^{\infty}\Bigg[\left[\left(-A_{na}^{(2)}S_{na}^{(2)} - P_{na}^{(2)}T_{na}^{(2)}\right)\right.$$

$$+\left(M_{na}^{(2)}B_{na}^{(2)} + R_{na}^{(2)}Q_{na}^{(2)}\right)\Big]\left(\frac{\cos\left(2\alpha_{nai}^{(2)}h\right)}{2\alpha_{nai}^{(2)}} - \frac{1}{2\alpha_{nai}^{(2)}}\right)$$

$$+\left(A_{na}^{(2)}T_{na}^{(2)} - M_{na}^{(2)}Q_{na}^{(2)}\right)\left(h - \frac{\sin\left(2\alpha_{nai}^{(2)}h\right)}{2\alpha_{nai}^{(2)}}\right)$$

$$+\left(P_{na}^{(2)}S_{na}^{(2)} - R_{na}^{(2)}B_{na}^{(2)}\right)\Big]\left(h + \frac{\sin\left(2\alpha_{nai}^{(2)}h\right)}{2\alpha_{nai}^{(2)}}\right)\Bigg] \tag{B8.7}$$

$$I_{ii}^{aa(3)} = \frac{a}{2}\sum_{n=1}^{\infty}\left(\left(A_{na}^{(3)}S_{na}^{(3)} - M_{na}^{(3)}B_{na}^{(3)}\right) t_1 + \left(A_{na}^{(3)}S_{na}^{(3)} + M_{na}^{(3)}B_{na}^{(3)}\right) \frac{\sin\left(2\alpha_{nai}^{(3)}t_1\right)}{2\alpha_{nai}^{(3)}} \right) \tag{B8.8}$$

where

$$M_{nai}^{(1)} = \frac{-1}{\alpha_n^2 + \beta_i^2}\left(A_{nai}^{(1)}\alpha_{nai}^{(1)}\alpha_n + j\beta_i k_0 B_{nai}^{(1)}\right) \tag{B8.9}$$

$$S_{nai}^{(1)} = \frac{-1}{\alpha_n^2 + \beta_i^2}\left(j\beta_i k_0 \varepsilon_{r1} A_{nai}^{(1)} + \alpha_{nai}^{(1)}\alpha_n B_{nai}^{(1)}\right) \tag{B8.10}$$

$$R_{nai}^{(2)} = \frac{1}{\alpha_n^2 + \beta_i^2}\left(\alpha_n \alpha_{nai}^{(2)} A_{nai}^{(2)} - j\beta_i k_0 B_{nai}^{(2)}\right) \tag{B8.11}$$

$$M_{nai}^{(2)} = \frac{-1}{\alpha_n^2 + \beta_i^2}\left(\alpha_n \alpha_{nai}^{(2)} P_{nai}^{(2)} + j\beta_i k_0 Q_{nai}^{(2)}\right) \tag{B8.12}$$

$$S_{nai}^{(2)} = \frac{-1}{\alpha_n^2 + \beta_i^2}\left(j\beta_i k_0 \varepsilon_{r2} P_{nai}^{(2)} + \alpha_{nai}^{(2)}\alpha_n Q_{nai}^{(2)}\right) \tag{B8.13}$$

$$T_{nai}^{(2)} = \frac{1}{\alpha_n^2 + \beta_i^2}\left(-j\beta_i k_0 \varepsilon_{r2} A_{nai}^{(2)} + \alpha_{nai}^{(2)}\alpha_n B_{nai}^{(2)}\right) \tag{B8.14}$$

$$M_{nai}^{(3)} = \frac{1}{\alpha_n^2 + \beta_i^2}\left(A_{nai}^{(3)}\alpha_{nai}^{(3)}\alpha_n + j\beta_i k_0 B_{nai}^{(3)}\right) \tag{B8.15}$$

$$S_{nai}^{(3)} = \frac{-1}{\alpha_n^2 + \beta_i^2}\left(j\beta_i k_0 \varepsilon_{r3} A_{nai}^{(3)} + \alpha_{nai}^{(3)}\alpha_n B_{nai}^{(3)}\right) \tag{B8.16}$$

The subscripts a and b refer to the respective regions, while the subscripts i and j denote the eigenmodes i and j. Note that these expressions are derived in the spectral domain to keep the computation consistent with the field solutions of the considered CPW. The same expressions for the I's can also be used for the double- and multiple-step CPW discontinuities.

Index

Printed and bound by CPI Group (UK) Ltd, Croydon, CR0 4YY
16/09/2026
14963414-0001